EAT FOR LIFE

The Breakthrough Nutrient-Rich Program
for Longevity, Disease Reversal, and
Sustained Weight Loss

Joel Fuhrman, MD

HarperOne

An Imprint of HarperCollins*Publishers*

This book contains advice and information relating to health care. It should be used to supplement rather than replace the advice of your doctor or another trained health professional. If you know or suspect you have a health problem, it is recommended that you seek your physician's advice before embarking on any medical program or treatment. All efforts have been made to assure the accuracy of the information contained in this book as of the date of publication. This publisher and the author disclaim liability for any medical outcomes that may occur as a result of applying the methods suggested in this book.

FIRST HARPERCOLLINS PAPERBACK EDITION PUBLISHED IN 2020

Designed by Kris Tobiassen of Matchbook Digital

Library of Congress Cataloging-in-Publication Data is available upon request.

ISBN 978-0-06-224930-2

23 24 25 26 27 LBC 11 10 9 8 7

This book is dedicated to all people suffering with serious medical issues who were never informed that they could recover their health with nutritional excellence. The lack of information has denied them their inalienable rights.

Contents

INTRODUCTION

The Nutritarian Diet

What should we call a diet designed to slow aging and maximize human lifespan? Which portfolio of foods best protects against cancer and dementia in later life? Despite all the confusion, is there one diet style that accurately and comprehensively addresses every nutritional feature that promotes longevity, that can be used as the gold standard of healthy eating for the modern world?

If so, such a diet cannot be judged on popularity. It cannot be judged by critics looking for mass appeal or making judgments based on limited knowledge, or bad science, or food preferences. It can only be judged accurately by knowledgeable individuals who have no preconceived dietary biases or food preferences. These individuals use impartial scientific thinking to review the supportive science to see whether the preponderance of evidence supports the conclusions.

After decades of rigor and scientific study, it is possible to conclude that the **Nutritarian diet** is that pinnacle of excellence in the nutritional world.

I coined the term "Nutritarian" to label and identify this nutrient-rich diet style and to set it apart from other diets. By paying attention not just to vitamins and minerals, but also to the thousands of other *phytonutrients*—that is, the beneficial chemicals found in plants—that are essential for maximizing immune function, such a diet style can have profound effects on extending healthspan (meaning the number of years we can expect to be healthy) and lifespan.

Thousands of people around the world now call themselves "Nutritarians" because they eat a nutrient-dense, plant-rich diet for better health and a better life. Simply put, a *Nutritarian* is a person who strives for more micronutrients—all the vitamins, minerals, and antioxidants essential for health—per calorie in their diet. Nutritarians seek to consume a broad array of micronutrients via their food choices because they understand that food has powerful disease-protecting and therapeutic effects.

The foods with the highest micronutrients per calorie are vegetables, and for optimal health and to combat disease, it is necessary to consume an adequate amount and variety of vegetables.

Nutritarians optimize their health potential by making food choices based on nutritional quality. This evaluation of the nutritional quality of what one eats should also include foods with documented effectiveness to improve immunity. A Nutritarian diet includes a full array of those foods with the most evidence for fighting cancer and prepares those foods in a manner that ensures and enhances the beneficial activity of their nutrients. While some foods may not have an elevated micronutrient density per calorie, they may still contain particularly effective phytonutrients that scientific studies have demonstrated to prevent cancer. Cooked mushrooms are a perfect example of a food that does not have high levels of vitamins and minerals compared with other foods; however, their unique compounds are powerfully protective against cancer, making them an integral part of a Nutritarian diet. As you can begin to see, this approach is all about understanding which foods best help support our bodies and minds and bringing those foods to our plates every single day.

Many people argue that there are various "best" diets depending on genetics and unique individuality. In other words, they believe that each individual has a different ideal diet that is based on specific genetic makeup. But does this accurately express the consensus of modern nutritional science compared with my argument that a Nutritarian diet is the most powerful life-extending diet style for the human species? This book examines with compassion, understanding, and insight all of the latest scientific research and dietary theories available to readers. It will help those interested in nutritional science, health, and longevity to clarify their thoughts about nutrition and be able to articulate the most critical and important elements to modify their own choices. In addition, this book provides readers with sound knowledge so they can help guide others who struggle with conflicting information, misinformation, and myths that make up much of the public debate about food and diet.

I have dedicated my life to investigating these issues, so I can be clear in stating that all the evidence points conclusively to the Nutritarian diet as being the single most powerful approach to sustain excellent health and a long life. That was the purpose of its design. Yet, this is not to say that minor adjustments should not be made for people with varying genetics, medical conditions, and food intolerances. These modifications, however, can be moderate and not impair the effectiveness of this dietary model as long as the adjustments to accommodate individual differences are consistent with the basic principles of the science regarding human longevity explained throughout this book.

For example, whether you are a vegan (you don't eat animal products) or for some biological reason you thrive better with some animal products in your diet, both approaches are governed by the same underlying biological and nutritional elements that shape a person's health destiny. The more the basic principle of eating a dietary portfolio with micronutrient excellence can be adhered to, while adjusting the food choices to fit individual constraints, the more predictable the slowing of the aging process and the protection against cancer.

It is only through an understanding of all the critical concepts of human nutrition that one can become enabled and empowered to most effectively make healthy food choices every day. This book is a guide to help you do just that.

YOU ARE THE CURE

I have been part of the medical community as a family physician for more than thirty years, and I can tell you that drugs and doctors cannot grant you excellent health or protection against disease and suffering. Almost every doctor knows this. The most effective health care is proper self-care. Reading this book, practicing the Nutritarian plan, and mastering its techniques will provide you with the best possible self-care: nutritional excellence.

The nutritional excellence I'm describing here can prevent and even reverse most medical problems within three to six months. This is a bold claim, but the facts, supported by scientific research and literature, show that most medical problems and medical tragedies we face in the modern world are the result of nutritional folly. Our standard American diet (SAD) has produced a sickly nation—one in which the majority of people are taking prescription drugs by the time they reach the age of 50. **Your body is made of the foods you eat, and when you eat the SAD, you get the diseases that most other Americans get.**

- Almost 40 percent of Americans die of heart attacks and strokes. *You don't have to be one of them.*

- Twenty-eight million Americans suffer from the crippling pain of osteoarthritis. *You don't have to be one of them.*

- Thirty-five million Americans suffer from chronic headaches. *You don't have to be one of them.*

- More than one hundred million Americans have diabetes or prediabetes. *You don't have to be one of them.*

- One-third of people older than 85 develop dementia. *You don't have to be one of them.*

- Thirty-eight percent of Americans are diagnosed with cancer. *You don't have to be one of them.*

You simply do not have to be sick.

We consider it normal to lose youthful vigor in our early 40s, carry 30 to 40 extra pounds, live with chronic illness in our late 40s and 50s, and live our last decades dependent on others. But all this should not be considered normal. These expectations are the result of a lifelong pattern of unhealthy living and misguided information. We should look forward to enjoying an active life well into our 90s. This seems like an outrageous expectation because most people spend a lifetime consuming an unhealthy diet. They don't make the connection that we are what we eat and ill health in the later years of our lives is the result of earlier poor choices.

I have cared for more than fifteen thousand patients; most of them first came to my office unhappy, sick, and overweight, having tried every dietary craze without success. After following the Nutritarian educational program for superior health and weight loss, they shed the weight they always dreamed of losing and kept it off easily. Most important, they were able to eventually discontinue their medications because they simply didn't need them anymore. When you learn and follow the Nutritarian program of eating, it is possible to

- Never have a heart attack or stroke

- Avoid dementia in later life

- Dramatically reduce your chance of getting cancer

- Prevent and heal digestive problems such as reflux, indigestion, constipation, and hemorrhoids

- Prevent and often resolve erectile dysfunction, high blood pressure, and other circulatory impairments

- Reverse and resolve autoimmune diseases such as psoriasis, lupus, and rheumatoid arthritis

- Prevent and reverse diabetes (type 2) and high cholesterol, eventually making drugs unnecessary

- Age more slowly, live longer, and maintain youthful vigor, intelligence, and productivity into your later years

Some people may be skeptical that I can make such radical claims, but these statements are supported by medical science as well as by thousands of clinical patient case histories. The reversal of diet-caused diseases occurs in a relatively short time and is easily observed by anyone who has a chronic disease who is following the Nutritarian program. However, don't make any decisions about starting the program yet. You first have to educate yourself by reading on. When you are ready to commit yourself to achieving superior health, I promise to make it almost impossible for you to fail.

THE *EAT TO LIVE* STORY

Eat to Live is the title of my bestselling book originally published in 2003 and revised in 2011. People often use that phrase to describe this micronutrient-rich eating style. For more than fifteen years now, and much to my surprise, I have received and continue to receive a huge volume of emails and letters from people expressing their gratitude and describing the miraculous changes that have occurred to their health, thanks to reading this book.

I wrote *Eat to Live* on the basis of what the scientific literature revealed to be the gold standard for health and longevity. I did not think it would be so popular, as I expected that the diet and nutritional advice in it would be too restrictive for the majority of people to consider its powerful message. But I wanted a book out there to represent this niche of dietary excellence and to make that knowledge public. Fortunately, over the years, mostly by word of mouth, millions of people have read *Eat to Live* and have gotten their health back. Eventually, it became a national best seller and was on the *New York Times* best-seller list for ninety weeks. I am fortunate to have been able to continue to write bestselling books—including *Super Immunity*, *The End of Diabetes*, *The End of Dieting*, and *The End of Heart*

Win the War on Disease Now

We can likely prevent more than 90 percent of all cancers and more than 95 percent of heart attacks and strokes with the advances in nutritional science discussed within these pages. The problem is that many Americans don't like the solution. Too many are still looking for a magic pill that can enable them to smoke and not get lung cancer or to consume ice cream, French fries, soda, and pizza and not get breast or prostate cancer. Life is not a fairy tale, and there is no magic in health care. The real world is unforgiving. Our bodies develop problems when they have been damaged by harmful influences.

But the good news is that we now know how to protect ourselves, and we have the capability to reverse disease and save lives. For those willing to open their minds and change their habits, lifelong excellent health is well within reach. And I am not talking about starvation diets or a horrible eating plan that can't be sustained. This Nutritarian diet approach is about putting the most flavorful, delicious, natural ingredients in the world in your body. It might take some time to get used to it, but I promise you that the benefits far outweigh any issues associated with changing what you eat.

Disease—that have had incredible life-changing effects on readers across the world.

Seeing the miraculous transformations brought about by my work enhanced my commitment and passion to making this knowledge as widely available as possible. It is too important to be ignored. Even though I have studied and utilized this dietary method as a medical therapy for more than three decades, seeing the large number of people whose lives have been transformed was thrilling. Thousands of people have lost dramatic amounts of weight—sometimes more than 100 pounds—without difficulty and never gained it back. More important, thousands have recovered from diseases such as diabetes, heart disease (high blood pressure, angina, cardio-myopathy), migraines, autoimmune conditions (psoriasis, lupus, rheumatoid arthritis, multiple sclerosis, Sjögren's syndrome, mixed connective tissue disease), fibromyalgia, allergies, asthma, acne, reflux esophagitis, kidney insufficiency, and many more. I have even seen dramatic improvements in chronic obstructive pulmonary disease (COPD), which most often occurs in smokers. I have also witnessed the usefulness of this approach in recovering from cancer. The results and success stories are astounding, and many will be featured throughout this book.

I have been blessed to have reached millions of people with this message of hope and healing. And I am grateful for the opportunity to make known the healing power of nutritional excellence and to motivate people to better care for their precious health.

Since writing *Eat to Live* I have written ten other books on health and nutrition. My goal in writing this book is to gather the most critical and important aspects of all my books into one volume and make these principles easier to understand and incorporate into your life. I also share the latest and most important new scientific discoveries and new lessons I've learned, including the obstacles people encounter when making dietary changes.

All the answers are here. You will see that people—sick or healthy, overweight or slim, young or old—can benefit from this dietary plan. It creates the environment necessary for our bodies to thrive and experience what amounts to a miracle in our modern world: a long life free of diseases such as heart disease, stroke, dementia, and even cancer.

THE BOOK YOU NEED TO READ

People often ask me, "If I could read just one book of yours that has the most comprehensive overview of your program and insights, which book should I read?" The answer is: this book. No dietary program has more scientific support regarding its principles and foods that slow aging, prevent disease, and prolong lifespan (with more than a thousand carefully vetted scientific references).

But there is another critical element here, and that is the therapeutic effects of the Nutritarian program to reverse chronic disease, including not just obesity, high blood pressure, high cholesterol, and obstructive coronary artery disease, but also asthma, migraines, and autoimmune diseases such as lupus, rheumatoid arthritis, scleroderma, psoriasis, and inflammatory bowel disease. It can even reverse early-stage cancers. I've also had beneficial results with people who have adopted this program while on chemotherapy and those with later-stage cancers. I tell people that there is always hope. At times, even I am astonished by what the body can do to heal itself when provided with the optimal nutritional environment to maximize self-healing.

How effective is this protocol therapeutically to reverse diseases? Over the past three decades my experience is that this eating style has been miraculously effective at enabling people to make complete recoveries from what are usually considered irreversible illnesses. A massive number of scientific studies corroborate my findings. My goal is for this book to provide all the information people need to adopt this program with confidence, to be able to modify it to their particular situations, and to feel secure that it will maximize both the quality and quantity of the days they live.

If you find yourself thinking that adopting a healthful diet means you have to give up the pleasure of eating, I am happy to reassure you that this is simply not true. Yes, it takes time to learn a new way to cook and prepare foods, and yes, it takes time for taste preferences to change and taste buds to become more sensitized to foods that have less salt and sugar. But if you stay with this plan, you will find that your taste buds and your sense of smell actually get stronger and adjust such that you *prefer* healthier foods. I am excited about the incredible menus and recipes you'll find in this book. You'll soon discover that not only will you like this way of eating—you will prefer it.

PREPARE FOR A NEW YOU

The information in these pages will change your life. Your decision to pursue superior health starts one of the most important journeys you can undertake. My life's work and expertise include a comprehensive and in-depth review of more than twenty thousand scientific studies on human nutrition over the past thirty years, and I have translated the science into recommendations that will help you alter your health destiny. Right now, I can state with certainty that this book and this conversation are where you begin your nutritional turnaround. I have seen the effects of this Nutritarian plan in action on many thousands of individuals with a wide range of diseases and health concerns. The bottom line: It works. Complete recovery from most chronic degenerative illnesses is possible because of the advances made in modern nutritional science.

Nothing illustrates the power of this way of eating more clearly than hearing from people who apply this knowledge and live it every day. Throughout this book you will find the real-life stories of people from across the country who have changed their lives by following this dietary plan. They are from different backgrounds, are different ages, and have different reasons for beginning this journey, but now they have one thing in common: excellent health.

When you read their success stories, you will gain a deeper understanding of the benefits that are possible when you make a commitment to your health and eat right. Nutritional excellence unleashes the miraculous human health potential hidden within all of us that is allowed to flourish when our bodies have an optimal exposure to nutrients—a fact that health authorities and those in the medical profession rarely consider.

The body is a miraculous self-healing machine when you supply it with an optimal nutritional environment. The information presented in this book provides the most effective way to create that environment for yourself. If you have high blood pressure, high cholesterol, diabetes, heart disease, indigestion, headaches, asthma, fatigue, body aches, or pain—or you want to prevent them—this plan is for you. You can reduce and eventually eliminate your need for prescription drugs. The Nutritarian diet can enable you to avoid angioplasty, bypass surgery, and other invasive medical procedures. By adopting this way of eating, you can make sure that you never have a heart attack, stroke, or dementia. These common events and diseases are not largely genetic, inevitable, or the consequence of aging; they are primarily the result of improper diet.

Many people are interested in this program because they want to lose weight. By adopting the Nutritarian plan, you will lose all the weight you want, even if diets have failed you in the past. I feel strongly that this is the most effective weight-loss plan ever, when you consider how many people keep the weight off. According to a study published in 2015, the nutritional program presented in this book is the most effective way to lose weight, especially if you have a lot of weight to lose. The subjects, followed for three years, lost and kept off more weight than the subjects of any other study that I could find.[1] Supporting these results is another 2015 study of more than 130,000 adults, which found that people who increased their intake of green leafy vegetables lost more weight over four years compared with people who didn't eat more of these healthy vegetables.[2] More and more, new medical studies are investigating and demonstrating that diets rich in high-nutrient plant foods have a suppressive effect on appetite and are most effective for long-term weight control.[3] Certainly, it is the healthiest way to lose weight.

THE EVIDENCE SUPPORTS

A Nutritarian diet demonstrates more long-term weight loss than any other diet.

A Nutritarian diet demonstrates more cholesterol-lowering than any other diet.[4]

A Nutritarian diet demonstrates more blood pressure–lowering than any other diet.[5]

A Nutritarian diet demonstrates more diabetes reversal than any other diet.[6]

A Nutritarian diet reduces hunger and food cravings more than any other diet.[7]

I have discovered—and reported in the scientific literature (the article is available online)[8]—that as people increase the micronutrient quality of their diets, their hunger lessens and their desire to overeat is curtailed. An important part of this book is the explanation of this scientific discovery, as it is an essential component that accounts for the long-term weight-loss success of this Nutritarian program.

There is no benefit if your weight loss is only temporary, and only a permanent commitment to healthful eating is effective for long-term weight control because it modifies and diminishes food cravings and feelings of hunger, enabling overweight individuals to be more comfortable eating fewer calories. Here you will learn about many natural foods that reduce appetite, slow the absorption of calories, and interfere with fat storage and fat-storage hormones. The foods that have powerful anticancer effects also support a favorable weight.

Many of my patients lose 20 pounds during the initial six weeks of the diet, and that is just the beginning. However, this book is nothing like your typical diet book because when the focus is on weight loss alone, the results are rarely permanent.

With the Nutritarian diet style, you don't have to count calories, measure portions, or weigh yourself regularly. You can eat as much food as you want, and over time you will become satisfied with fewer calories. This is an eating style that you will learn to enjoy forever.

This book presents you with logical, scientific information that explains the connection between diet and health. Let these facts change the way you think about food. Incorporate the information into your life by using the Nutritarian meal plans and great-tasting recipes included in Chapter 9.

As you adopt this diet style, you will find that you truly enjoy and prefer this way of eating more and more the longer you do it. This book will guide you through your transition as you step up to better health. If you need to lose weight, you will shed pounds naturally and miraculously, merely as a side effect of eating so healthfully.

This Nutritarian program works well for those people who gain an in-depth knowledge and understanding of the science and logic supporting it, and it takes time and effort to learn this body of knowledge. However, once you have learned and put into practice this information, you will become a nutritional expert. The key to achieving your ideal weight will be in your hands—and in your mind.

Weight loss is important, but it should not be your main goal. Rather, your primary objective should be to obtain long-lasting, excellent health. Achieving and maintaining an ideal weight is just part of the whole healthy new you. A person can be at a healthy weight and still not be healthy. **As a Nutritarian, reaching a healthy weight will be a pleasant and automatic by-product that occurs naturally on the road to maximizing your health.**

Applying the information in this book to your life will help you achieve long-term success. It will create new, healthy behaviors that will eventually become effortless. It is so highly effective that it enables you to never be concerned about your body weight again as you take control of your health destiny.

BE ONE OF THE FEW, NOT ONE OF THE MANY

I'm fortunate to say that millions of people have read my books, and many of them consider themselves Nutritarians. If you are "new" to the idea of eating nutrient-dense foods, you may not even know what nutrient density means and what phytonutrients are. I hate to call you "victims," but that is what we all have been. We are all products of our upbringings. Nearly all of us were brought up on the standard

American diet (SAD)—meats, fried foods, dairy products, white flour and lots of sugar, among other unhealthful things. Like classic victims, we grew to love the things that were killing us—in this case, the food.

In addition, for people who use food as emotional comfort (and you most likely know who you are), this has led to more than just cravings; it has led to an addiction. If you habitually turn to food for comfort, it may be hard for you to hear that you are a food addict, or you may be relieved to hear what sounds so right. It explains why you've yo-yo dieted, why you can't stop yourself from continuing the food battles. Every time you try to stop eating the way you do, you can't. This addiction has to be taken quite seriously, because food addiction, like any other addiction, can be dangerous and deadly. However, if you are reading this, it is not too late to change. I can show you how to get rid of the food addiction that is sabotaging your health.

I want to congratulate you for purchasing this book, because you are now ahead of the game. You are interested in the relationship between what you eat and your health. You've now successfully jumped one of the biggest hurdles. You may even have attempted to eat differently and more healthfully than people around you. The secret to achieving spectacular results is your willingness to discard old beliefs for new and wonderful realizations. I offer you this simple idea: If you adopt this Nutritarian program, you will achieve your ideal weight, slow down the clock of aging, and prevent and even reverse disease, all at the same time. While doing this, you will also discover a level of enjoyment in eating that you have never imagined. It all begins with a state of mind, and reading this book means that you are ready!

Your success depends on your willingness to learn. It is okay if you are doubtful; just continue being curious. The way to get the most out of this book is to read it thoroughly and completely. I recommend that you suspend decision-making at first. It is best if you wait until you have learned more so that you are in an educated position to decide whether you disagree with something and why.

Try not to let what you have learned in the past cloud your learning here; nutritional science has changed dramatically in recent years.

I designed this scientific, nutritional program after years of studying thousands of articles from the scientific literature, seeing the most effective nutritional interventions used by physicians around the world, and observing and testing these methods with patients and in scientific studies.

The preponderance of evidence points in the same direction: This program works, and it works best for those people who understand it best. Read, underline, practice the recommendations, and ask questions. Those who learn and understand this program typically find they achieve remarkable results.

BECOME A NUTRITARIAN

Go at your own pace. I advise you to master the material in each chapter before moving on to the next one. The goal of this program is to teach you to eat healthfully, enjoy it, and learn to prefer a healthy eating style instead of a disease-promoting one. The "Quick Summary" at the end of each chapter will help make sure you understand the main points.

As you learn more and eat this way, your taste buds will gradually realign themselves and become stronger and more discriminating. As you become healthier, you will lose your psychological dependence on unhealthy foods. One way of implementing the Nutritarian diet is to gradually increase the micronutrient content of your diet, thus allowing your taste buds to acclimate to the change. So, you have a choice. You can go cold turkey on certain foods and jump right in to the highest level of nutritional excellence, or you can make smaller changes first. As you increase your knowledge about and preference for healthy foods, you can choose to move forward—at your own pace.

Over a period of time, as you increase your intake of nutrient-dense foods and replace unhealthful foods with better choices, you will reset your internal taste preferences and hunger drive. Once this happens, you will be amazed at how easy it becomes to follow this way of eating and maintain your ideal weight forever, without dieting.

To help you see your progress, I have given foods Nutrient IQ Scores. These numbers are a way for you to quantify the nutritional content of various foods and will help you learn which foods you should be choosing to eat more of on a daily basis.

A DIFFERENT DIET WORLD

I know how frustrating it is to dive into a promising diet only to meet with failure. You blame yourself, which sets up a vicious cycle of guilt and punishment. This expresses itself through self-defeating behavior. Most people who have been through this cycle have given up on the idea of reaching their ideal weight. It seems impossible, and their failure in the past only reinforces this belief.

That will not be the case here. Mediocre expectations yield mediocre results, so you must, right now, raise your expectations. Give this program a true test and follow it as directed, and I am confident you will get results that are different from those you have had in the past.

Many people find their hope for the future, enthusiasm, and creativity blossom as their health improves. You may even become more impressed with the wonder of organic foods, which can be real things of beauty with almost magical gifts to our health. Think about the people who worked to have them come to life; the joy of growing clean, unpolluted, and nonpolluting foods; the protection of the soil quality and the clean water that contribute to these gifts we enjoy. Supporting the farming of organic plants also better supports our planet and our planet's people. I see food not as an expense, but more as an investment. Nourishing your body with the best-quality food is the best investment you can make for your life, your future, and your happiness.

At the beginning and end of each chapter and throughout this book, I include case histories of real people who each represent thousands of people with similar health transformations from having adopted a Nutritarian diet. These people and their stories are both motivating and instructional. They bring life to the science, and it is easy to see how enthusiastic and grateful these individuals are for having learned this message. These stories are important to the process of changing the way you eat.

I'm so excited to be on this extraordinary life-changing journey with you, so let's begin!

CHAPTER ONE

The Science of Longevity

Whatever your goals are for health—disease reversal, longevity, weight loss, or simply feeling stronger and more energetic—the information in the pages of this book will empower you to achieve and sustain the results you are searching for. If you are reading this book because your goal is to effectively lose weight and keep it off, you still have to eat for health and longevity, not just for an optimal weight. This Nutritarian program works because it isn't simply about short-term benefits; it is a comprehensive lifestyle change that will transform your health once and for all.

Lots of diet theories, weight-loss gimmicks, and fads have come and gone. But none of these diets should be continued over the long term unless they are lifespan-favorable—in other words, unless they promote overall well-being and longevity. It is not healthful or even really possible to cut out all carbs or drink just smoothies for the rest of your life. A diet program must be evaluated on the basis of its long-term benefits, not short-term weight reduction or any other short-term measurement.

There are lots of crazy diet programs, and many of them actually work in the short term by tricking the body's natural mechanisms.

Sandra Goodson McClanahan
Lost 125 Pounds and Got Rid of Her Diabetes

I was a real food addict. As a high school teacher, I found myself feeling drained physically and emotionally—and financially—every year by the school's candy drive. I used to eat all of my band fundraising candy in school, and then I'd have to pay for it myself with the money I earned giving piano lessons. At 5 feet, 2 inches tall and 252 pounds, and with a family history that included diabetes, heart disease, and cancer, I knew it was time to face the music: I had to break my addiction to junk food to save my life.

Born in 1961, I grew up in Florida surrounded by relatives who took pride in preparing delicious Southern dishes. I was always the chubby one, but active and outgoing, and the other kids rarely teased me about my weight. As I got older, my weight gain accelerated, and I developed health problems. After I became an adult and started having kids, I became morbidly obese and lived that way for years. At my heaviest, I weighed 252 pounds and had a body mass index (BMI) of 46. I suffered from migraines, heart palpitations, extreme sweating, exhaustion, high cholesterol, and high blood pressure. What scared me the most was the onset of "a ravenous thirst." This, I knew, was a flashing neon sign that signaled "Diabetes Ahead."

One day, in October 2015, I took a big swig of water in front of my class, but when I tried to speak, my mouth had gone cotton dry. It scared the heck out of me! I knew I had out-of-control diabetes, but I was too afraid to even find out about it. I went home that night and started researching "diabetic diet" online. That's where I discovered Dr. Fuhrman and his books *The End of Diabetes* and *Eat to Live*. As I was reading online reviews, my resolve almost faltered when the first comment noted that this way of eating emphasized vegetables (including lots of green veggies), plus fruits, nuts, and seeds. The only vegetables I ate were potatoes, mashed or French fried—but I knew I had to change that way of eating. In fact, I had to change everything and internalize Dr. Fuhrman's teachings so that the changes I made would be permanent.

I immersed myself in those teachings. I listened to his audiobooks night and day. I listened in the shower, while prepping food, while driving to work, even while I was falling asleep at night. In the beginning, I allowed myself to "cheat" occasionally—but found that this stalled my

weight loss and left me with constant cravings for foods common in the standard American diet. Determined to change my life, I took a leap of faith. In his books, Dr. Fuhrman promised that when you adopt the Nutritarian way of eating, your taste buds will change and you will no longer crave SAD foods, so I committed fully to a Nutritarian diet—and he was right! After six weeks, I didn't even think about cheat meals—and the weight started to come off again.

After three years of following the Nutritarian diet, I feel better and more energetic than I ever have before. I no longer have migraines, bloating, or stiffness in my joints. My blood pressure is textbook normal now. My cholesterol, calcium, blood sugar, protein—all perfect. I am filled with energy and a zest for life. I had to chase after a student down the sidewalk recently, and it felt good to run—even at age 56!

The Nutritarian diet has become my way of life—and I plan ahead to make sure I have healthy food choices available wherever I go: field trips, work conferences, even vacations. On a recent trip to Peru, I took cans of unsalted beans, plus premeasured bags of raw nuts, oatmeal, and more. I also requested fruits and veggies at restaurants. I came home lighter than when I left.

I recently celebrated a milestone: having lost half my body weight. I now weigh 126 pounds and fit into size 7 skinny jeans. I feel incredible! When my students see old pictures of me when I was heavy, they think that I've Photoshopped them! And I am so proud to share what I have learned and to be able to help others too.

The only problem is that over time, these programs will shorten your lifespan. If you do not maintain a diet program in the long term, the results will be only temporary, and you will regain the weight that you initially lost. Losing weight and then gaining it back again down the road is not only worthless; it is harmful.

When you lose weight and put all of it back on again, whether slowly or quickly, you can be regaining more visceral fat, the fat surrounding your organs, than you had to begin with. It is especially harmful to regain weight quickly because it will increase the amount of visceral fat in your body. Visceral fat poses far more significant health risks than subcutaneous fat. Weight regain also means you add more plaque in your blood vessels and coronary arteries, and buildup of that can lead to heart issues. In other words, yo-yoing your weight is not good for you, and it can jeopardize your health. These crash-and-burn diets are worse for you than maintaining a stable weight—even one that is overweight.

NEVER PUT ON WEIGHT RAPIDLY, AS IT PROMOTES ACCUMULATION OF VISCERAL FAT.

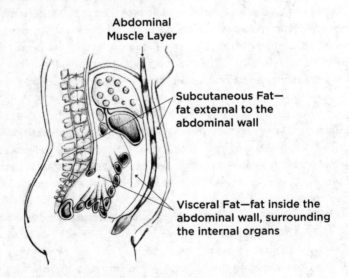

Abdominal
Muscle Layer

Subcutaneous Fat—
fat external to the
abdominal wall

Visceral Fat—fat inside the
abdominal wall, surrounding
the internal organs

A Nutritarian diet is different from these unsustainable weight-loss schemes. When you follow it, you are encouraged to make PERMANENT changes in your eating style to slow aging, prevent cancer and dementia, and extend your life. This program identifies the most lifespan-promoting foods and eating patterns, and then focuses on how to make this optimal eating style become a part of your lifestyle, complete with food you want to eat. With enough knowledge, skills, and practice, you will find that this diet can become a way of life that you enjoy immensely, and you can maintain the benefits forever. This program is designed using creative recipes that will teach you how to prepare these lifespan-promoting foods in a way that maintains their anticancer nutrients and also makes them taste great.

There are four principles of a Nutritarian diet. I share them all here and then unpack them one by one as we discuss them throughout the book.

1. The First Principle
   The only proven strategy for slowing aging and prolonging lifespan is moderate calorie restriction in the environment of micronutrient excellence.

2. The Second Principle
   A diet has to be hormonally favorable to enable maximal lifespan.

3. The Third Principle
   Optimal exposure to all macronutrients and micronutrients humans require is needed to maximize health and lifespan—this is called Comprehensive Nutrient Adequacy (CNA).

4. The Fourth Principle
   Synthetic chemicals, toxins, pathogenic bacteria, parasites, and other disease-causing substances should be avoided.

THE FIRST PRINCIPLE OF A NUTRITARIAN DIET

The only proven strategy for slowing aging and prolonging lifespan is moderate calorie restriction in the environment of micronutrient excellence.

Before we go any farther, I want you to read the sentence above carefully. Say it out loud a few times; memorize it. Write it down on an index card and tape it to your fridge if that will help. This statement forms the basis of a Nutritarian diet and is its first and most important principle.

If we don't establish a foundation of good health first, then it is pointless to focus on obtaining and maintaining a favorable weight. Though not everyone or even most people will embrace this overriding premise—that your health and longevity must be preserved through what you eat—it forms the foundation of this program and the reason for the development of the Nutritarian diet.

Food supplies us with *macronutrients* and *micronutrients*. There are four macronutrients—fats, carbohydrates, protein, and water. Micronutrients—vitamins, minerals, antioxidants, and phytochemicals (chemical compounds found in plants, also called phytonutrients)—do not contain calories. A Nutritarian diet strives for micronutrient excellence by including a wide variety of micronutrient-rich foods and avoiding foods that are not nutritious. Excess fat, excess carbohydrates, and excess protein all shorten lifespan. To put it another way: Consuming too many calories will shorten your life.

The higher the nutritional quality of your diet, the less you will desire to overconsume calories.

When you eat unhealthfully, it becomes almost impossible for you to not crave and consume excess calories. Most people do not consume enough micronutrients daily. Because their micronutrient needs aren't met, they can't control food cravings and overeating behavior.

What you put in your mouth turns into your living self. The food you eat makes you who you are, both physically and emotionally.

Everything you eat positively or negatively affects the structure and elasticity of your cell membranes, the tensile strength of your cartilage, the elasticity of your blood vessels, and the interior machinery of the cells in your body. Outside of unusual circumstances or very rare genetic defects, cells function normally when exposed to a diet with sufficient colorful, natural plant material. Conversely, they behave abnormally, and sometimes in a bizarre fashion, when these plant-derived nutrients are lacking.

My health equation, which represents the first principle of the Nutritarian diet, describes this important phenomenon of human existence with three letters:

$$H = N \, / \, C$$

Healthy Life Expectancy = Nutrients divided by Calories

This means that your healthy life expectancy is determined by the nutrients per calorie, or the *micronutrient density* of your diet. Or said in another way, the more low-nutrient calories you eat, and the more lacking your diet is in micronutrients, the more miserable your later years will be, and the shorter your life. The term "healthy life expectancy," or *healthspan*, means the period of life spent in good health, free from the chronic diseases and disabilities associated with aging. This formula reminds you about this critical foundational principle—that everything you put in your mouth counts toward determining your future health.

But while food may be the most important and overriding influence on your health, it is not the sole determinant of healthspan and lifespan. Exercise or activities; chronic stress; happiness; exposure to drugs and chemicals in our environment, air, and water; and genetic and other factors all play a role too.

Macronutrients are fats, carbohydrates, protein, and water.

Micronutrients are vitamins, minerals, and phytonutrients.

To apply this concept, it helps to consider your **micronutrient density exposure**, which is your lifetime exposure to micronutrients (per calorie) that you have consumed. The more empty calories you have eaten and continue to eat, the higher your health risk; conversely, consuming more phytonutrients reduces your health risks.

Nature is dependable and consistent in terms of biological causes determining outcomes. You don't get a free ride for smoking cigarettes for thirty years; it damages you and shortens your life. The sooner you quit, the better. It is the same with food. If you eat more nutritionally barren foods, such as white bread, oil, and sugar—which supply calories but no significant micronutrient load—you age faster, create disease, suffer needlessly, and shorten your life. Every bite of white bread, pita, bagel, candy, ice cream, cookie, pizza, and croissant enhances your suffering and takes time and quality off the end of your life—in proportion to the amount you consume. Empty calories come with a cost that you have to pay three times:

- First, you pay in the present by feeling ill and being overweight and sickly.

- Second, you pay in the future, with poor health and loss of brain function and memory as you age.

- Third, you pay again down the road with the years of life taken away from you.

The popular idea that our health is predetermined by genetics, putting good health out of our control and determined by access to medical care, is not only **not** supported by evidence, but it is absurd.

The foods we consume throughout our lives are the largest factor determining our health destiny.

The Nutritarian diet radically reduces your consumption of low-nutrient foods and radically increases your intake of high-nutrient foods. The program is not about counting calories. One of the benefits of eating foods with a high nutritional bang per caloric buck is that it leads to a lower caloric exposure naturally and less desire for excessive calories, so you don't have to count them. That's right. No calorie counting.

Adequate consumption of vitamins, minerals, and phytochemicals is essential for a healthy immune system and for empowering your body's detoxification and cellular repair mechanisms that protect you from cancer and other diseases. Not surprisingly, the foods that have a high nutrient density come straight from nature, meaning plant foods, mostly vegetables. Nutritional science in the past twenty years has demonstrated that plant foods, especially from colorful plants, contain a huge assortment of protective compounds, most of which are still being discovered and studied by scientists. We are learning that these compounds work in fascinating ways to

- Detoxify carcinogens

- Repair DNA damage

- Reduce free radical formation

- Facilitate the removal of toxins from the body

Only by eating an assortment of nutrient-rich, natural foods can we obtain the diversity of elements we need to protect ourselves from common diseases. By eating a diverse array of these colorful, nutrient-rich plants, we can make the human body resistant to cancer, even if we have a genetic predisposition or have been exposed to causative agents.

THE STANDARD AMERICAN DIET (SAD) IS DRAMATICALLY DEFICIENT IN MICRONUTRIENTS

The SAD is composed predominantly of low-nutrient foods. Animal products, including meat, eggs, dairy, and fish, are relatively low in vitamins and minerals per calorie compared with plants. However, the overriding deficiency in animal-based foods is the complete lack

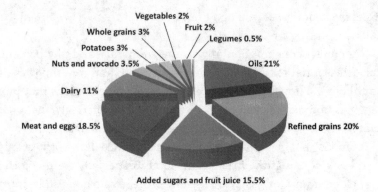

Vegetables 2%
Fruit 2%
Whole grains 3%
Legumes 0.5%
Potatoes 3%
Nuts and avocado 3.5%
Oils 21%
Dairy 11%
Meat and eggs 18.5%
Refined grains 20%
Added sugars and fruit juice 15.5%

Food Consumption Data: How Most People Eat

Source: US Department of Agriculture. Economic Research Service. Food Availability (per Capita) Data System. https://www.ers.usda.gov/data-products/food-availability -per-capita-data-system/food-availability-per-capita-data-system/. Last updated 26 Aug 2019.

of antioxidants and phytonutrients. Refined grains, oils, and sweeteners also lack these essential micronutrients.

We require a continuous exposure to phytonutrient-rich foods throughout life to maintain a normal-functioning immune system and to prevent inflammation. People cannot have disease-free lives if they are not regularly consuming these delicate nutrients contained in whole plants, particularly vegetables. Beans/legumes, nuts, and seeds and, to a lesser degree, whole grains are also rich in life-enhancing phytonutrients.

The excess poor-nutrient calories consumed by the vast majority of individuals lead to weight gain and disease, including cancer. A typical bagel is similar to a piece of chicken in that they both contain calories, and neither has a significant micronutrient load or phytonutrient content.

Shockingly, the SAD contains only about 2 percent of calories from vegetables—the food group most linked to a longer lifespan and protection from heart disease and cancer. For a diet to be considered healthful, at least 90 percent of its calories should come from vegetables, fruits, legumes, nuts, seeds, and intact whole grains. This eating style is consistent with that found in the Blue Zones around the world where rates of cancer and heart attacks are very low (see more on Blue Zones in Chapter 6). Like other primates, we are dependent on green vegetables and other colorful plants for health and longevity.

WHAT'S YOUR NUTRIENT IQ?

Prevailing food myths largely contribute to our overweight population and to poor health for many people. Lifestyle-related diseases are the most common cause of death, but according to a 2011 poll by *Consumer Reports on Health*, 90 percent of Americans believe that they eat a healthy diet. In addition, 43 percent of Americans polled reported that they drink at least one sugar-sweetened drink each day, 40 percent said that they eat "pretty much everything" that they want to eat, and 33 percent of overweight and obese individuals reported that they were at a healthy weight.

This highlights the nutritional misinformation that abounds in our society. Americans have not yet grasped the concept of nutrient density and its importance for health and longevity. Most Americans do not understand that whole plant foods are the best foods for their health. Instead, they are led to believe that processed foods labeled "low-fat" or "low-carb," artificially sweetened beverages, pasta, grilled chicken, and olive oil make up a healthful diet.

The nutrient density in your body's tissues is proportional to the nutrient density of your diet. To illustrate which foods have the highest nutrient-per-calorie density, about ten years ago I created the Aggregate Nutrient Density Index, or ANDI. The ANDI ranks the nutrient value of many common foods on the basis of how many nutrients they deliver to your body for each calorie consumed. Unlike food labels, which list only a few nutrients, ANDIs are based on thirty-four important nutritional parameters. The following nutrients were included in the evaluation:

FIBER, CALCIUM, IRON, MAGNESIUM, PHOSPHORUS, POTASSIUM, ZINC, COPPER, MANGANESE, SELENIUM, VITAMIN A, BETA-CAROTENE, ALPHA-CAROTENE, LYCOPENE, LUTEIN AND ZEAXANTHIN, VITAMIN E, VITAMIN C, THIAMIN, RIBOFLAVIN, NIACIN, PANTOTHENIC ACID, VITAMIN B6, FOLATE, VITAMIN B12, CHOLINE, VITAMIN K, PHYTOSTEROLS, GLUCOSINOLATES, ANGIOGENESIS INHIBITORS, ORGANOSULFIDES, AROMATASE INHIBITORS, RESISTANT STARCH, RESVERATROL, PLUS THE ORAC SCORE (ORAC STANDS FOR OXYGEN RADICAL ABSORBANCE CAPACITY AND IS A MEASURE OF THE ANTIOXIDANT OR RADICAL SCAVENGING CAPACITY OF A FOOD)

The ANDI ranks food on a scale of 1 to 1,000, with the most nutrient-dense cruciferous leafy green vegetables (such as kale, collards, and mustard greens) scoring 1,000. You can find a comprehensive list of ANDIs in my book *Nutritarian Handbook & ANDI Food Scoring Guide* (2010). The ANDI was used by Whole Foods Market for years and increased the company's sales of healthy, nutrient-rich produce because when consumers were able to recognize that certain foods were powerfully protective and had a high nutrient content, they were encouraged to buy more of them.

Recently, I developed an updated food-scoring system, **Nutrient IQ Scores**, that is based on portion sizes of foods eaten rather than calories, so it can be more practically utilized to track the quality of one's diet. (See full chart on pages 32 and 33.)

Not surprisingly, the foods that have a high Nutrient IQ Score are straight from nature—primarily vegetables, legumes, and fruits. Because phytochemicals are largely unnamed and unmeasured, these rankings may underestimate the healthful properties of colorful, natural plant foods, so the comparative nutrient density of many of these whole foods may be even higher than these scores indicate.

Even though attention should be placed on nutrient-rich foods, it is also important to achieve micronutrient diversity and to eat an adequate assortment of lower-ranked plant foods in order to obtain the full range of human dietary requirements. I need to stress that you do NOT have to score foods and keep track of points to eat a Nutritarian diet. This scoring system encourages you to make better food choices, but mostly it demonstrates the nutritional value of colorful plants compared with that of animal products, oils, and processed foods. It is a motivational tool to help you eat more healthfully; however, you do not have to keep track of your calories, and you do not have to keep track of your nutritional scores—you only have to understand how to lay out a healthful and nutritionally balanced menu, and then stick to it.

Clearly, some foods are more protective than others. The foods with the strongest evidence supporting their anticancer and longevity-promoting effects are green vegetables, particularly leafy green cruciferous vegetables, followed by the other **G-BOMBS**, which is an acronym for **G**reens, **B**eans, **O**nions, **M**ushrooms, **B**erries, and **S**eeds. These are the foods that have documented benefits to prevent cancer and other diseases, as discussed in detail in the next chapters.

THE THEORIES OF BIOLOGICAL AGING AS THE CAUSE OF ALL CHRONIC DISEASE

Aging is not biologically or genetically programed in advance, as has been thought by some people in the past.[1] Today, the leading theories of aging involve understanding the cumulative effects of multiple pathways of damage occurring concurrently.[2] This cumulative damage comes both from environmental influences and from programmed immune system decline.

As we are exposed to more waste products and toxins, we age faster, if we don't supply the micronutrients our bodies need to neutralize and remove them. The toxins and waste products build up, causing slow damage. This damage includes cross-linking of tissue proteins, DNA damage (including methylation defects), an increase in reactive oxygen species (ROS) including superoxide radicals, and the glycation of proteins and buildup of advanced glycation end products (AGEs)—all of which play a role in accelerating aging.

Nutrient IQ Scores Based on Typical Serving Sizes

	Serving Size	Nutrient IQ Score*
Kale, cooked	1 cup	112
Collards, cooked	1 cup	112
Mustard greens, cooked	1 cup	112
Turnip greens, cooked	1 cup	112
Watercress, cooked	1 cup	112
Arugula, cooked	1 cup	112
Swiss chard, cooked	1 cup	112
Bok choy	1 cup	90
Broccoli	1 cup	90
Broccoli rabe	1 cup	90
Brussels sprouts	1 cup	90
Cauliflower	1 cup	90
Cabbage	1 cup	90
Kohlrabi	1 cup	90
Radishes	1 cup	90
Turnips	1 cup	90
Endive, cooked	1 cup	82
Escarole, cooked	1 cup	82
Spinach, cooked	1 cup	82
Arugula, raw	1 cup	79
Collard greens, raw	1 cup	79
Kale, raw	1 cup	79
Mustard greens, raw	1 cup	79
Turnip greens, raw	1 cup	79
Watercress, raw	1 cup	79
Asparagus	1 cup	64
Artichoke	1 item	64
Cucumber	1 cup	64
Endive, raw	2 cups	64
Escarole, raw	2 cups	64
Fennel	1 cup	64
Green beans	1 cup	64
Green pepper	1 cup	64
Okra	1 cup	64
Romaine or other lettuces	2 cups	64
Snap peas	1 cup	64
Snow peas	1 cup	64
Spinach, raw	2 cups	64
Zucchini	1 cup	64
Bean sprouts	1 cup	60
Eggplant	1 cup	60
Mushrooms, cooked	¼ cup	60
Onions, raw	¼ cup	60
Radicchio	1 cup	60
Red pepper	1 cup	60
Tomatoes	1 cup	60
Tomato sauce or pasta sauce, low sodium	1 cup	60
Yellow squash	1 cup	60
Beans, lentils, or split peas, cooked	½ cup	52
Edamame	½ cup	52
Lima beans, cooked	½ cup	52
Bean pasta, cooked	1 cup	52
Tempeh	1 cup	45
Beets	1 cup	45
Carrots	1 cup	45
Corn	1 cup	45
Green peas	1 cup	45
Parsnips	1 cup	45
Pumpkin	1 cup	45
Rutabaga	1 cup	45
Sweet potato	1 cup	45
Winter squash (butternut, acorn, spaghetti)	1 cup	45
Blackberries	½ cup	45
Blueberries	½ cup	45
Raspberries	½ cup	45
Strawberries	½ cup	45
Cranberries, fresh	½ cup	45
Gooseberries	½ cup	45
Goji berries	¼ cup	45
Walnuts	¼ cup	45
Cherries	⅔ cup	41
Chia, flax, or hemp seeds	2 Tbs	41
Pomegranate, juice or kernels	¼ cup	37
Sunflower, pumpkin, or sesame seeds	¼ cup	34
Onions, cooked	¼ cup	30
Cashews, almonds, pistachios, pecans, hazelnuts, or Brazil nuts	¼ cup	26
Pine nuts	¼ cup	26
Barley, cooked	1 cup	26
Buckwheat, cooked	1 cup	26
Farro, cooked	1 cup	26
Steel cut oats, cooked	1 cup	26
Teff, cooked	1 cup	26
Wild rice, cooked	1 cup	26
Quinoa, cooked	1 cup	26
Turmeric, ground	1 tsp	25
Avocado	¼ cup	23
Vegetable juice, 100% vegetable	1 cup	22
Apricots, fresh	2 items	19
Figs, fresh	2 items	19
Grapefruit	1 cup	19
Grapes	1 cup	19
Kiwi	2 items	19
Kumquats	4 items	19
Mango	1 cup	19
Melons (watermelon, honeydew, cantaloupe)	1 cup	19
Orange	1 item	19
Papaya	1 cup	19
Peaches or nectarines	1 item	19
Pears	1 item	19

	Serving Size	Nutrient IQ Score*		Serving Size	Nutrient IQ Score*
Pineapple	1 cup	19	Cold cereals made from 100% whole grains, <15 g sugar/serving	1 cup	5
Plums	2 items	19	Plain yogurt, fat-free or low-fat, no added sugar	6 oz	5
Tangerines or clementines	2 items	19	Milk, skim or 1%	1 cup	5
Old-fashioned oats, cooked	1 cup	19	Eggs	1 item	4
Tofu	1 cup	15	Higher-mercury seafood (tuna, shark, swordfish, grouper, red snapper, mahi mahi, halibut, orange roughy)	4 oz	4
Soy, hemp, or almond milk, unsweetened	1 cup	15			
Green tea	1 cup	15			
Cashew or almond butter	1 Tbs	13	Couscous, cooked	1 cup	4
Sunflower butter	1 Tbs	13	Quick or instant oats, cooked	1 cup	4
Sesame seed butter (tahini)	1 Tbs	13	Black tea	1 cup	4
White potatoes	1 cup	12	Poultry	4 oz	3
Celery	½ cup	11	Plain yogurt, full-fat, no added sugar	6 oz	3
Iceberg lettuce	2 cups	11	Milk, 2% or whole	1 cup	3
Apple	1 item	11	Cheese	2 oz	2
Banana	1 item	11	Coffee	1 cup	2
Peanuts	¼ cup	11	Whole wheat bread products, not 100% whole grain		0
Cold cereals, made from whole grains or nuts, no added sugar	1 cup	11	White bread products		0
Basil, fresh	2 Tbs	10	White pasta		0
Dill, fresh	2 Tbs	10	White rice		0
Parsley, fresh	2 Tbs	10	Cold cereals, not 100% whole grain		0
Cilantro, fresh	2 Tbs	10	Beef, pork, lamb		0
Gingerroot, fresh	1 Tbs	10	Hotdogs, sausage		0
Cinnamon, ground	1 tsp	10	Cold cuts		0
Orange juice, 100% fruit	1 cup	7	Dried meat products (jerky)		0
Salmon, wild	4 oz	7	Yogurt with added sugar		0
Apricots, dried	¼ cup	7	Frozen yogurt		0
Coconut, dried	¼ cup	7	Ice cream		0
Dates	¼ cup	7	Pizza		0
Figs, dried	¼ cup	7	French fries		0
Raisins	¼ cup	7	Potato, corn, or other chips		0
100% whole grain bread, wraps, or pita	1 item	7	Crackers		0
Brown rice, cooked	1 cup	7	Milk chocolate or dark chocolate <65% cocoa		0
Pasta, whole wheat, cooked	1 cup	7	Cookies, pie, cake		0
Cocoa powder, unsweetened	2 Tbs	7	Apple or grape juice		0
Dark chocolate, 85–100% cocoa	1.5 oz	7	Carbonated drinks		0
Dark chocolate, 65–80% cocoa	1.5 oz	6	Energy drinks		0
Peanut butter	1 Tbs	6			
Salmon, farmed	4 oz	5			
Lower-mercury seafood (scallops, clams, mussels, oysters, shrimp, lobster, tilapia, mackerel, cod, flounder, haddock, crawfish, catfish, black sea bass, hake, sole, squid, sardines)	4 oz	5			

Note: Unless indicated, vegetable scores are for either raw or cooked vegetables.

*The Nutrient IQ Scores are for adult men. Women should multiply scores by 1.2, and children younger than 12 should multiply scores by 1.75.

Scientific research suggests that biological aging is a treatable condition. Some amount of accumulated damage is inevitable with chronological age, but not most of it. If we work to prevent the processes that drive biological aging, we reduce our risk of developing diabetes, Alzheimer's disease, cancer, and heart disease—that is, we reduce the risk of all the diseases that rob Americans of their healthspan.

Theoretically, if we could specifically prevent cancer, then many of the people who would suffer from cancer still would suffer from heart disease or Alzheimer's. But since the same biological process is the underlying cause of all of these diseases, preventing the cause also prevents all of these diseases, adding healthy years to our lives. If our aim is to target and inhibit biological aging, we have to address this collection of interrelated processes, which include

- Chronic inflammation

- Damage to DNA and proteins (including oxidative damage)

- Cellular senescence (cumulative DNA damage leading to dysfunction and loss of ability to divide)

- Reduced capacity for tissue repair (stem cell dysfunction)[3]

A Nutritarian diet is designed to address the most important causes of inflammation, which drives obesity, heart disease, diabetes, cancer, dementia, and most other chronic diseases. Chronic inflammation and oxidative damage to DNA and proteins are major mechanisms of biological aging, and they are preventable with the right diet and lifestyle.[4] **Caloric restriction with adequate micronutrient intake is the most consistent intervention that slows or hinders these hallmarks of aging, therefore extending lifespan.** We can trace the process of biological aging to cellular signals that respond to nutrient and calorie availability.

The overall message is that excess calories and low-nutrient foods turn off longevity signals. Phytochemicals, caloric restriction, and exercise turn on these signals. Importantly, these signals are common to all of the age-related chronic diseases.

SLOWING AND MEASURING BIOLOGICAL AGING

Leukocyte telomere length is one indicator of biological aging. Telomeres are portions of DNA at the ends of our chromosomes that protect our genetic material and are shortened with each cell division. Telomere shortening is essentially cellular aging. Excess production of free radicals—molecules that can cause damage to our cells—from a poor diet creates inflammation that is associated with telomere shortening.[5] Excess free radicals cause oxidative stress—chronically creating premature aging and disease. A healthful diet enables us to control and contain free radical production and has been shown to protect the length of our telomeres, therefore decreasing our biological age.[6] Slower biological aging is associated with healthy lifestyle variables, such as lower body mass index (BMI) and higher circulating carotenoids (from colorful plant foods and exercise), whereas having more body fat ages you more quickly and increases mortality.[7]

One of the most important and overlooked factors that can effectively slow aging and enhance lifespan is reducing our metabolic rate with moderate calorie restriction. We see in the media and on the internet promises about supplements and diets that can speed metabolism. So can the right food, supplement, or breakfast boost our metabolism for the rest of the day, aiding weight loss? Not really. No decent evidence backs the theory that taking agents to enhance metabolism will be effective for long-term weight reduction. With lots of misinformation floating around advising methods to enhance metabolism, it's no surprise that people equate a higher metabolism with good health, when the opposite is in fact true.

SLOW METABOLISM IS LINKED TO LONGEVITY

When we use the word "metabolism," we are usually referring to *resting metabolic rate*, which is the amount of energy (calories) that the body requires per day for its basic functions at rest. Most people believe that it is desirable to raise their metabolism because they will burn more calories and consequently lose weight. Having a slightly lower resting metabolic rate is thought to predispose some individuals to weight gain, especially in our obesogenic culture— that is, a food environment that leads to obesity.[8] However, there are unfavorable consequences to running your body at faster than normal speed, and raising your metabolism is not the key to weight loss.

The chemical reactions of normal everyday functioning produce by-products. In particular, cellular energy production produces reactive oxygen species (ROS) as a by-product, which can damage DNA, proteins, and lipids. Although we have natural antioxidant defenses, oxidative damage can still occur, especially if we don't take in adequate antioxidants from our diet.[9] Oxidative damage accelerates aging.[10] In other words, the rate of living slowly wears out our cellular machinery, so if we function at a faster rate—that is, with a faster metabolism—the body will "wear out" more quickly. In animals, energy expenditure is indeed inversely related to lifespan.[11] A faster metabolic rate means faster energy turnover and greater production of free radicals, which leads to increased oxidative damage and more potential for disease.

Human studies measuring resting metabolic rate directly come to the same conclusion. In one, metabolic rate was measured by two different methods at the start of the study. Subjects were followed for eleven to fifteen years, and deaths from natural causes were recorded. Researchers observed that for each 100-calorie increase in twenty-four-hour resting metabolic rate, the risk of premature death increased by 25 to 29 percent.[12]

Resting metabolic rate is partially genetically determined, but we can control our calorie intake.[13] Caloric restriction and mild negative energy balance have been shown to reduce resting metabolic rate, and in contrast, overeating increases resting metabolic rate.[14] Furthermore, caloric restriction has been consistently shown to prolong maximal lifespan by up to 60 percent in animals.[15]

Keep in mind that although exercise raises total calorie expenditure, it does not raise the body's basal metabolism, that is, the number of calories needed to meet your energy expenditure at rest.

Exercise is the only safe way to "raise metabolism" because it activates the peripheral tissues to utilize more calories and increases muscle mass, which in turn increases total calorie expenditure.[16] It has been shown that exercise promotes longevity.[17]

The goal is to eat so healthfully that your desire to overeat is reduced, as well as your metabolism a bit. You will comfortably desire less food, without getting too thin. My Nutritarian recommendations actually make you feel more satisfied with less food and give you the ability to enjoy food more without overeating.

Studies support what I have said here: Having a fast metabolism does not mean that you are healthier—in fact, it means that you are aging more quickly. Instead of trying to increase your metabolism with thermogenic stimulants, such as green tea, caffeine, herbs, and medicines to aid weight loss, try to slow your metabolism and become slim with a lower-calorie, high-nutrient diet for a longer, healthier life.

THREE FACTORS MOST POWERFULLY SLOW AGING

The trifecta to slow aging is a *Nutritarian diet*, which is rich in phytochemicals, along with *moderate calorie restriction* and *exercise*. These three factors slow all the processes associated with biological aging. When your body fat is lower and you do not overeat, your metabolism slows a bit, and as we have just seen, a slower metabolism is one of the components of aging more slowly.[18]

Judging from the current trends of popular diets (paleo, ketogenic, and others), most Americans are inappropriately trying to *raise* their metabolism so they can eat more food and not gain weight. The secret to the real fountain of youth, however, is to moderately lower your metabolism so you can comfortably eat less food and not get too thin. When we eat somewhat less and stay slim, our body temperature lowers, our respiratory quotient (calories burned via breathing) decreases, and even our thyroid slows. Consequently, the body does not need as many calories and does not become too thin with a moderately lower caloric intake.

Together, these three metabolic modulations (moderately lower thyroid hormone production, respiratory quotient, and body temperature) effectively slow the aging process in all body systems. In particular, the mild reduction in thyroid function is a well-studied but otherwise unknown healthspan and lifespan asset.[19] For example, a recent study reported that low-normal thyroid function was associated with 3.5 years of extra life and a decreased incidence of cardiovascular disease.[20] While it is well-known that abnormally high thyroid activity or giving a bit too much thyroid hormone to treat low thyroid function increases the risk of irregular heartbeat, it is not generally known that higher thyroid activity, with blood results in the high range of normal, significantly increases heart attacks and sudden cardiac death even in people not taking any thyroid medication.[21] Furthermore, a study presented at a 2018 American Heart Association scientific session confirmed earlier studies showing a 40 percent increased risk of atrial fibrillation with higher thyroid activity, even within the normal range.[22]

Since people with atrial fibrillation are five times as likely to have a stroke, one could accurately say that eating less and eating right decreases the risk of stroke by lowering the risk of atrial fibrillation, as it lowers the metabolic rate and slows the thyroid a bit. A faster metabolism does not just age you faster, it also increases your risk of stroke, heart attack, cancer, and overall mortality.[23]

By balancing the mild drop in metabolism that naturally occurs as we age with body building and osteoporosis-preventing exercises, we can maintain muscle mass and muscle strength as we grow older. The art of maximally pushing the envelope of healthspan and lifespan involves maintaining a slim body with lower body fat throughout life, while striving to maintain sufficient muscle and bone strength as we age. Exercise is the only safe way to enhance the utilization of calories by the body as it does not accelerate aging (unless done to excess).

AMERICANS MISS OUT ON AT LEAST
TEN YEARS OF HEALTHY LIFE

The life expectancy in the United States is approximately 81 for women and 76 for men;[24] however, healthy life expectancy is estimated to be 70 and 67, respectively.[25] This means that, on average, Americans spend the last ten years of their lives in a state of poor health in which they are unable to live a full, quality life.

Blue Zones are areas in the world that have produced cultures whose members live the longest. The diets in these regions all share similar traits: They are high in plant-based foods and extremely low in animal products. As a result, in these locations people eat more home-grown vegetables and legumes and generally have longer lifespans. Typically, they have a higher percentage of centenarians (individuals who live past age 100). It is important to note that the dietary patterns in Blue Zones were not methodically designed on the basis of a comprehensive review of more than twenty-five thousand nutritional studies, as is the case here with the Nutritarian diet. Instead, they represent the availability of foods in that particular area. The diets in the Blue Zones are life-enhancing for sure, but they are not optimal. The reason is because they have not been designed to be as nutritionally complete or micronutrient diverse as a Nutritarian diet.

Comparing diets from various regions and evaluating their health-promoting effects is useful, but a Nutritarian diet is different in that it takes the dietary practices from the Blue Zones a step farther—it utilizes the important lessons learned from the Blue Zones in conjunction with the most important recent scientific discoveries on aging, lifespan, and the power of particular foods and food ingredients to protect against cancer and later-life cellular senescence (deterioration) that leads to premature aging and chronic disease. A Nutritarian approach improves upon the Blue Zone diets, making it even more lifespan-favorable.

A good example is one of my longtime patients, John Pawlikowski. John came to me in 1994 with triple-vessel coronary artery disease and was taking multiple medications to control his blood pressure. John's cardiologist had just recommended a stent placement at age 72. I persuaded John not to have the stent placed and instead to follow the Nutritarian program. He did so and reversed his heart condition, normalized his stress test, and resolved his high blood pressure and cholesterol, thus getting off all his medications. He is alive today and thriving without the need for these medications (or cardiologists) at age 98.

A more well-known example is Scott Nearing, who wrote with his wife, Helen, the 1954 book *Living the Good Life: How to Live Sanely and Simply in a Troubled World*. Scott and Helen lived on their homestead in Maine, growing almost all their own food. He celebrated his 100th birthday in 1983 and gracefully passed away soon after. Helen lived into her 90s and passed away after a car accident.

Taking advantage of modern nutritional science, which is not available to Blue Zones, enables the potential for an extended healthspan and lifespan. Not only can we design a more optimal menu, but in the wealthier areas of the world, because of modern transportation and refrigeration, we have all-year access to fresh vegetables, leafy greens, sprouts, frozen berries, nuts, seeds, onions, and fresh and dried mushrooms. Simply put, we have a range of health-promoting, anticancer foods that was not available to our ancestors or in most Blue Zones today. Most importantly, people in the Blue Zones eat more vegetables but do not generally eat a sufficient amount of green cruciferous vegetables, which is the most powerful longevity-promoting food.[26]

RECENT ADVANCES IN NUTRITIONAL SCIENCE HELP US TO LIVE LONGER

We are in an era of nutritional breakthroughs that give us a unique opportunity to be super healthy; or, we can eat the worst diet imaginable. It is up to you.

Robust evidence now supports the fact that a wide complement of carotenoids is strongly linked to longevity and chronic disease prevention.[27] More than six hundred carotenoids are synthesized by

April Bromiley
Lost 40 Pounds (from 155 to 115) and Resolved Multiple Sclerosis

After a lifetime of eating what I considered a "healthy" standard American diet, I was faced with a sudden—and devastating—health crisis at the age of 48: half my tongue became numb, and the right side of my body followed; I lost control of my bladder; and I slurred like a drunk when I spoke. I dragged my right leg and had a hard time gripping a pen. It was astonishing to lose control of my body so suddenly.

After an MRI and a diagnosis of multiple sclerosis, I was shocked and could not imagine that this was possible. But I remembered my childhood diet, which included few vegetables and was heavy on animal products. And this was made even worse because I had lived next to the boardwalk, which meant lots of saltwater taffy and chocolate molasses paddle pops.

I was faced with a treatment plan that included daily injections of immune-suppressing drugs to shrink three large MS-caused lesions in my brain, so I sought desperately for a safer alternative. On the advice of a friend, I visited an acupuncturist, who recommended a plant-based nutritional approach and lent me a copy of Dr. Fuhrman's *Eat to Live*, which I read cover to cover.

Continued on next page

Armed with new knowledge about the healing power of a high-nutrient, plant-based diet, I made an appointment with the head of the multiple sclerosis division at Philadelphia's Thomas Jefferson University Hospital. I was searching for encouragement and hope, but what I was told was terrifying. I mentioned Dr. Fuhrman's *Eat to Live* protocols and asked about using nutritional methods to treat multiple sclerosis, but the doctor informed me that diet had no effect on MS. He said the lesions in my brain were permanent, and only injections could slow their growth. I was unwilling to give up and submit to a treatment plan that would suppress my immune system and increase my risk of cancer, so I stood my ground. I told him that I wanted to try the Nutritarian diet first and, if in six months my lesions were the same or worse, I would submit to injections. He was not happy with me.

My first order of business was to clear my home of "bad" foods, and with my husband's support, we gathered our favorite treats into a laundry basket and gave them away. I actually sobbed as I handed over that basket. But after just six weeks of eating a Nutritarian diet, I saw dramatic improvements. I was driving in my car, and I started to feel the other side of my tongue! I could wiggle it around. I was joyous; this meant that my body was repairing itself and not ripping me apart.

When I went for a follow-up MRI a few months later, I was told that my three brain lesions had shrunk. But the doctor shrugged it off, saying that I had just been "especially inflamed" during the original MRI. I was confident that Dr. Fuhrman's nutritional protocols were working, so I thanked the doctor—and never went back.

I never looked back, either, and got totally well. Five years ago, we moved to Boquete, Panama, and I am thriving without MS and following a Nutritarian diet. I joke that I use more local fruits and vegetables in my cooking than my Central American neighbors—I incorporate the produce into smoothies, salads, and other dishes. So many of our Panamanian friends ask me for the recipes that I learned from Dr. Fuhrman.

I'll never go back to my old eating habits. I remember how it felt to wet my pants in public and feel my body rip itself apart from the inside out. Thanks to Dr. Fuhrman and this way of eating, I am lean, strong, and healthy. I am at a perfect weight, and I have more energy and vitality than I ever did. I feel joy!

plants and act as antioxidant pigments. Carotenoids are found in green vegetables; tomatoes; and yellow, red, and orange fruits and vegetables such as carrots, sweet potatoes, papaya, and red peppers. Typical American consumption of these foods is quite low: 0.14 cups per day of green vegetables and 0.4 cups per day of all red, yellow, and orange vegetables combined.

We also know of newly discovered substances with identified age-delaying functions, such as the mushroom-derived antioxidant ergothioneine and the bacterial metabolites pyrroloquinoline quinone (PQQ) and queuine. PQQ is synthesized by soil bacteria and is present in vegetables, and queuine is produced by favorable gut bacteria. These substances protect against the degradation of cellular protein transcription and of mitochondria.

A Nutritarian diet with its emphasis on high-nutrient plant foods is carefully designed to include the full symphony of nutritional instruments and to assure, through the conservative use of supplements, that no nutrition insufficiencies exist. A Nutritarian diet checks off every box that enhances playspan, healthspan, and lifespan, and in doing so also becomes therapeutically effective to reverse disease.

CHAPTER ONE: QUICK SUMMARY

Advances in nutritional science have given us an unprecedented opportunity in human history to live longer, better, and even happier than ever before. Choosing an eating style by focusing only on weight loss leaves you open to fad diets, gimmicks, and questionable eating habits. So you'll need to shift your focus from eating solely for weight loss to eating for health and longevity. This means finding a diet that is sustainable, emphasizes whole plant foods, is hormonally favorable, offers the full portfolio of anticancer superfoods, and maximizes the number of micronutrients per calorie. In other words: the Nutritarian diet.

THE FIRST PRINCIPLE OF THE NUTRITARIAN DIET

The only proven strategy for slowing aging and prolonging lifespan is moderate calorie restriction in an environment of micronutrient excellence. All animals (including humans) can age more slowly and achieve their maximal lifespans when they eat a diet that avoids the overconsumption of macronutrients and supplies adequate micronutrient exposure. A Nutritarian diet includes a wide variety of micronutrient-rich foods and avoids foods that are not nutritious. Excess fat, excess carbohydrates, and excess protein shorten lifespan.

Macronutrients (Calories)
Fat
Carbohydrates
Protein

Micronutrients (No Calories)
Vitamins
Minerals
Antioxidants
Phytochemicals

THE HIGHER THE NUTRITIONAL QUALITY OF YOUR DIET, THE LESS YOU WILL DESIRE TO OVERCONSUME CALORIES

When people's micronutrient needs are not met, they are over-whelmed by food cravings and the desire to overeat. These deficiencies also leave people susceptible to critical diseases and serious medical conditions.

YOU ARE (LITERALLY) WHAT YOU EAT

What you put in your mouth makes you who you are, both physically and emotionally. The structure of the cell membranes, their elasticity, the tensile strength of your cartilage, the elasticity of your blood vessels, and the interior machinery of the cell—everything you put in your mouth affects these things and counts toward determining your future health. Other factors that play a part include exercise, sleep, stress, happiness, and exposure to drugs and chemicals in our environment, air, and water.

MEMORIZE THIS HEALTH EQUATION: H = N / C

Your healthy life expectancy (H) is determined by the nutrients (N) per calorie (C), or the micronutrient density of your diet. The term "healthy life expectancy" (or healthspan) is a measurement of how healthy you are when you are older and combines this with how long you live. Your lifetime exposure to micronutrients (per calorie) is a critical factor here. The more empty calories you consume, the higher your risk of developing health problems, lowered brain function, and loss of memory; while high phytonutrient consumption is associated with lower risk across the board.

YOUR HEALTH IS IN YOUR HANDS

As tempting as it may be to blame our current health challenges on fate, genetics, or cellular damage from poor eating habits we learned from our parents, the truth is that diet is the largest factor that determines our health destiny. By eating an assortment of colorful, nutrient-rich plants, we can make our bodies resistant to cancer, diabetes, heart disease, autoimmune disease, and other chronic ailments—even if we have a genetic predisposition or have been exposed to causative agents.

ANIMAL PRODUCTS, OIL, FLOUR, AND SUGAR = SAD

The standard American diet is a dangerous, low-nutrient eating style that promotes excess weight, inflammation, cancer, and other chronic diseases. What makes the SAD particularly dangerous is that only about 2 percent of calories come from vegetables. For a diet to be considered healthful (consistent with the Blue Zones around the world, where cancer and heart attack rates are very low), at least 90 percent of its calories should come from vegetables, fruit, legumes, nuts, seeds, and intact whole grains.

WHAT'S YOUR NUTRIENT IQ?

Nutrient IQ scores illustrate which foods have the highest nutrient-per-calorie density. This scoring system makes it easy to track the quality of your diet and to encourage you to eat a wide variety of colorful, whole-plant foods, including the **G-BOMBS** (Greens, Beans, Onions, Mushrooms, Berries, and Seeds).

AGING AND CHRONIC DISEASE

Calorie restriction with adequate micronutrients is the most consistent intervention that reduces the risk of diabetes, Alzheimer's disease, cancer, and heart disease, therefore extending lifespan. Conversely, having more body fat, insulin resistance, and triglycerides ages us faster and increases mortality from multiple causes. You'll discover the real fountain of youth by following a phytochemical-rich Nutritarian diet, practicing moderate calorie restriction, and getting some exercise. These three actions will allow you to moderately lower your metabolism so you can eat less food and not get too thin.

SLOW METABOLISM IS LINKED TO LONGEVITY

Resting metabolic rate is partially genetically determined, but we can control our calorie intake. Calorie restriction and mild negative energy balance have been shown to reduce resting metabolic rate. In contrast, overeating increases resting metabolic rate. When we eat less and stay slim, our body temperature lowers, our respiratory quotient decreases, and even our thyroid slows, so the body does not need as many calories, and we don't get too thin and lose muscle.

Most people believe that having a "slow" metabolism will cause them to gain weight and that "speeding up" their metabolism will help them burn more calories and lose weight. However, a faster metabolism means we age faster and "wear out" more quickly. A faster metabolic rate increases your risk of cancer and overall mortality.

HEALTHY LIFE EXPECTANCY: SAD EATERS, BLUE ZONES, AND NUTRITARIANS

On average, Americans spend the last ten years of their lives in a state of poor health. In Blue Zones around the world, we generally see longer lifespans and a higher percentage of centenarians than in the United States. But while this eating style is beneficial when compared with the SAD, it is still not optimal. A Nutritarian diet takes the Blue Zone diet a step farther, utilizing the power of superior nutrition to prevent cancer, heart disease, and other chronic illnesses—making it even more lifespan-favorable.

Your Hormones and Your Health

Hormones—particularly insulin, insulin-like growth factor-1 (IGF-1), and estrogen—are major factors in determining the rate at which you age, modulating your longevity and your risk of developing cancer, particularly breast and prostate cancers. The good news is that you can achieve favorable levels of these self-produced hormones by following a Nutritarian diet.

THE SECOND PRINCIPLE OF A NUTRITARIAN DIET

A diet has to be hormonally favorable to enable maximal lifespan.

A Nutritarian diet is designed to be hormonally favorable because eating adequate micronutrients is not sufficient for optimal health if your diet is increasing hormones to levels that are detrimental to your long-term health. Excessive production of and exposure to certain hormones accelerates aging and promotes cancer development and spread.

David Palk
Lost 155 Pounds and Resolved Diabetes and Depression

Growing up in the Midwest, I was always active and enjoyed water sports and outdoor activities every day. I even thought I ate a healthy diet—80 percent vegetarian meals—but I grew increasingly overweight, and my health deteriorated. By the time I was 41 years old, I weighed 330 pounds.

This lifetime struggle with my weight took on a new urgency with the sudden onset of symptoms associated with type 2 diabetes: severe thirst, exhaustion, brain fog, and more. I was sick and depressed and spent most days in bed. I knew I couldn't go on like this. I suffered from migraines, sleep apnea, depression, and anxiety. I even had severe cramping in my legs, especially at night, which prevented sleep and caused issues with walking. I also had digestive problems, with constant gas, bloating, and rectal bleeding. I had severe boils, acne, skin tags, and skin infections.

When I finally visited my doctor for blood tests, they found out-of-control type 2 diabetes, with a hemoglobin A1c level of 11.4. The doctor prescribed 500 milligrams of metformin and 10 units of insulin per day. Determined to find a different path to wellness, I made an appointment at Dr. Fuhrman's Wellness Center. I knew that I had to change what I was doing. I had tried everything else over the years to lose weight and get healthier, and I had failed every time. I was committed to following every element of the Nutritarian diet.

As I followed Dr. Fuhrman's eating style, I gradually lost my cravings for unhealthy foods. Believe it or not, I started craving greens and salads instead! I dropped an incredible 155 pounds in a year and my health was transformed. My digestion was better, and my energy levels returned. I no longer use my sleep apnea machine, which is huge! I used to suffer from three to four migraines a week, but now I have zero. My skin is much better, almost all of my skin tags are gone, and I'm off all medication.

I now weigh 175 pounds. Every day is an adventure—it's like I've been given a new lease on life. I started exercising when I started this way of eating, and now I do fifteen to thirty minutes of cardio and thirty minutes of weights every day; and every night after dinner, I take a

thirty-minute walk. My health transformation—including my 155-pound weight loss—has inspired other members of my family to begin their own health journeys. Two have joined the Nutritarian Women's Health Study, and others have started to incorporate the Nutritarian eating style into their lives.

My advice to anyone contemplating making the change is: Don't wait—do it now! If I had known the outcome, I would have done this decades ago. I constantly think about what successes I would have had in my life if I had done this when I was younger. It's worth every bite of greens!

When I returned to my doctor's office for a follow-up visit a year after my diabetes diagnosis, the physician and his staff were astonished by the dramatic changes to my health. And given my professional career in IT, I got a chuckle out of the fact that even the computer system couldn't believe my transformation. When the nurse typed in my weight and numbers, the computer kept alerting her that something was incorrect, because the numbers were so different. Everybody laughed.

It has been well-studied and confirmed scientifically that excessive exposure to self-produced estrogen, insulin, and IGF-1 increases angiogenesis and cell proliferation and growth, with higher levels significantly increasing the risk of prostate and breast cancers.[1] These hormones can be regulated and the risks mitigated through a Nutritarian diet. *Angiogenesis* means to promote the growth of new blood vessels and expand the blood vessel distribution tree to feed new tissue growth. Angiogenesis promotion is typically associated with the growth of fat on the body and the growth and spread of cancer, as both fat cells and cancer cells secrete angiogenesis-promoting factors to facilitate their growth.

Excess fat on the body is a risk factor for developing many diseases, particularly the hormonally sensitive cancers (such as breast or prostate cancer) via the enhanced production of estrogen from fat cells. Fat cells live in a low-oxygen environment (hypoxia) and as a result produce and shed more free radicals, creating an environment of chronic inflammation. Because body fat is poorly perfused with blood vessels, chronic inflammation within the fat cells further increases insulin resistance, which in turn increases insulin requirements and insulin levels in the bloodstream. Since insulin is a growth hormone, which enables fat storage and cellular replication, higher levels further promote tumor growth and angiogenesis.

The increase in inflammatory compounds, particularly reactive oxygen species (ROS), cytokines, and lipokines, also stimulates aromatase activity, increasing estrogen levels in the body and even higher levels within the breast tissue. Estrogen is the collective name for a group of hormones that promote and regulate the development of sex characteristics. In females, this includes the breasts, the endometrium (lining of the uterus), and the menstrual cycle; in males, estrogen regulates the maturation of sperm and the sex drive. Three types of estrogen occur naturally in the body: estradiol, estriol, and estrone. A fourth, estetrol, is produced only during pregnancy. Increased estrogen is a major factor connecting excess body fat with breast and prostate cancers.[2] Androgens produced by the adrenal

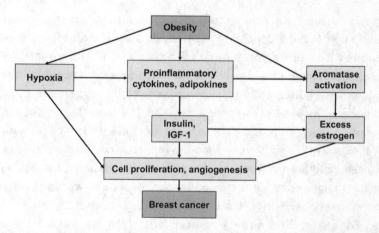

Note: Mushrooms contain natural aromatase inhibitors and suppress estrogen production in breast tissue, radically reducing the risk of breast cancer. Mushrooms also contain angiogenesis inhibitors.

cortex and the postmenopausal ovary are converted into estrogens by aromatase. This mechanism of estrogen production in a person with excessive body fat can lead to local estrogen levels in breast tumors that are as much as tenfold higher compared with a person of normal weight.[3] Even though ovaries are the main source of estrogen in premenopausal women, if a woman is significantly overweight, the excess estrogen added on by body fat stores can elevate the levels further, thus increasing cancer risk in younger women too. This trio of breast cancer–causing elements—inflammatory compounds, excessive estrogen, and excessive insulin and IGF-1—creates precancerous changes and then permits cancer cells to proliferate.[4]

THE COMPLICATED RELATIONSHIP AMONG INSULIN, FAT, AND CANCER

Excess insulin is another critical factor contributing to the development and progression of cancer. For decades, epidemiological studies have reported an increased risk of cancer in people with obesity and type 2 diabetes, especially those taking insulin.[5] So elevated insulin levels do not merely increase the risk of diabetes and heart disease, but they have proangiogenic and strong cancer-promoting effects.[6]

When you eat high-glycemic carbohydrates such as white rice, white flour, honey, and maple syrup, your blood sugar levels spike, triggering the pancreas to produce insulin. The term "high-glycemic" refers to the high speed by which glucose enters the bloodstream. A food with a high glycemic load is broken down to simple sugar quickly, and those glucose calories enter the bloodstream rapidly. The more glucose entering the blood quickly, especially within the first hour after consumption, the higher the glycemic load.

For example, in a meta-analysis of thirty-nine studies, a diet supplying a high glycemic load was associated with increased risk of colorectal and endometrial cancers, and a meta-analysis of ten prospective studies demonstrated a link between higher glycemic load and breast cancer.[7] Another study demonstrated that for every 100 grams of white rice consumed per day, breast cancer risk increased 19 percent, whereas the same amount of whole grains, brown rice, or beans had almost the opposite effect.[8]

Not only do high-glycemic foods enter the bloodstream rapidly, but so do the fat calories coming from oils and food high in animal fat. When too many calories enter the bloodstream rapidly, they enhance body fat storage and stimulate addictive centers in the brain. What makes these foods even more fattening and unhealthy is that oils and saturated animal fats in the diet can worsen insulin resistance, increasing insulin secretion further in response to ingested carbohydrates.[9]

The surfaces of our cells are covered in cave-like structures, formed by cholesterol and proteins, called "caveolins" and "cavins," that are involved in the binding and uptake of insulin. Dietary saturated fat, largely from animal products but also from excess oil consumption, distorts these structures, reducing the response to insulin and worsening glycemic control. This distortion increases the demand on the pancreas to secrete more insulin. When both dairy

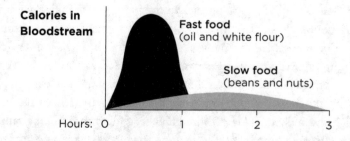

fat (such as from butter or ghee) and high-glycemic carbohydrates (such as white rice or honey) enter the bloodstream, they provoke an excessive insulin response, which promotes fat storage and excess cell replication and growth. Excess growth promotion in adults is a major contributor to cancer development.

It is clear that dietary oils and animal-based fats are fattening foods. Excess fat on the body increases the stored fat in muscle, or intramyocellular lipids, which interferes with the efficient utilization and uptake of insulin into cells. Rapidly absorbed fats are stored quickly as body fat, but when you get your fat from eating nuts and seeds, for example, the fat enters the bloodstream so slowly that appetite is suppressed for many hours. Plus, the fat calories from nuts and seeds are not all absorbed; some are lost in the stool, but the remaining absorbed calories go into the bloodstream gradually, so they can be preferentially burned for energy rather than stored. Thus it is not as easy to gain weight from eating them.[10]

Dietary oils, on the other hand, quickly and efficiently increase body fat accumulation, enhancing insulin resistance. The more overweight and insulin-resistant you are, the higher your insulin response to ingested carbohydrates. *Insulin resistance*, also known as "impaired insulin sensitivity," is a condition in which the body's fat, liver, and muscle cells are no longer responding appropriately to circulating insulin, and the beta-cells of the pancreas are forced to produce more and more insulin in order for the body to remove, utilize, and store the glucose in the bloodstream. The more insulin-resistant one is, the higher one's risk of heart disease and cancer.[11]

The ability of the body (specifically, the pancreas) to increase insulin production means that insulin resistance typically has no symptoms at first, because the pancreas can produce a huge amount of insulin in order to maintain relatively normal blood sugar levels. Over time, however, the overworked beta-cells of the pancreas can "poop out" and lose their ability to produce a high level of insulin, so eventually higher blood glucose levels (prediabetes) and ultimately type 2 diabetes occur. With high insulin resistance, the blood sugar levels can be up in the diabetic range even though the beta-cells, those responsible for making and releasing insulin, are still pumping out much higher levels of insulin than a person of normal weight would require.

Insulin, as the chief fat-storage hormone, also prevents you from losing body fat. Besides allowing uptake of glucose into cells for use and storage as fat, insulin also inhibits lipolysis, that is, fat breakdown and utilization. It makes people fatter and makes it hard to lose weight.

> When you eat white flour, white rice, and sweets,
> you don't lose weight even with lower caloric intake.

THE MORE CALORIES YOU CONSUME, THE MORE YOU PRODUCE FREE RADICALS

For most of human history, access to adequate nutrients and calories was the main concern for eating behavior. Now, our low-micronutrient, high-calorie modern diet drives oxidation from excess free radicals, causing inflammation. Excess calories drive free radical production, without the corresponding ability of the body to control and remove them because of the lack of phytonutrients in processed foods and animal products. All excess calories are proinflammatory, and this inflammation interferes with effective insulin function, enhancing insulin resistance and increasing insulin production further. Excess dietary fatty acids, especially saturated fats, cause inflammation in the part of the brain called the hypothalamus and disrupt satiety signaling (which tells us when we have eaten enough), which

Cancer Cases Attributable to Obesity

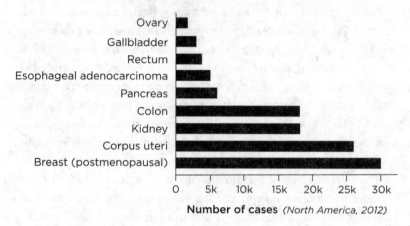

Number of cases (North America, 2012)

Source: World Health Organization. International Agency for Research on Cancer.
"Cancer Attributable to Obesity." https://gco.iarc.fr/causes/obesity/home.

enhances appetite. This perpetuates the vicious cycle of overeating,
inflammation, and more insulin resistance—promoting disease and
type 2 diabetes.

THE MICROBIOME AND INSULIN METABOLISM

Any discussion of the cause and repair of altered glucose and insulin
metabolism would not be complete without addressing the microbial composition of the gut—or the gut's *microbiome*. The beta-cells of
the pancreas are also prone to toxicity from proinflammatory foods,
increasing their destruction and the development of type 2 diabetes.
That is, a poor diet with more animal products and processed foods
and less colored plant food favors a proinflammatory microbiome,
which also contributes to insulin resistance. An unfavorable microbial composition is another source of low-grade inflammation from
both trimethylamine oxide (TMAO) and endotoxemia from excess
gram-negative bacteria.

It is also well-known that dietary excellence with calorie restriction can rapidly reduce insulin resistance before significant fat loss occurs.[12] The Nutritarian program typically resolves type 2 diabetes in weeks or a few months, well before the individual has obtained his or her favorable weight. Flooding the body with antioxidants and other phytonutrients along with moderately restricting calories can immediately start to reduce inflammation and undo some of the damage. Steadily losing weight in a healthy manner reduces angiogenesis, reduces estrogen production, and reduces insulin levels even when there is still substantial weight to lose.

Excess secretion of insulin also leads to increased IGF-1 expression. Both increased insulin and increased IGF-1 are associated with tumor growth in animal models and in human studies.[13] The ability of high-glycemic carbohydrates to promote cancer involves many mechanisms, including the promotion of angiogenesis and cellular replication and the suppression of *apoptosis*—that is, the capacity of a damaged cell to stop itself from replicating.[14]

Glucose and insulin surges actually have a dual negative effect on apoptosis. Insulin resistance and high circulating insulin levels increase fatty liver and the death of liver and pancreatic cells, creating a loss of organ function and accelerating the progression of diabetes.[15] At the same time, immune system surveillance is reduced, and apoptosis of precancerous cells is halted.[16] Epidemiological studies that investigated the association between insulin and the risk of certain cancers in humans without diabetes document that higher insulin levels were associated with prostate, breast, colorectal, and endometrial cancers.[17]

The hormonal pathways, particularly the insulin/IGF-1 signaling combination, also play a key role in regulating aging affecting SIRT1 genes. These genes are involved with repairing DNA damage and gene silencing, that is, suppressing genetic defects. The reduction in gene silencing from excess hormone production leads to the wrong genes being activated, which can result in an increase in cancer, heart disease, and dementia.

Let's remember that it is the triad of exposure to adequate phyto-nutrients, moderate reduction of calories, and exercise that elevates longevity proteins and reduces cancer-promoting hormones, which are the only real fountain of youth, enabling us to age more slowly.

THE CARBOHYDRATE HIERARCHY

Some high-carbohydrate foods promote longevity and prevent body fat accumulation and diabetes, and others do the opposite. It all depends on the presence of fiber and resistant starch, glycemic load, phytochemical content, and other factors.

The scientific literature has been very clear that high-glycemic, processed foods are so disease-causing that they promote heart disease even more than saturated fats from animal products. This is a chilling concern, considering all the white bread, white rice, and white potato products people consume in this country and around the world.[18] The undisputed reality is that some carbohydrate-rich

Glycemic Load of Common Foods

	Glycemic Index		Glycemic Index
White potato	29	Apples	9
White rice	26	Kiwi	8
White pasta	21	Green peas	8
Chocolate cake	20	Butternut squash	8
Corn	18	Kidney beans	6
Sweet potato	14	Black beans	6
Grapes	14	Watermelon	6
Rolled oats	13	Oranges	4
Whole wheat	11	Cashews	2
Mango	11	Strawberries	1
Lentils	9		

Sources: Atkinson FS, Foster-Powell K, Brand-Miller JC. International tables of glycemic index and glycemic load values: 2008. *Diabetes Care*. 2008;31(12):2281-83; Foster-Powell K, Holt SH, Brand-Miller JC. International table of glycemic index and glycemic load values: 2002. *Am J Clin Nutr*. 2002;76(1):5-56.

foods are healthier than others, and we can't paint them with the same brush. Discussing the merit of starchy foods or carbohydrates in the diet "in general" is almost irrelevant, as the individual food in question has to be considered with awareness of all its qualities.

Consider white potatoes as an example of a high-glycemic carbohydrate with a favorable caloric density, because they contain only about 120 to 160 calories each. The Nurses' Health Study demonstrated an 18 percent increased risk of diabetes for each potato eaten daily by overweight women. Notably, it was the glycemic load of the potatoes themselves that was implicated, rather than any added or associated butter or oil.[19]

Ominously, data from the North Carolina Colon Cancer Study, with more than one thousand cases of colon cancer compared against more than one thousand controls, demonstrated a more than 50 percent increased risk of rectal cancer in those eating three potatoes a week compared with one, and more than an 80 percent increase comparing 5.6 servings a week to 1 serving.[20] These findings are quite shocking, but note again that these heightened risks were observed only for non–whole grains and potatoes, not the discretionary fat intake, such as sour cream or butter. This signifies that the association is demonstrating glycemic risk and not merely an association with a dietary pattern or the fatty toppings used on a potato. The disease risk becomes more pronounced as body weight and insulin resistance increase, which specifically identifies the potato's glycemic effect, as it becomes a more significant stimulator of excess insulin release as a person's body weight increases. The fat-promoting effects of insulin make high-glycemic foods more fat-promoting and make it more difficult for a person who is overweight to lose weight.[21]

Overall, though, it is neither the inclusion nor the exclusion of one high-glycemic food, such as a potato, that will make a diet good or bad. Nonetheless, care should be taken to eat only a limited amount of higher-glycemic carbohydrates, such as white rice and white potatoes, and to eat them with a meal that includes plenty of greens, beans, nuts, and seeds. Moderate portion size is also important, such as eating half of a medium potato. A healthier diet would utilize a variety of starches with a more valuable nutritional profile, such as turnips, rutabaga, winter squash, butternut and acorn squash, chestnuts, parsnips, carrots, peas, corn, and intact whole grains.

Commercially baked goods made from white flour are worse than white potatoes and white rice. Even worse, when you eat these foods, especially while consuming oils, meats, and cheeses, the body can store fat even more effectively than it could if you ate just the meat alone with greens or the potato and rice alone with greens. Nevertheless, though it is not as bad as sugar and white flour, white potato is still not the most favorable high-starch vegetable to choose, and white rice is not a preferred form of grain.

Virtually everyone interested in nutrition knows that white flour products are a poor carbohydrate source for a longevity-enhancing or disease-reversing diet. This can be understood by comparing the nutritional qualities of white flour products with those of more healthful carbohydrate sources such as peas, corn, and beans. We can do the same with white potatoes, comparing them to peas, corn, and beans.

When comparing the nutritional profiles of high-carbohydrate plant foods, we must consider

1. Fiber content

2. Percentage of slowly digestible starch

3. Percentage of resistant starch

4. Micronutrient content

5. Caloric density

6. Glycemic index/load

7. Beneficial qualities of other foods in the diet that may be displaced to allow room for this food

Once we have this information, we can place carbohydrate-rich foods in a hierarchy based on their nutrient levels, fiber content,

and amount of resistant starch. As the amounts of fiber and resistant starch increase, one can note more dramatic benefits for people
who are overweight and/or diabetic. Resistant starch is counted as
calories on labels and charts, but during digestion, 90 percent of
resistant starch calories are lost, thus acting more like a type of fiber.
Beans, starchy vegetables, intact grains, and other nutritious, high-
carbohydrate natural foods are generally the most heavily emphasized in my recommendations because of all the criteria above.

Vegetables, legumes, nuts, seeds, and fruits are all rich in fiber,
and, besides the phytonutrients, the fiber itself has some cancer-
fighting actions. Indigestible polysaccharides—fiber and resistant
starch—help keep blood glucose and insulin levels down and also
act as prebiotics, promoting the growth of beneficial gut bacteria.[22]
Higher fiber intake is associated with a reduced risk of breast, colorectal, esophageal, pancreatic, gastric, and endometrial cancers.[23]

For preventing breast cancer in particular, fiber has the additional benefit of lowering estrogens by facilitating the excretion of
estrogen.[24] A comprehensive 2019 study looked at dietary fiber content over 135 million person-years and corroborated the commonly
accepted notion that more fiber in the diet is beneficial to weight loss,
protects against chronic disease, and lengthens lifespan. Those people with the most fiber in their diets lived the longest, and the more
fiber, the better the outcomes.[25]

WHAT IS RESISTANT STARCH?

Resistant starch is present in natural carbohydrate-containing foods.
While most starch is broken down by digestive enzymes, converted
into simple sugars, and absorbed in the small intestine, resistant
starch is more like a fiber—it resists enzymatic degradation in the
small intestine and travels to the colon for degradation by bacterial
action into short-chain fatty acids. However, only a small percentage
of those fatty acids are absorbed and utilized for energy. Public health
authorities in recent years have accepted the food value of resistant

starch because it has anti-diabetic and weight-loss benefits. A committee of the United Nations and World Health Organization stated that the discovery of resistant starch is "one of the major developments in our understanding of the importance of carbohydrates for health in the past 20 years."[26]

Resistant starch is satiating, yet its calories do not get absorbed well. It also promotes health and weight loss by other mechanisms, such as

- Encouraging the growth of beneficial bacteria, which reduce intestinal pH, bile acids, and ammonia

- Producing, when fermented by bacteria, short-chain fatty acids that reduce body fat storage

- Reducing the glycemic effect of other foods, even when eaten at separate meals

	Resistant Starch (%)	Resistant Starch (%) + Fiber (%)
Black beans	27	70
Navy beans	26	62
Lentils	25	59
Split peas	25	58
Corn	25	45
Brown rice	15	20
Rolled oats	7	17
Whole wheat flour	2	14
Pasta	3	9
Potato	3	5

Source: Bednar GE, Patil AR, Murray SM et al. Starch and fiber fractions in selected food and feed ingredients affect their small intestinal digestibility and fermentability and their large bowel fermentability in vitro in a canine model. *J Nutr.* 2001;131(2):276–86.

The hierarchy of carbohydrate quality includes not just the factors above but also the amount of slowly digestible starch and a food's nutrient density. We can use this information to devise a dietary protocol that reduces exposure to the highest-glycemic carbohydrates.

Note that beans run away with the prize for the healthiest carbohydrate choice. You achieve dramatic glycemic (that is, glucose-moderating) benefits from

- Using beans in place of other carbohydrate-rich foods

- Using more green vegetables, both raw and cooked, with other low-glycemic vegetables

- Using more nuts and seeds in place of carbohydrate-rich foods

As an example, a two-group controlled trial encouraged one group to increase legume intake by 1 cup a day and another to increase their intake of whole grains by the same amount. A clear benefit occurred for adding more whole grain, but more dramatic benefits occurred for the addition of beans, as noted in the chart below.[27]

	Whole Grain Group	Bean Group
Fiber increase (g/1,000 cal)	1.9	10.0
Glycemic load reduction	−5	−48
HbA1c (%)	−0.3	−0.5
Body weight (lb)	−4.4	−5.7
Fasting glucose (mg/dL)	−7	−9
Triglycerides (mg/dL)	−9	−21
Cholesterol (mg/dL)	−2	−9
Systolic blood pressure (mm Hg)	0	−4
Diastolic blood pressure (mm Hg)	0	−3

In addition to beans being glycemically favorable themselves, their fermentation and prebiotic effects lower the glucose absorption of other foods in the diet. These glucose-lowering benefits occur not merely in the meal eaten with the beans, but later as well when no beans are eaten. This has been called the "second-meal effect."[28] Beans have multiple benefits for favorable glycemic response, weight reduction, and anticancer effects.

Across a broad spectrum of regions and ethnicities, beans/legumes have been found to be the most consistent and reliable predictor of

Brown Rice Is No Longer Recommended

Even though brown rice is a whole grain, I no longer recommend its consumption because of significant arsenic contamination of most of the brown rice available for purchase in the United States—even organic brown rice and wild rice. Arsenic is a causative factor in many cancers, and it also promotes heart disease.

The United States is the world's leading user of arsenic. Since 1910 about 1.6 million tons of it have been used for agricultural and industrial purposes. Lead-arsenate insecticides were used for many decades, and the residue still lingers in farm soils today, even though these pesticides were banned in the 1980s. Other arsenic-containing ingredients in animal feed to prevent disease and promote growth are still permitted. Fertilizer made from poultry waste can contaminate crops with arsenic as well.

In the United States as of 2010, about 15 percent of rice acreage was in California, 49 percent in Arkansas, and the remainder in Louisiana, Mississippi, Missouri, and Texas. That south-central region of the country has a long history of producing cotton, a crop that was heavily treated with arsenic pesticides for decades in part to combat the boll weevil. Since the hull of rice takes up arsenic so readily, it is safer to eat other mild-tasting grains such as amaranth and quinoa.

longevity. An 8 percent reduction in death was reported for every 20 grams (2 tablespoons) of beans eaten daily.[29] Beans, nuts, and seeds have numerous anticancer compounds, including phytic acid and inositol pentakisphosphate, which has been shown in animal studies to inhibit tumor growth, migration, and invasion and also augment natural killer (NK) cell activity.[30] Eating more beans as a replacement for other foods simply aids all metabolic parameters that enhance cardiovascular health.[31]

IGF-1: THE MASTER LIFESPAN REGULATOR

Insulin-like growth factor-1 (IGF-1) is a hormone with a structure similar to that of insulin. It is a growth-promoting signal that is important during childhood, contributing to brain development and muscle and bone growth; IGF-1 levels peak during the teen years and twenties and then decline with age. Growth hormone from the pituitary gland stimulates IGF-1 production in the liver. Circulating IGF-1 is regulated predominantly by dietary protein intake, especially animal protein. Animal protein—which is higher in essential amino acids and more biologically complete compared with plant protein—increases IGF-1 much more than plant protein.[32] High-glycemic, refined carbohydrates also raise IGF-1 levels.[33]

The link between animal protein enhancement of IGF-1 production and increased death from cancer is well-established in the scientific literature.[34] A good example is a study published in 2014 that tracked animal product consumption in six thousand individuals between the ages of 50 and 65 for more than eighteen years.[35] Researchers found a fourfold increase in cancer mortality when comparing those consuming more than 25 percent of calories from animal products with those consuming less than 10 percent. There was a 75 percent increase in overall mortality over that eighteen-year period in the group eating more animal protein.

Interestingly, the high-protein group were consuming even less than the average animal protein level most Americans consume. This means that individuals who eat a paleo or ketogenic diet, which encourages much higher levels of animal product consumption, may experience higher levels of cancer. Many of these individuals are consuming 50 to 80 percent of their calories from animal products.

Plus, there was a seventy-three-fold increased risk of developing diabetes in the higher protein group, and a twenty-three-fold increased risk in the moderate protein group compared with the lower protein group. This increased risk of diabetes with higher protein was consistent at all ages.

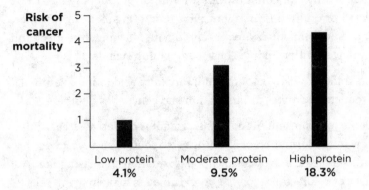

Animal Protein Intake and Risk of Cancer Mortality

Source: Levine ME, Suarez JA, Brandhorst S et al. Low protein intake is associated with a major reduction in IGF-1, cancer, and overall mortality in the 65 and younger but not older population. *Cell Metab.* 2014;19(3):407–17.

Every long-term study using hard endpoints (death, heart attack, or cancer) demonstrates increased premature mortality with increased consumption of animal products.

This study is only one of several long-term studies using various naturally raised animal products from all over the world and looking at the "hard endpoints" of death, heart attack, or cancer that show the same thing: High consumption of animal protein means higher IGF-1 levels, accelerated aging, and higher premature death rates.[36] High IGF-1 is associated with higher rates of breast, prostate, colorectal, and lung cancers as well as lymphomas.[37] Among animal proteins, dairy protein is the most potent for increasing IGF-1;[38] accordingly, greater consumption of dairy products is linked to increased risk of prostate cancer.[39] Red and processed meats are also associated with an increase in the risk of other common cancers.[40] Unquestionably, this is an important reason why consumption of animal products needs to be curtailed for optimizing healthspan and lifespan.

There are multiple other reasons, besides elevating IGF-1 levels too much, why the consumption of animal products needs to be restricted or omitted from a longevity-promoting diet. For instance, carcinogenic compounds are present in both red and white meat and are enhanced with cooking. These compounds include:

- Arachidonic acid, a proinflammatory fatty acid found in animal products, promotes inflammation and cancer development.[41]

- Heme iron, present in red meat, promotes oxidative stress and aids in the formation of N-nitroso compounds and cytotoxic and genotoxic aldehydes by lipoperoxidation, increasing the risks of colorectal cancer.[42] Cyotoxic means toxic to living cells and genotoxic means toxic to DNA molecules in genes, causing mutations, tumors, etc.

- Trimethylamine oxide (TMAO), another inflammatory compound that is produced from eating animal products, contributes to cancer and heart disease.[43]

- The association between red meat and cancer is thought to be largely due to heterocyclic amines (HCAs) and polycyclic aromatic hydrocarbons (PAHs), which are formed when meats are cooked; this occurs with white meats as well. For processed meats, the major factor appears to be that nitrite used as a preservative forms carcinogens called nitrosamines (or N-nitroso compounds) during production and storage, and more during the digestive process.[44]

Obviously, red meat is dangerous. The World Health Organization (WHO) has declared processed meats to be carcinogenic and red meat to be "probably carcinogenic."[45] Twenty-two scientists from ten countries forming an International Agency for Research on Cancer (IARC) Working Group, who considered more than eight hundred different studies on cancer in humans, more than seven hundred epidemiological studies on red meat, and more than four hundred epidemiological studies on processed meat, came to these conclusions about meat and cancer.

However, many people still wonder whether such results mean that even small amounts of pasture-raised meats can increase the risk of death. A study specifically looked carefully at this question. Seventh-Day Adventists are interesting to study because many are vegans or vegetarians and others eat animal products only occasionally or only in small amounts. Results from the Adventist Health Study-2 showed a significant increase in death even in those eating fewer than 2 ounces a day of red meat, demonstrating that even a small amount is worse than eating none at all. The authors of this study believe their work supports previous conclusions. "Our findings give additional weight to the evidence already suggesting that eating red and processed meat may negatively impact health and lifespan," comments Michael Orlich, coauthor of the study and codirector of the Adventist Health Study-2.[46]

Consuming eggs has also been linked to prostate cancer and gastrointestinal cancers. Eating at least two and a half eggs per week, compared with eating less than half an egg per week, was linked to an 81 percent increase in the risk of lethal prostate cancer in men.[47] An analysis of forty-four studies of participants who consumed fewer than three, three to five, or more than five eggs per week found that as egg consumption increased, the risk of gastrointestinal cancers (esophagus, stomach, colon, and rectum) increased as well. Compared with no egg consumption, increases in risk were 13 percent, 14 percent, and 19 percent for the three groups, respectively.[48] Eggs are high in cholesterol and choline, and high blood cholesterol and choline have each been linked to increased risk of prostate cancer.[49] Cholesterol contributes to cellular processes important to cancer development, such as proliferation and migration.[50] Eggs are particularly rich in choline, which gets metabolized by gut bacteria into the proinflammatory compound TMAO, which contributes to the development of cancer and heart disease.[51]

Eggs may cause TMAO levels to rise the most, but this compound also increases after you consume meat and fish, because diets rich in animal products support the growth of the *Acinetobacter* bacteria species, which is needed to convert dietary choline and carnitine to TMAO, which then promotes cholesterol deposits in blood vessels, higher blood pressure, and endothelial inflammation. A 2017 meta-analysis found that higher circulating TMAO levels were associated with a 23 percent higher risk of cardiovascular events and a 55 percent higher risk of mortality.[52]

A longevity-promoting diet has certain characteristics, such as being vegan or low in animal products and high in fiber and phytochemicals. The foods that are well-documented to slow aging and prevent cancer that contain these protective phytonutrients include many carbohydrate-containing plants, such as fruits. Diets that try to restrict or omit carbohydrates fuel unfavorable bacterial growth in the gut and are also dangerously low in protective phytonutrients. These carbohydrate-restricting diets are not lifespan-favorable.

The last nail in the coffin of low-carb diets such as paleo and ketogenic came in August 2018 with the conclusion of a massive investigation into the health effects of such diets for weight loss. One of the study authors, Maciej Banach, of the Medical University of Lodz, Poland, said: "We found that people who consumed a low-carbohydrate diet were at greater risk of premature death. Risks were also increased for individual causes of death including coronary heart disease, stroke, and cancer. These diets should be avoided."[53]

This study prospectively examined a nationally representative sample of 24,825 participants of the US National Health and Nutrition Examination Survey (NHANES) from 1999 to 2010. As a press release from the European Society of Cardiology announced: "Compared to participants with the highest carbohydrate consumption, those with the lowest intake had a 32% higher risk of all-cause death over an average 6.4-year follow-up. In addition, risks of death from coronary heart disease, cerebrovascular disease, and cancer were increased by 51%, 50%, and 35%, respectively."[54]

Interestingly, the relationship between a low-carbohydrate diet and total mortality was more pronounced in nonobese (48 percent increase) versus obese (19 percent increase) people. Normal-weight individuals eating lots of animal products die young. In other words, this means that people controlling their weight by restricting carbohydrates are paying a severe price—their life. This heightened danger was shocking to researchers, who went on to strengthen this data by performing a meta-analysis of seven prospective cohort studies with 447,506 participants and an average follow-up of 15.6 years. This also found clear increased risks for all causes of mortality with carbohydrate avoidance, further confirming the dangers of low-carbohydrate diets and the avoidance of fruit.[55]

We always need to look to long-term data to ascertain the usefulness of a dietary approach and the claims made about it. Eating a "Twinkie diet" may have short-term advantages (if overall calories are reduced), but of course those benefits are worthless if you die much younger. Clearly, the reduced intake of fiber, micronutrients, and phytonutrients as a result of these trendy ketogenic and paleo diets takes a serious toll.

IDEAL LEVELS OF IGF-1 ARE BETWEEN 100 AND 160 NG/ML

A US study categorizing participants as following a diet with a moderate or high amount of animal products reported typical IGF-1 levels between 200 and 250 ng/mL in middle-aged adults.[56] Two studies comparing adults on a vegan diet with about 10 percent of calories from protein (no animal protein) to those on a Western or SAD diet with about 17 percent of calories from protein (fewer animal products than the average American consumes) found the average IGF-1 level to be 200 ng/mL, with the vegan average lower than 150 ng/mL. IGF-1 levels in endurance runners averaged 175 ng/mL.[57]

In the Nurses' Health Study, premenopausal women with IGF-1 levels higher than 207 ng/mL had a higher risk of breast cancer.[58] In the Physicians' Health Study, prostate cancer risk increased significantly once IGF-1 levels increased above 185 ng/mL.[59]

All of this information leads to the conclusion that for most adults, keeping IGF-1 levels below 175 ng/mL is likely important, with less than 150 ng/mL being even more protective. However, serum IGF-1 levels less than 80 ng/mL are likely detrimental, especially after the age of 75. Restricting animal protein during most of adult life to maintain a relatively low IGF-1 is an important objective for people who desire superior health and life extension.

A meta-analysis looked at ten studies investigating IGF-1 levels and all-cause mortality. The authors found a U-shaped association, meaning that IGF-1 levels on both the low end and the high end of the spectrum were associated with increased risk of premature death.[60] The lowest risk was at the 55th percentile of serum IGF-1 and increased in both directions for all-cause, cancer, and cardiovascular mortality. These data suggest that we should aim for an IGF-1 level near the middle, between 100 and 160 ng/mL, for healthy people. A few studies, primarily in European populations, have attempted to define average IGF-1 levels for healthy people in different age ranges.[61]

Age	Average Serum IGF-1 (ng/mL)
21–30	158–230
31–40	135–220
41–50	121–193
51–60	98–150
61–70	85–140
71–80	85–95
81+	85–90

The European Prospective Investigation into Cancer and Nutrition (EPIC) study reported average serum IGF-1 levels of approximately 210 ng/mL, suggesting that this is a typical level for adults on a Western or SAD diet.[62]

AFTER AGE 75 PROTEIN REQUIREMENTS INCREASE

Older adults have somewhat different nutritional needs than younger adults and require more protein. On average, muscle mass peaks between ages 20 and 30 and then decreases. It is estimated that muscle mass falls 8 percent per decade from age 30 to 80, and then 15 percent per decade. Muscle strength declines even more quickly, approximately 3 percent per year.[63] Older adults require adequate protein intake coupled with strengthening exercises to counteract this muscle loss and prevent functional decline and falls.

The US recommended dietary allowance (RDA) for protein (0.8 g/kg body mass) is adequate for younger and middle-aged adults. However, studies providing older adults with the RDA of protein have reported negative nitrogen balance or a reduction in thigh muscle—indicating that this level of protein intake is inadequate for maintaining muscle mass.[64]

This research suggests that a more appropriate protein intake for older adults is more than 1 g/kg per day (more than 70 grams for an average 150-pound male).[65] Since very healthy eaters, such as Nutritarians, age more slowly, their protein needs would not likely increase until years, even decades, later. The increasing need for protein in the elderly is a very gradual process that occurs in different decades in different people, depending on their health and rate of aging. Typically, the need for increased protein occurs after age 70 in people eating a standard Western diet, and after age 85, later, or not at all, in those who have been following a Nutritarian diet for many years.

This is an important issue, because preventing loss of muscle mass and maintaining brain function help elderly people to avoid disability and to continue to live independently longer. In fact, a study published in 2018 evaluated functional tasks in older adults who had completed two dietary surveys over a twelve-year follow-up period. Subjects who consumed more than 1.2 g/kg per day of protein compared with those who consumed less than the RDA of 0.8 g/kg per day were more likely to be able to complete a number of tasks independently, including walking, going up and down stairs, kneeling or crouching, and lifting heavy objects. Higher physical activity levels and lower BMI also contributed to independence in these everyday tasks.[66]

As we age, the body produces less growth hormone, leading to lower IGF-1 levels. For a small percentage of people, their ability to adequately digest and assimilate protein decreases enough with aging to make it difficult for them to get enough protein on a vegan diet. Even after a vegan diet is optimized for protein adequacy (by adding, for instance, greens, beans, quinoa, hemp, soybeans, tempeh, and Mediterranean pine nuts), these individuals (whose IGF-1 is too low) may thrive better and live longer if a plant protein supplement (such as a pea and hemp protein powder) and/or a small amount of animal products is added to their daily diet to raise IGF-1 above 90 ng/mL.

Adequate IGF-1 levels are required for the maintenance of bone mass, muscle mass, and brain function in the elderly.[67] Low IGF-1 levels associated with an increased risk of disease or mortality are generally 80 ng/mL or lower.[68] At the same time, many elderly people still have dangerously elevated IGF-1 levels, contributing to their increased risk of cardiovascular events and deaths from cancer.[69] It is always important to determine where we fall on the spectrum of circulating IGF-1 levels to help us decide how much protein is advisable as we age.

Eating greens, seeds, and beans supplies sufficient protein in a diet, therefore preventing excessive lowering of IGF-1 in the elderly, which is more often seen with other plant-based diets (such as high-starch vegan diets or macrobiotic-type diets, which may not provide enough protein for some individuals over the age of 75).

When this dietary adjustment is necessary to prevent muscle wasting and also to prevent IGF-1 levels from getting too low, we still need to be aware of the dangers of higher amounts of animal products. Their use should be minimized to less than 10 percent of total calories, or just enough to raise IGF-1 above 90 ng/mL. Too much animal products can be dangerous at any age, even when IGF-1 levels drop too low.

Anyone who does not thrive on a vegan or near-vegan diet should still be cautious about adding animal products. If animal product use is deemed a necessary choice for some individuals, they should still utilize a maximum of only 2 ounces a day for females and 3 ounces a day for males and attempt to increase their intake of high-protein plant foods to minimize the need for animal products. I caution against diets higher in animal protein because they are solidly linked to negative health outcomes, such as cancer, heart disease, and premature death. As noted above, studies of large populations show that even small amounts of red meat in one's diet increase risk of death. Plus, in studies that make the comparison, plant protein is consistently linked to health advantages over animal protein. **The public tends to see protein as a "super" nutrient and strives to consume more, when in reality most people get either more than enough or too much in their diets.**

Protein Content of Selected Foods

Plant-Based Product	Serving Size	Grams of Protein/ Serving	Plant-Based Product	Serving Size	Grams of Protein/ Serving
Pea protein powder	1 oz	22.1	Oats, uncooked	½ cup	5.3
Pumpkin protein powder	1 oz	18.7	Flaxseeds, ground	¼ cup	5.1
Edamame	1 cup	18.5	Corn	1 cup	5.1
Lentils, cooked	1 cup	18.0	Brazil nuts	¼ cup	4.8
Tempeh	½ cup	16.8	Walnuts	¼ cup	4.6
Soy protein powder	1 oz	15.8	Pine nuts, regular	¼ cup	4.6
Kidney beans, cooked	1 cup	15.3	White rice, cooked	1 cup	4.4
Black beans, cooked	1 cup	15.2	Whole wheat bread	1 slice	4.0
Bean pasta, cooked	1 cup	14.7	Broccoli, cooked	1 cup	3.7
Chickpeas, cooked	1 cup	14.5	Kale, cooked	1 cup	3.5
Hemp protein powder	1 oz	12.8	Macadamia nuts	¼ cup	2.7
Hemp seeds	¼ cup	12.6	Pecans	¼ cup	2.5
Tofu, firm	½ cup	11.2	White potato, baked	1 cup	2.4
Pine nuts, Mediterranean	¼ cup	10.6	French fries	2.5 oz	2.4
Pumpkin seeds	¼ cup	9.7	Romaine lettuce	2 cups	1.2
Peanuts	¼ cup	8.9			
Chia seeds	¼ cup	8.6	**Animal Products**		
Wheat germ	¼ cup	8.2			
Quinoa, cooked	1 cup	8.1	Whey protein powder	1 oz	21.1
Almonds	¼ cup	7.6	Chicken, white, cooked	2 oz	17.6
Sunflower seeds	¼ cup	7.3	Ground beef, 85%, cooked	2 oz	14.7
Pasta, whole wheat, cooked	1 cup	7.0	Salmon, cooked	2 oz	14.4
Peas, frozen	1 cup	7.0	Steak, porterhouse, cooked	2 oz	14.1
Sesame seeds	¼ cup	6.4	Yogurt, plain, low-fat	4 oz	11.9
Pistachio nuts	¼ cup	6.2	Eggs	1	6.2
Cashews	¼ cup	6.2	Milk, nonfat	½ cup	4.1
Spinach, uncooked	1 cup	5.3	Milk, whole	½ cup	3.8
			Ice cream, vanilla	½ cup	2.3

Meeting the higher protein needs of the elderly (as well as of serious athletes) primarily with plant protein sources is preferable in order to minimize the risks associated with animal protein. Adding a small amount of a plant-based protein supplement—in addition to regularly eating seeds, nuts, and beans—is likely beneficial for older

Janet Marchegiani
Lost 100 Pounds and Recovered from Multiple Sclerosis

I have multiple sclerosis (MS) and had to take oral steroids for many years. I was very overweight and had high cholesterol and hypertension. I was in a car accident and my heart went into a dangerous rhythm. I was so sick because of the shots, steroids, other medications, and MS. Everything was getting worse. I have been a vegetarian since I was 18 years old, but when I saw Dr. Fuhrman on my local PBS station, I knew I could change to a better, healthier vegetarian diet. I decided to give it a go.

In thirteen months, I lost over 100 pounds and stopped taking the steroids. I have no migraines, no high blood pressure, and have not had any further problems with the MS—it disappeared. I have not felt this spectacular in more than twenty years!

We recently lost my father after his twenty-one-year battle with ALS. I am committed to health and happiness after witnessing his incredible struggle. I follow Dr. Fuhrman's lifesaving advice: I eat G-BOMBS (Greens, Beans, Onions, Mushrooms, Berries, and Seeds) every day, make a salad the center of my meals, and follow Dr. Fuhrman's Nutritarian recipes. Thank you, Dr. Fuhrman, for developing this healthy lifestyle—it is saving my life.

adults to optimize muscle protein synthesis, negating the need for animal products as a supplemental protein source.

A Nutritarian diet is already designed for adequate protein intake throughout life. However, it is more important after age 75 that we consume high-protein plant foods such as hemp seeds, sunflower seeds, Mediterranean pine nuts, edamame, tempeh, water-cooked dried soybeans and other beans, and high-protein greens such as broccoli. For the elderly and others with higher protein needs, a wide variety of supplemental plant proteins, such as hemp protein, pea protein, and pumpkin protein, are available today. These plant proteins are healthier choices than animal proteins. Soy protein isolates may elevate IGF-1 levels excessively. In other words, even if you need to increase your protein intake, it is unlikely you would have to use animal products to do so.

CHAPTER TWO: QUICK SUMMARY

Hormones—specifically insulin, IGF-1, and estrogen—are major factors in determining the rate at which you age, modulating your longevity and your risk of developing cancer, particularly breast and prostate cancers. The good news is that you can achieve favorable levels of these self-produced hormones by following a Nutritarian diet.

The second critical principle of a Nutritarian diet is that one's diet must be hormonally favorable to enable maximal lifespan. In order to keep these hormones from getting dangerously high, we must keep the diet low enough in both animal protein and sugar (and other sweetening agents).

ESTROGEN

What you need to know about estrogen:

- Excess fat on the body stimulates excess production of estrogen.

- In postmenopausal women, body fat is the main source of estrogen biosynthesis, which is mediated by aromatase—a complex of enzymes that is found in the breast and other body fat stores and also within some types of tumor tissue.

- Mushrooms have natural aromatase inhibitors and suppress estrogen production in breast tissue, radically reducing the risk of breast cancer. They are best eaten cooked.

- The inflammatory environment created by excess fat increases the activity of aromatase, which further enhances the production of estrogen.

Androgens produced by the adrenal cortex and the postmenopausal ovaries are converted into estrogens by aromatase. This can lead to local estrogen levels in breast tumors that are as much as tenfold higher compared with levels in the circulation. Even though ovaries are the main source of estrogen in premenopausal women, if a woman is significantly overweight, the excess estrogen added by body fat stores can elevate the level further, thus increasing cancer risk.

INSULIN

The main function of insulin, which is secreted by the pancreas, is to supply your body's cells with glucose to use for energy. In a sense, insulin "opens the gate" to your cells, allowing glucose to enter. Insulin is the fat-storage hormone blocking fat breakdown and turning sugar into fat, promoting weight gain. Problems occur when

- You have excess fat on the body, because the fat on cells' surface membranes blocks the receptors, making it harder for glucose to get into the cells—called insulin resistance. This spurs the pancreas to produce more and more insulin.

- Your diet contains high amounts of high-glycemic carbohydrates, oils, and saturated animal fats. High-glycemic foods cause glucose to enter the bloodstream rapidly and your blood sugar levels to spike—which triggers your pancreas to produce excessive insulin. Excess consumption of saturated fat disrupts insulin receptor function.

Insulin resistance increases insulin production, which then creates further problems, especially inflammation and angiogenesis, which in turn results in increased risk of colorectal, endometrial, prostate, and breast cancers. Insulin resistance and the resultant pancreatic "poop out" are the main factors that cause type 2 diabetes. High circulating insulin (and glucose) promotes fatty liver and apoptosis (death) of liver and pancreatic cells, creating loss of organ function and accelerating the progression of diabetes.

INSULIN-LIKE GROWTH FACTOR-1
IGF-1 has a structure similar to that of insulin; it is a growth-promoting signal that contributes to brain development and muscle and bone growth in childhood. Growth hormone from the pituitary gland stimulates IGF-1 production in the liver. IGF-1 levels peak during your teens and 20s and then decline with age. What you need to know:

- Animal protein increases IGF-1 levels much more than plant protein because it is higher in essential amino acids and therefore more biologically complete.

- High-glycemic, refined carbohydrates also raise levels of circulating IGF-1.

- Elevated IGF-1 levels are associated with accelerated aging; higher rates of breast, prostate, and colon cancers as well as lymphomas; and higher premature death rates.

Animal Protein, IGF-1, and Cancer Deaths
The link between animal protein enhancement of IGF-1 production and increased death from cancer is well-established in the scientific literature. Every long-term study using hard endpoints (death, heart attack, or cancer) demonstrates increased premature mortality with increased consumption of animal products.

IGF-1 Levels in the Elderly and Higher Protein Requirements

As we age, the body produces less growth hormone, leading to lower IGF-1 levels. Both excessively low or high IGF-1 levels could lead to serious health problems in the elderly: Low levels can lead to frailty, immune system dysfunction, cognitive problems, and premature death, and elevated levels increase the risk of cardiovascular events and deaths from cancer. Adequate IGF-1 levels are required for the maintenance of bone mass, muscle mass, and brain function in the elderly. Older adults require adequate protein intake coupled with strengthening exercises to counteract muscle loss and prevent functional decline and falls.

Use Caution When Meeting Higher Protein Needs

Meeting higher protein needs primarily with plant protein sources is preferable to animal products in order to minimize the risks associated with animal protein. However, a small number of people may thrive better and live longer when a plant protein supplement and/or a small amount of animal products is added to their daily diet to raise IGF-1 levels above 100 ng/mL (the ideal level of IGF-1 is likely between 100 and 160 ng/mL). If animal product use is deemed necessary, it should be limited to 2 ounces a day for females and 3 ounces a day for males—and be coupled with an increase in the consumption of high-protein plant foods.

It is important to restrict the consumption of animal protein during your adult life to maintain a relatively low IGF-1 level. The greens, seeds, and beans in the Nutritarian diet most often supply sufficient protein to prevent excessive lowering of IGF-1 in the very young and in the elderly (people age 75 and older). This is an important issue, as preventing loss of muscle mass and falls and maintaining brain function help elderly people to avoid disability and continue to live independently longer.

CHAPTER THREE

It Is All About the Plants

I hope you are starting to see a pattern with the science in these pages thus far. Our everyday health and energy and our lifespan can be maximized by a diet rich in the critical micronutrients supplied by plants. The Nutritarian diet takes into consideration all the underlying health factors that lead to possible disease and premature death to offer a solution like no other. It is backed by science, and the results continue to be astonishing.

Nutritarians age more slowly and live longer because their diets have adequate micronutrients for all stages of life. In other words, Nutritarians aim to obtain the ideal level and diversity of micronutrient intake while at the same time preventing the excessive consumption of calories.

Since neither animal products nor processed foods contain a significant amount of antioxidants and phytochemicals, deficiencies of these important micronutrients are far too common. For your metabolism to run well, you not only need basic macronutrients—fat, carbohydrates, and protein—as caloric fuel but you also need fifteen or so vitamins that are coenzymes, fifteen or so minerals that are required in enzymes, omega-3 and omega-6 fatty acids, and nine essential

Scott MacLean
Lost 90 Pounds, Reversed Severe Coronary Artery Disease, Retinal Occlusion, and Stage 4 Melanoma

My doctors at the cancer treatment center would be astonished to know that not only have I surpassed their five-year survival prognosis, but now, eight years later, I am cancer-free, having recovered from Stage 4 melanoma. They would be even more surprised to know that I reversed severe cardiovascular disease, lymphedema, vision loss, sunlight allergy, high blood pressure, and Achilles tendinitis—all through a Nutritarian diet.

I was raised on a diet that was heavy in meat, bread, dairy, and potatoes—with dessert a "must" after every meal. I was fit and active through school, but my unhealthy eating took its toll as the years passed. By the time I reached my mid-30s, my weight had shot up past 250 pounds. I developed severe Achilles tendinitis, and I had trouble sleeping. Plus, I had 80 percent vision loss in my left eye due to retinal occlusion and getting injections of medications into my eye. And I suffered from very high blood pressure and was facing heart bypass surgery for obstructive coronary artery disease. Then I was diagnosed with cancer: Stage 4 melanoma.

I underwent four surgeries and had several lymph nodes removed. The prognosis after surgery was that I had only a 25 percent chance of living the next five years. My body was swollen with lymphedema (a condition caused by damaged lymph nodes that makes your arms and legs swell), causing me to lose feeling on the left side of my body. I also couldn't go outside because I was too allergic to the sun—even with SPF 110 sunblock, my skin would still burn severely. I knew my life was in grave danger. I also realized that traditional medical protocols couldn't offer me the answers and help I so urgently needed.

I felt alone and scared, thinking that I was going to die. Desperate to find a cure, I tried a variety of diets, cleanses, and fasts, with little success. Then one day, I came across Dr. Fuhrman's website and discovered the Nutritarian diet and its benefits. I was skeptical at first but decided to give it one last try. The results astonished and delighted me.

Within months, this new way of life reversed every single condition I had! The weight started to melt off immediately, and in one year, I went

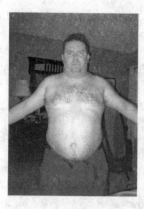

from 250+ pounds to 165 pounds and have remained at this weight ever since. My Achilles tendinitis went away; whereas before I could hardly walk, now I could run with no symptoms. My vision loss was reversed—a development that baffled my eye doctor, but I was delighted I could see again.

Even more dramatic was the reversal of my high blood pressure and cardiovascular disease. My coronary arteries are now 100 percent clear of blockages, which astonished the doctors and radiologist. And my lymphedema resolved, and I regained all feeling in my left arm and the left side of my body.

The most dramatic change in my health is my recovery from Stage 4 melanoma. The cancer is completely gone now, and I'm pretty sure the doctors and staff at the cancer center believed I would have been dead by now. I'm sure I would be if I had kept up with their treatment recommendations.

Within a few months, my sun allergy also disappeared, as did my sleep problems. I sleep like a baby now, and all the other little aches and pains are gone, too.

Since becoming a Nutritarian, I "fired" all my doctors and physio-therapists. They all reached out to me several times and tried their very best to persuade me to keep taking my meds and to come back for more procedures. They even offered to help me enroll in drug trials. But I had witnessed the incredible healing power of the Nutritarian diet firsthand. I politely refused their treatments and stopped taking their calls. Eventually, they stopped calling.

Continued on next page

In the years since I began my health transformation, I forged a close bond with nature, and I'm also eager to spread the message that the health benefits of the Nutritarian eating style are profound and far-reaching. There is healing and light energy in plants born of good soil; once harnessed, the power of plant-based fuel for our bodies is endless.

I have shared this life-saving message with friends and family. My mother, who suffered from severe lupus, adopted the Nutritarian eating style and also recovered. She has also regained her ability to walk and now has her energy back. My wife, Sonia, also switched to this way of eating and has lost weight and feels terrific.

I have some simple advice for anyone who's thinking about starting a Nutritarian style of eating: Just get it done. I tell people, "Put your plants on and take your weight off!"

---

amino acids. All in all, this is a total of roughly forty substances that are the bare minimum for basic bodily function. However, there are thousands of lifespan-extending phytochemicals, many of which are newly discovered. No one compound is essential for life, but the full symphony of nutrients and phytonutrient exposure is essential for a long healthy life. This brings us to the third principle of the Nutritarian diet.

THE THIRD PRINCIPLE OF A NUTRITARIAN DIET

> Optimal exposure to all macronutrients and micronutrients humans require is needed to maximize health and lifespan— this is called Comprehensive Nutrient Adequacy (CNA).

Sustained health requires micronutrients. It is just that simple. Virtually every metabolic pathway functions better with a full assortment of micronutrients present. Deficiencies in micronutrients may not be severe enough to create immediate clinical symptoms, but in the long term, suboptimal intake leads to an increased risk of the diseases associated with aging. Because of our population's inadequate consumption of fruits, vegetables (especially green vegetables), beans, nuts, and seeds, practically every American is deficient at some level in these protective nutrients.

Our lack of nutritional diversity impacts our DNA. Whenever human cells are studied in a nutritionally insufficient environment, we see increased DNA damage.[1] For example, a vitamin B12 deficiency, which is common worldwide, does not just lead to anemia and nerve damage, it also damages the DNA of our cells.[2] Likewise, the DNA damage in someone who is chronically deficient in folate is equivalent to the damage from ten times the allowable annual exposure to ionizing radiation.[3] In other words, the lack in the diet of greens and beans (which are high-folate foods) has an impact on our DNA that is similar to that of smoking or radiation exposure.

As you can see in the graph, insufficient intake of vitamins and minerals is common in Americans, even with widespread use of multivitamin and mineral supplements.

Insufficient Intake of Vitamins and Minerals in the US

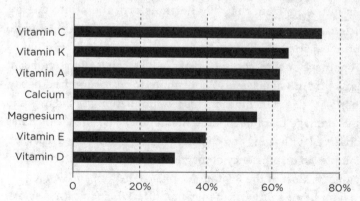

Percent of Americans meeting recommendations

Source: Fulgoni VL, Keast DR, Bailey RL, Dwyer J. Foods, fortificants, and supplements: where do Americans get their nutrients? *J Nutr.* 2011;141:1847–54.

More than a decade ago, Dr. Bruce Ames first came up with the *triage hypothesis*, which states that even moderate insufficiencies of micronutrients can shorten one's life.[4] When faced with insufficiency, the body uses whatever nutrients are available to ensure that the most basic and pressing metabolic functions are fueled first. If a needed nutrient is not available, the body compromises long-term health to ensure short-term critical function. When our stores of a vitamin or mineral are limited, our body prioritizes the most essential and immediate functions of the vitamin—like a doctor in an emergency room using triage to decide which patients need the most immediate attention. The body will always support survival and reproduction over long-term protective functions. Over time, this prioritizing damages our long-term health and longevity. Stabilization and maintenance of long-term survival proteins and stem cells will be sacrificed for immediate needs when the level of nutrients is insufficient. This means that outward effects of vitamin and mineral insufficiencies are invisible until we age, but are likely even more dangerous than typically thought.

Insufficiencies can lead to DNA damage and mitochondrial dysfunction. The mitochondria are organelles found in cells that help process nutrients from the cell and transform them into energy. Insufficient amounts of vitamins A, C, and E; folate; niacin; vitamin B6; biotin; calcium; iron; magnesium; and zinc have all been linked to DNA damage or chromosome breaks, which can lead to significant problems.[5] Insufficiencies also contribute to mitochondrial decay. Mitochondrial decay includes increased production of oxidants and oxidative damage to mitochondrial membranes and DNA, which has been observed with aging and contributes to age-associated diseases. For example, an insufficiency of magnesium results in mitochondrial DNA damage. Most of the B vitamins, plus zinc, iron, and copper, are essential for heme synthesis in mitochondria, which in turn is essential for the mitochondrial reactions that produce energy.

When nutrients are scarce, the body will triage which nutrients are most important for survival to sustain physiological function and reproduction; it will move away from the regulation and repair of cellular DNA and proteins that increase longevity. At that point, the stage is set for the development of chronic disease and accelerated aging, especially DNA breakage, which can lead to cancer.[6] When an adequate supply of necessary nutrients is available, however, both short-term and long-term health are preserved.

In the past, we thought that as long as one consumed enough of the essential nutrients, all of the other nutrients would be built out of the essentials. Vitamin deficiency led to primary diseases associated with vitamin shortage, such as scurvy, pellagra, and beriberi. If the deficiency was not corrected, plenty of problems occurred. However, now we know that there are subclinical deficiencies of many micronutrients. For most of us, this lack of micronutrients is not severe enough to cause immediate symptoms or death, but it will accelerate aging and impact our lives as we age.

In 2009, Ames and his team at Children's Hospital Oakland Research Institute published an article that explored the triage theory and asked the question: Is micronutrient inadequacy linked to diseases of aging?[7] First, they showed that Ames's triage theory for micronutrients was indeed correct. Second, they clearly demonstrated that the current recommendations for vitamin K intake need to be increased to prevent heart disease, osteoporosis, arterial and kidney calcification, and even cancer. Vitamin K is concentrated in dark green plants such as kale, collards, and spinach; it is present in only small amounts in most multivitamins. Very simply, if you don't eat greens and get enough folate and vitamin K, it's as if you are being exposed to radiation in terms of health effects. The SAD is grossly deficient in vitamin K and the reference daily intake (RDI) is ridiculously low too. Four weeks of Nutritarian menus contains an average daily vitamin K intake of 1,400 mcg—more than ten times the RDI of 120 mcg.

In summary, the foods you choose to regularly consume should contain optimal amounts of all nutrients that humans require to thrive throughout life. The major deficiencies that lead to chronic

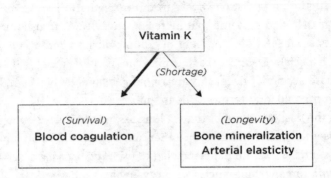

illnesses and premature death are the result of a scarcity of antioxidants and phytochemicals in green vegetables and other colored plants. This is perhaps the easiest thing we can do to live longer and healthier lives: Eat more plants and vegetables of all colors.

THE VEGAN QUESTION: SOME SUPPLEMENTS ARE IMPORTANT

Eating a diet rich in natural plants can raise health issues if it is not supplemented properly. Though a vegan diet may be the most lifespan-favorable diet, there are drawbacks and risks from lack of exposure to beneficial nutrients, such as vitamin B12, zinc, and docosahexaenoic acid (DHA), which are more readily available in animal products. Deficiencies and even chronic insufficiencies of these three micronutrients can have life-threatening consequences.

All the kale and wild berries in the world won't do you much good if you lose brain function or become unable to walk because of a vitamin B12 deficiency. As a person's diet includes more whole plant foods and fewer animal products, the risk of having suboptimal levels or even deficiencies of vitamin B12, zinc, and DHA may increase. This is true in some individuals more than in others. This suboptimal exposure to these micronutrients may also still be present when someone is eating small amounts of animal products. We can get the full benefit from plant-based eating only when we understand the risks of nutrient deficiencies and pay attention to these few but necessary nutrients that are not as easily accessible from plants.

A Nutritarian diet includes a wide portfolio of protective plant foods to supply a full spectrum of phytonutrients. It also includes supplemental ingredients to ensure all nutrients are accounted for in the sweet spot for health optimization and prevention of serious consequences of chronic insufficiency, even if not severely deficient.

ZINC AND PROTECTING AGAINST PNEUMONIA AS WE AGE

Particularly for the elderly, infectious diseases such as pneumonia are preventable causes of sickness, hospitalization, and death. Pneumonia is an infection of the lungs that can be caused by bacteria, viruses, or fungi. Influenza is a common cause of viral pneumonia, which in some cases can lead to bacterial pneumonia. Influenza and pneumonia combined are the seventh leading cause of death in adults 65 and older.[8]

In our 70s, 80s, and 90s, taking nutritional care of the immune system becomes increasingly important, as immune function declines as we age, making influenza and pneumonia greater threats. This dampened immune function, for most SAD-eating Americans beginning around ages 60 to 65, increases our susceptibility to developing pneumonia, the risk of complications, and the time it takes to recover. Declining immunity increases the risk of cancers as well.

Research suggests that maintaining adequate intake of zinc is particularly important for preventing pneumonia in the elderly.[9] This is an important consideration because the antioxidants found in plant food, called *phytates*, limit the absorption of zinc from plants such that only about 20 percent of the zinc present in beans and greens is absorbed by the body. So even if the content of a plant-based diet appears to meet our zinc requirements, the levels supplied are still suboptimal, and with absorptive capacity decreasing further with aging, this potentially places the aging vegan at increased risk.[10]

Zinc has many different functions in the body; it drives hundreds of chemical reactions, reduces oxidative stress, plays a structural role by stabilizing proteins, and regulates the expression of many genes. It is also required for neurotransmitter release and insulin packaging and secretion.[11] Zinc is essential for DNA synthesis and cell proliferation, and for this reason, highly proliferating cells—like immune cells—depend on an adequate supply of zinc. Furthermore, growth or function of macrophages, neutrophils, natural killer cells, T cells, and B cells is impaired by zinc deficiency.[12] In short, zinc plays a critical role in the body, and we must ensure that we maintain adequate levels of it throughout our lives.

The characteristics of immune system decline with age are similar to those of zinc deficiency, implying that diminishing levels of zinc could be a major factor in the age-associated decline of immune function. Zinc needs are estimated to be about 50 percent higher for people who follow a completely plant-based diet. Also, other minerals interfere with zinc; iron and calcium compete with zinc for absorption, and copper competes with zinc for binding proteins inside the body's cells.[13] Several studies have found zinc deficiency to be prevalent in the elderly. This is not a surprise, given that fewer than half of older adults in the US consume adequate zinc.[14]

In a study of older adults living in nursing homes, those with normal serum zinc levels had a lower incidence of pneumonia and required half as many antibiotic prescriptions compared with those who had low zinc levels.[15] Several studies on supplementation suggest that improving zinc status in the elderly improves immune system function and resistance to infection. In a study published in 2007, adults 55–87 years old had lower plasma zinc levels and higher oxidative stress and inflammatory markers compared with younger adults. Half of the subjects took zinc supplements for twelve months, and the other half took a placebo. The rate of respiratory infections and markers of inflammation and oxidative stress were lower in the zinc group than in the placebo group.[16]

As mentioned, pneumonia is a major cause of premature, largely preventable death in the elderly. Eating a Nutritarian diet, with its attention to the broad array of immune-supporting phytochemicals, slows the aging process and improves immune function, lowering the risk of pneumonia and other infections. However, maintaining adequate levels of zinc is a valuable adjunct to help prevent later-life deaths by further promoting the optimal functioning of the immune system.

Zinc also is helpful in preventing mood disorders. A meta-analysis of seventeen studies measuring zinc levels demonstrated that zinc concentration averaged 1.85 mmol/L lower in depressed subjects versus controls, and greater depression severity was associated with greater relative zinc deficiency.[17] So, zinc is an important nutrient to pay attention to for anyone following a healthful, plant-centered diet, especially older adults.

FISH AND LONG-CHAIN OMEGA-3 FATTY ACIDS FOR HEALTHY AGING

"Healthy aging" means getting old without chronic disease or cognitive and physical dysfunction. Despite some confusion and disagreement among studies on the health effects of omega-3 fatty acids derived from fish, there is overwhelming evidence that higher levels of long-chain eicosapentaenoic acid (EPA) and DHA (both are omega-3 fatty acids) measured in the blood over time are associated with a lower risk of unhealthy aging and reduced brain function.[18]

A review and meta-analysis published in 2012 set out "to assess the role of omega-3 supplementation on major cardiovascular outcomes" and found "no definite benefits."[19] An even larger 2018 review showed largely the same thing, with only an insignificant advantage when heart patients took fish oil supplements.[20] As a result of these studies and the conclusion of the large-scale randomized VITAL trial (**VIT**amin D and Omeg**A**-2 Tria**L**), many in the vegan community were quick to shout out, "Fish oils do NOT prevent heart attacks." However, the lead investigators of these studies explain that the results are complicated and that is not what the study showed.[21] In fact, it showed that 2,000 IU of vitamin D did reduce cancer deaths even though the subjects were tracked for only five years. It also showed that omega-3 fish oils did reduce the risk of heart attacks in people eating little or no fish. Supplements had a negligible effect only in those who had an adequate intake of EPA and DHA—exactly what we would expect. If you are getting these nutrients from fish, extra supplementation will have little impact on your health.

If we look at the African Americans in the VITAL study who were not eating lots of fish, heart attack rates were reduced by 77 percent in those who took supplements compared with those who received a placebo. In the entire study population, considering those who ate fewer than one and a half servings of fish per week, heart attacks were reduced by 28 percent and the risk of all cardiovascular events (stroke included) was reduced by 19 percent. We shouldn't expect that omega-3 supplements would be beneficial for those people who eat more fish, so the results actually did not prove anything extraordinary, nor did they determine that supplemental DHA and EPA was worthless, as is so often claimed.

Still, the benefits need to be tempered and put in perspective. The SAD causes heart attacks and premature death. Taking fish oils, aspirin, or even statins is not going to change that; we are still going to see lots of heart attacks and strokes in people who rely on the SAD. We have to face the reality that the quality of our diets is too powerful to overcome the effects of an unhealthy diet with a few pills.

All the contradictory studies on the long-chain omega-3 fatty acids DHA and EPA are helpful to analyze in depth. Remember that these studies do not identify and treat only people found to be deficient in DHA; they treat everybody. My in-depth review of the details of hundreds of such studies indicates that *both* deficiency and excess of long-chain omega-3 fatty acids can be harmful.

The idea that fish oil supplementation may not be a great idea first came to light in 2003, after the Diet and Angina Randomized Trial (DART-2) was published. It was the first study on fish oil that followed people with heart disease for many years. It randomized 3,114 men, younger than age 70 and with angina, to one of four groups:

1. those advised to eat two portions of oily fish weekly, or to take three fish oil capsules daily (3 grams),

2. those advised to eat more fruit, vegetables, and oats,

3. those given both the above types of advice, and

4. those given no specific dietary advice.

Mortality was ascertained after three to nine years. The researchers determined that the group advised to eat more fruits and vegetables did not follow the instructions and thus did not make enough dietary modifications for the results to be valid. The group advised to eat more fish and take fish oil did follow the instructions, and that group was found to have a 25 percent increased risk of death and more than a 50 percent increased risk of sudden cardiac death.[22] The excess risk was largely found among the subgroup given fish oil capsules. There is clearly a problem with 3 grams of fish oil a day—that is a lot of fish oil!

One problem with fish oil is that it is vulnerable to oxidative damage, or rancidity (and furthermore, some supplements are more rancid and contain more contaminants than others). Another problem is the high amount of fish oil supplementation that many people recommend, such as in this study. My review of all the studies on this issue, and the tremendous number of differences found between the studies, demonstrates that exposure to some fish or a small amount of fish oil seems to offer benefits. However, the minute you start taking high doses (more than 1 gram daily) of fish oil, problems seem to erupt and negate many of the benefits.

It seems that a deficiency of the long-chain omega-3 fatty acids EPA and DHA is not healthspan- or lifespan-favorable, but too much may be not favorable as well. But remember: Just because an excess of some supplemental ingredients may have risk does not mean we permit and accept deficiencies. When we identify a deficient population, then supplementation becomes clearly beneficial.

Consider a study published in 2013 that I consider to be one of the best to evaluate the risks and benefits of DHA and EPA adequacy.[23] It evaluated the risk of death in 2,692 individuals between ages 70 and 80, following the levels of these beneficial fatty acids in their blood. These individuals did not have a history of heart disease at baseline. Total mortality, as well as fatal and nonfatal heart attacks and strokes, was tracked for sixteen years. This study was particularly informative because it relied on blood tests to determine the DHA and EPA levels, rather than just on self-reported dietary histories. The study concluded that higher blood levels of omega-3 EPA and DHA were associated with lower total mortality, especially deaths from heart disease. The result was more than two years of longer life in the highest quintile.

Another study from 2013 corroborated this finding—that an adequate, but not excessive, amount of EPA and DHA was lifespan-favorable.[24] In a multiethnic cohort (whites, Hispanics, African Americans, and Chinese Americans) of 2,837 US adults, blood levels and dietary intake were examined. For the following ten years, cardiovascular events, deaths, and strokes were monitored. This variable-controlled analysis found that both dietary and circulating EPA and DHA levels were inversely associated with cardiovascular events.

The conclusion is clear. We need the long-chain omega-3 fatty acids EPA and DHA. But it is important to recognize that both too-low levels and too-high levels can place us at increased risk of cardiovascular events. The need to determine safe and beneficial levels holds true, in most cases, with other nutrients that the body requires.

Given all the findings, I recommend supplementing EPA and DHA to prevent insufficiency only if you do not eat fish regularly. That said, you must not overdo supplementation. It is ideal to use an algae-based DHA and EPA supplement in a relatively low dose (such as 200 to 300 milligrams total) to prevent deficiency—and also to protect yourself from exposure to the environmental contaminants potentially found in fish and fish oil.

FATTY ACID DEFICIENCY, DEMENTIA, AND DEPRESSION

Even though the risk of heart disease for a vegan would be exceedingly low, and even lower for someone eating a Nutritarian-style diet, there is still the potential risk of depression and dementia from fatty acid deficiency, especially from chronically low EPA and DHA levels if these omega-3 fatty acids are not supplemented.[25]

More than a dozen epidemiological studies have reported that reduced levels of long-chain omega-3 fatty acids are associated with increased risk for age-related cognitive decline or dementia such as that found with Alzheimer's disease.[26] DHA promotes the growth and maintenance of nervous tissue and improved cognition. In animal models, DHA adequacy prevents the amyloid accumulation commonly seen in Alzheimer's. Most critically, a study on participants in the Women's Health Initiative showed increased brain shrinkage with aging in women with lower blood levels of DHA evaluated eight years after initial DHA measurements.[27] The hippocampus area plays an important role in memory and usually begins to atrophy before symptoms of Alzheimer's appear. This area of the brain was found to be particularly vulnerable to shrinkage associated with a low omega-3 index (red blood cell EPA plus DHA) comparing the lowest quartile with the highest. The mean index of the lowest quartile was 3.4 percent and the mean index of the highest quartile was 7.5 percent.

This is of particular concern considering the large number of unsupplemented vegans who demonstrate deficiency as measured by the omega-3 index. In this study, 166 healthy vegans showed a wide range in DHA-EPA levels that did not correlate with the amount of short-chain omega-3 intake in their diets.[28] This means that more alpha-linolenic acid (ALA, a plant-based omega-3 fatty acid) from flaxseeds and walnuts did not translate into higher levels of EPA and DHA, suggesting that the major variation was primarily due to genetic differences in conversion enzymes.

A significant percentage of the vegans had levels below 3 percent (as indicated in the graph), suggesting a serious risk of later-life brain compromise. This study also demonstrated that a low-dose, algae-derived supplement of only 254 milligrams of EPA (82 milligrams) plus DHA (172 milligrams) was sufficient to normalize the omega-3 index results on subsequent blood tests.

Omega-3 Index in Vegans

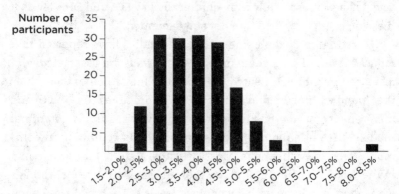

My more than thirty years of clinical experience treating vegans who have some failure to thrive on a vegan diet has repeatedly demonstrated individuals whose lack of DHA and EPA resulted in an episode of depression or anxiolytic depression. These observations are consistent with studies demonstrating that patients with depression commonly have lower EPA and DHA levels.[29] DHA deficiency in depressed patients relative to healthy control subjects is linked to anxiety and depression. The low omega-3 fatty acid status commonly observed in patients with major depressive disorder may reduce the effectiveness of some antidepressants.[30] Supplementation with DHA and EPA, especially EPA, has been shown to be helpful for people who are depressed.[31]

Other common deficiencies might additively increase depression risk. Low vitamin D (also found in fish), best measured with a 25-hydroxy vitamin D blood test, can also exacerbate the risks of dementia and depression and has been associated with diminished global cognitive function and greater decline over a four-year period. In one study, patients with 25-hydroxy vitamin D values higher than 30 ng/mL showed insignificant decline, those with levels of 20 to 29 ng/mL showed moderate decline, and those with levels less than 20 ng/mL showed severe decline. These findings were matched almost perfectly in another study as well.[32]

The Nutritarian diet is effective because it relies on the latest science and because it accommodates the varied needs of individuals. We must all pay attention to these factors to assure that people counting on longevity-promoting dietary advice are not left with some irreversible loss of memory and brain function. I see many patients who eat healthfully and follow the advice of dietary leaders, yet they still have low levels of these important nutrients, with some of them developing dangerous and even irreversible brain pathology and thousands of others placing themselves at needless risk.

WHY SOME VEGAN DIETS FAIL AND HOW TO REMEDY THAT PROBLEM

A diet style that most effectively reverses heart disease is vegan or near-vegan. But we must make sure, in our valiant efforts to win the war against excess caloric consumption and the overconsumption of animal products, that we don't ignore the potential nutritional risks of a vegan diet. It has become apparent from extensive research and clinical experience that a very-low-fat vegan diet (that excludes nuts and seeds and omega-3 supplementation) should ignite a spark of caution. That eating style is not natural to humans' genetic heritage and has never been pursued for generations. Its promotion as the "perfect" diet for all can result in failure to thrive in children and can be unsafe for pregnant and nursing women, increasing the risk of postpartum depression. In addition, an unsupplemented vegan diet can increase the risk for major depression and later-life dementia. Furthermore, as shown in the Adventist Health Studies and other prospective studies (see Chapter 7), such a diet can increase the risk of life-threatening arrhythmias for people with advanced heart disease.

We must always err on the side of caution when we face controversy or inadequate definitive studies. Very little in the world of nutritional science is known with 100 percent certainty. We always need to consider the preponderance of the evidence within a conservative framework, along with broad clinical experience.

Susan Walters
Lost 84 Pounds and Now Is Free of Diabetes and Thyroid Disease

I am a public-speaking coach, and rarely at a loss for words, but in 2016, a diagnosis of type 2 diabetes left me speechless. At 52 years of age, I weighed 261 pounds with a BMI of 36, and I had been feeling awful for a long time before the diagnosis. I had high blood pressure, I was thirsty all of the time, and I was experiencing brain fog and tightness in my chest. Finally, I visited my doctor, who confirmed the bad news: I had type 2 diabetes, with a hemoglobin A1c (HbA1c) level of 8.9 percent—as well as thyroid disease and likely heart disease too. Much of what happened next was a blur.

I couldn't say anything—I started crying. With one part of my brain, I could hear the doctor rattling off instructions: I'd have to stick my fingers regularly to test my blood sugar, start taking this drug and that one, and on and on. But I wasn't paying attention—all I could think about was whether I would lose my eyesight or a limb, have a heart attack or a stroke, or any of the other things I associated with diabetes. I felt like I was in a waking nightmare.

Even in this panic, I recognized that this diagnosis had been the work of a lifetime. Growing up in a suburb outside Chicago, I had enjoyed a typical American childhood—fueled by the SAD. Breakfast would usually be cereal, toast, pancakes, eggs, or oatmeal; lunch usually consisted of sandwiches—deli meat with butter and cheese, accompanied by potato chips. Dinner was almost always meat and potatoes, with pizza as an occasional change of pace. Desserts and sugary treats weren't just part of my mealtimes—in my family, birthdays, good grades, and good behavior were celebrated with ice cream. Chocolate ice cream, chocolate sauce, chocolate morsels—these snacks were part of my daily food pyramid.

This diagnosis of diabetes forced me to take action. I had gotten myself into this situation—so I was determined to get myself out. I left the doctor's office and went straight to a bookstore and proceeded directly to the health section. That's where I found Dr. Fuhrman's book *The End of Diabetes*—and the knowledge that would change my life. There it was, in writing. I didn't have to just treat my diabetes—I could *reverse* it, I could get RID of it! I read that book cover to cover and learned how eating the right foods could completely transform my health. So I did everything Dr. Fuhrman said to do.

I focused on myself and made changes to my lifestyle; I followed the Nutritarian recipes and meal plans to the letter. I lowered my meat consumption to 6 ounces a week and added beans, nuts, seeds, and more greens to my diet. I also ate onions, mushrooms, and berries. I started walking every day to get exercise. I lost more than 30 pounds in three months! At the end of those three months, new blood work showed that my HbA1c level had dropped dramatically, to 5.8 percent—I was not diabetic anymore! I have continued to get even healthier and lose more weight since then; now, two years later, I have lost 84 pounds and have been able to get off seven medications.

And not only am I not diabetic, I'm not even prediabetic! All of my symptoms are gone! My blood pressure is normal, my thyroid normalized, and I am feeling great. The brain fog and chest pain have disappeared. My husband and I are both sleeping through the night, I can keep up with my kids, and I have lots of energy. I have more patience with everyone and am actually having fun in my life.

My husband, Jeff, and our two daughters, Jenna, 15, and Jessica, 11, have also embraced this incredibly healthy way of eating. It's especially important to us because Jessica is a leukemia survivor. I make sure that the anticancer superfoods G-BOMBS are included in our meals every day—we have fun trying to find new and different ways to do that. We all feel so much better and happier now that we are eating so healthfully.

I plan to use my public-speaking skills to share this great way of living with as many people as possible. I want to tell them: You will love how you feel. You *CAN* take control of your health. What you put into your body *DOES* matter. Food is medicine. If I can do it, you can, too!

My thirty years of a very busy clinical practice, which caters to this population of vegans and health seekers, may place a different light on some of these issues. Many people sought me out after they had developed problems eating a vegan diet. I have communicated with other medical professionals with similar practices who have reported similar or identical findings, shedding light on these potential risks of a vegan diet that is not appropriately supplemented to ensure nutritional adequacy. The most likely observed deficiencies found to cause problems are vitamin B12, zinc, iodine, DHA-EPA, and vitamin D.

Certainly, vitamin B12 is the most important supplement to use with a vegan or near-vegan diet or as we age, but it is also important to assure a source of iodine in the diet since so many of us do not use iodinated salt or eat seaweed regularly.

Without a predetermined philosophy or agenda, we must turn to a clear, unbiased evaluation of today's science, in conjunction with assessment of the broad spectrum of individual needs, to ensure favorable dietary and health outcomes. The Nutritarian diet does just that to ensure optimization and safety for all.

CHAPTER THREE: QUICK SUMMARY

There are two reasons why people who follow a Nutritarian diet age more slowly and live longer: First, their diet contains adequate levels and a diversity of micronutrients for all stages of life, and second, Nutritarians consume this broad range of micronutrients without consuming excess calories.

THE THIRD PRINCIPLE OF THE NUTRITARIAN DIET

A diet must include all of the macronutrients and micronutrients that humans need to maximize their health and lifespan—this is Comprehensive Nutrient Adequacy (CNA).

Because the animal products and processed foods that make up the bulk of the SAD lack significant levels of important micronutrients, people who eat this way are almost always deficient in antioxidants and phytochemicals. Humans require sufficient macronutrients (calories) plus approximately forty micronutrients for their metabolisms to run normally. Long-term deficiencies inevitably lead to an increased risk of diseases associated with aging.

THE TRIAGE HYPOTHESIS

"Triage" is the process of determining the order of medical treatment on the basis of the urgency of need and available resources. In 2006, Dr. Bruce Ames introduced the triage hypothesis to describe the body's response to nutritional insufficiency. When nutrients are scarce, the body is programmed to choose immediate survival over long-term health. Basic physiological functions and reproductive needs are given priority, while processes that are less critical in the moment, but that affect long-term health, get fewer resources.

THE IMPORTANCE OF YOUR DIETARY PORTFOLIO

Throughout your life, it is essential for you to eat a diet that contains optimal amounts of all the nutrients that humans require to thrive. Though a vegan diet may be the most lifespan-favorable diet when supplemented appropriately, serious risks are associated with lack of exposure to the beneficial nutrients that are more readily available in animal products, such as vitamin B12, zinc, and DHA. A Nutritarian diet utilizes supplements (judiciously) to assure that all nutrients are included in adequate amounts.

THE ROLE OF ZINC

Zinc drives hundreds of chemical reactions in the body, reduces oxidative stress, plays a structural role by stabilizing proteins, and regulates the expression of many genes. Zinc is also required for neurotransmitter release and insulin packaging and secretion. It is essential for DNA synthesis and cell proliferation, and for this reason, highly proliferating cells—such as immune cells—are dependent on an adequate supply of zinc. Diminishing levels of zinc could be a major factor in the age-associated decline of immune function. Several studies have shown zinc deficiency to be prevalent in the elderly. Fewer than half of older adults in the United States consume adequate zinc.

THE IMPORTANCE OF LONG-CHAIN OMEGA-3 FATTY ACIDS

Eating fish or taking fish oil supplements raises concerns because many fish contain contaminants and fish oil can become rancid. I recommend taking a conservative dose of algae-based DHA and EPA to prevent deficiency in these long-chain omega-3 fatty acids. When it comes to EPA and DHA blood levels, it is important to strive for adequacy. If levels are too high over time, we can be placed at increased risk of cardiovascular events and cancer (possibly prostate cancer for men). On the other hand, if levels are too low over time, we are at increased risk of depression, brain shrinkage, and dementia, in addition to an increased risk of cardiovascular events. Adequate levels of omega-3 fatty acids, however, have been shown to have important brain health and longevity benefits. Supplementing the diet with omega-3 fatty acids is especially important for vegans, because fatty acid deficiency due to chronically low EPA and DHA puts them at increased risk for depression and dementia.

CHAPTER FOUR

Your Health Is in Your Hands

WITHOUT NUTRITIONAL STRESS WE LIVE LONG, HEALTHY LIVES

Living healthfully to 100 years is all about eating much more of the healing superfoods—the ones that have the most scientific evidence demonstrating their lifespan-enhancing and anticancer effects. In the race to get to the moon, we landed a man there and planted a flag for all humanity. In the race to win the war on cancer, we also have planted a flag: We know how to prevent the vast majority of cancers.

The only problem is that people don't like the answer we have found: vegetables. Most are looking for a magic pill or potion that can enable them to eat pizza, burgers, and fried chicken and not get breast and prostate cancers. Unfortunately, our health trajectory is not a fairy tale based on dreams and pixie dust. Instead, it is the unyielding consequence of all the choices we have made during our lives. Some people may feel threatened and upset when told that their behavior is responsible for the health tragedies they incur down the road; but the fact is, we are in more control than we think, and we

Sue Kehoe
Now Optimistic About Life

I was diagnosed with estrogen-positive invasive breast cancer (Stage 4) in May 2011 at the age of 46. I had a very large mass (4 by 2 inches) in one breast, and the cancer had metastasized to three places in my liver. The prognosis from the oncologist was very poor, and when pushed, doctors said they expected me to live less than two years.

Obviously, I was discouraged with the diagnosis but listened to the doctors and immediately began six months of treatment: aggressive chemotherapy to possibly shrink the tumors, followed by a lumpectomy and radiation. The first chemo treatment lasted more than four hours and made me so weak and ill that after being driven home I had to be carried into the house. Knowing I had six months more of this debilitating treatment, I began to feel hopeless. When I was at my lowest point and felt like giving up, I happened to turn on the TV for a distraction and there was Dr. Fuhrman starting his PBS show "3 Steps to Incredible Health!" He said, "I'm going to give you the knowledge that can set you free from chronic disease." That got my attention! I now believe a higher spiritual power put me in front of the TV to hear Dr. Fuhrman's message at the perfect time.

I never even considered that maybe the food I ate had anything to do with cancer or that it would have any effect on my cancer prognosis. But I listened intently to Dr. Fuhrman's talk and immediately started doing research online and reading his books. The science made complete sense. Of course the food you eat has everything to do with your immune system and your body's ability to fight cancer! That same day my husband, Steve, and I started eating a 100 percent Nutritarian diet with a focus on G-BOMBS. When I shared this newly found knowledge with my oncologist, he insisted that diet and exercise wouldn't have any effect on the cancer at this stage.

Steve and I had eaten very poor diets that consisted of about 90 percent restaurant food. We were very busy running a small business and didn't have any time to cook or even consider our health, for that matter. Over the years, we both added weight and felt sicker, just like most people in our society. But within weeks of starting to eat a Nutritarian diet, both of us started to feel more energized. The next chemo

treatment was difficult, but I got through it much better than the first round. Amazingly, each treatment got easier rather than progressively harder. After the chemo treatments ended, I kept following Dr. Fuhrman's advice and began feeling better and stronger. I sensed that I was healing, but doctors cautioned me not to be too hopeful because the cancer was so advanced.

Once my hair grew back after chemotherapy, Steve and I were able to take a trip to Italy and spend a week at a retreat with Dr. Fuhrman. I left Italy with the one thing I really needed . . . HOPE! I felt empowered that I could control my future health destiny, including the cancer. I continued to learn and attended numerous Dr. Fuhrman nutritional seminars and retreats. The more I learned and met others who had reversed major chronic diseases, and were cancer-free for decades, the more my confidence grew.

That was more than eight years ago, and I have no signs of cancer and I have lost 100 pounds and kept if off. Steve's cholesterol also dropped from almost 300 to less than 150. We are both healthier and happier than we have ever been and are forever grateful to Dr. Fuhrman and the tremendous work that he does.

have the freedom and security to be proactive so we can live without fear and know we can protect ourselves and our loved ones.

Of course, there are always exceptions to every rule, as well as to the general rule that you can only be as healthy as the health-giving properties of the foods that you put into your mouth. We do not live in a pristine bubble, and there can be a degree of randomness in the location and severity of toxin-induced DNA damage. However, those rare exceptions do not change the fact that nutritional excellence can prevent the vast majority of cancers. The most common cancers would be very rare in a population that followed these healthful Nutritarian dietary guidelines. And by "healthful," I mean a diet with more than 90 percent of calories coming from unrefined, unprocessed, organic, whole plants. This diet is even more protective when those plants are nutrient-rich and chock-full of anticancer phytochemicals.

THE FOURTH PRINCIPLE OF A NUTRITARIAN DIET

Synthetic chemicals, toxins, pathogenic bacteria, parasites, and other disease-causing substances should be avoided.

Obviously, we should avoid exposure to dangerous bacteria, parasites, and worms and known chemical carcinogens such as smoke, chemicals, and asbestos. However, more typical toxic exposures come from cooked meats and other cooked animal foods containing dangerous compounds such as heterocyclic amines, polycyclic aromatic hydrocarbons, and lipid peroxides formed during high-temperature cooking. All of these are linked to an increased risk of cancer.[1] Then we have exposure to packaging chemicals such as bisphenol-A (BPA), dioxin, food additives, arsenic, toxic metals, and even glyphosate (a weed killer) and other chemicals used on commercial crops.

ACRYLAMIDE AND AGES

Acrylamide is a relatively newly discovered dietary toxin first reported in 2002. The primary dietary sources of acrylamide are potato chips, French fries, and other dry-cooked starchy foods such as bread products and breakfast cereals. Coffee is also a source of acrylamide.[2]

Acrylamide is one of hundreds of chemicals known as Maillard reaction products (MRPs) formed when foods are heated at high temperature with dry heat. The Maillard reaction occurs when sugars react with amino acids in foods, producing the characteristic browning effect of frying, baking, and roasting. Acrylamide is not produced during boiling.[3] Acrylamide has been classified as a Group 2A carcinogen (probably carcinogenic to humans) by the International Agency for Research on Cancer (IARC).

A meta-analysis of four prospective studies found that high dietary acrylamide intake was associated with a 39 percent increase in endometrial cancer risk in women who had never smoked. In meta-analyses of multiple studies, an increase in the risk of ovarian cancer among never-smoking women has also been reported, as well as an association with kidney cancer.[4]

A related family of chemicals known as advanced glycation end products (AGEs) are created when acrylamide and other MRPs bind to proteins in foods or proteins in the body. AGEs, also known as glycotoxins, are a diverse group of highly oxidant compounds that accumulate in the body and cause cumulative pathogenic damage leading to disease, premature aging, and death. AGEs can induce cross-linking of collagen, increasing vascular stiffening and entrapment of low-density lipoprotein (LDL) particles in the artery walls. Its oxidation effects can further oxidize LDL cholesterol (the "bad" cholesterol), accelerating atherosclerosis (the accumulation of plaque on the artery walls) and heart disease risk.[5] Though these toxins are associated with baked, high-glycemic carbohydrates, cooked animal products such as barbecued chicken and pan-fried bacon have the most AGEs.[6]

AGES AGE US. THEY

- Increase vascular permeability[7]
- Increase arterial stiffness[8]

- Oxidize LDL, accelerating atherosclerosis[9]

- Enhance oxidative stress and inflammation in endothelial cells[10]

- Create diabetic complications, damaging the kidneys, eyes, and nerves[11]

To reduce your exposure to acrylamide and AGEs, cook starchy foods like turnips and sweet potatoes mostly in stews or soups because cooking in water prevents acrylamide formation. When you bake or roast vegetables, do not allow them to brown. Increasing the water content helps—soaking starchy foods in water before roasting reduces acrylamide formation.

PERSISTENT ORGANIC POLLUTANTS

Some pesticides and other industrial chemicals remain in the environment even though they have not been produced or used in decades. The list includes now-banned substances such as the insecticide DDT (dichlorodiphenyltrichloroethane) and the chemicals known as PCBs (polychlorinated biphenyls). Dioxins also fall under the category of persistent organic pollutants (POPs), which are released into the environment from waste incineration, burning of fuels, and forest fires. They are called "persistent" because they persist in the environment for a very long time; they are easily transported by water or wind and do not break down readily. Because they are fat-soluble, POPs accumulate in the fatty tissues of animals, so our main route of exposure to them is via fatty animal–based foods, such as fish, butter, and ground beef.[12]

According to data compiled by the Environmental Working Group, farmed salmon may be the biggest concern for PCB exposure. Seven of ten farmed salmon samples from grocery stores were found to be contaminated with PCBs. Furthermore, the group found that farmed salmon had sixteen times more PCB content than wild-caught salmon and four times more than beef.[13] Sport-caught fish or shellfish are often high in PCBs and DDT. Commercial fish that are high in these

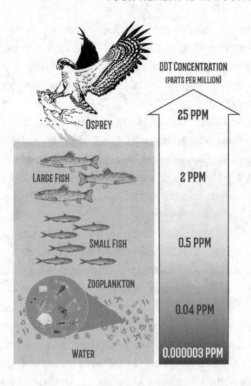

DDT CONCENTRATION
(PARTS PER MILLION)

OSPREY — 25 PPM

LARGE FISH — 2 PPM

SMALL FISH — 0.5 PPM

ZOOPLANKTON — 0.04 PPM

WATER — 0.000003 PPM

pollutants include Atlantic or farmed salmon, bluefish, wild striped bass, white and Atlantic croaker, blackback or winter flounder, summer flounder, American eel, and blue crab.

METHYLMERCURY

Methylmercury is a heavy metal that humans are exposed to primarily by eating predatory fish—that is, fish that eat other fish. It is a known neurotoxin and is harmful to pregnant women and children because methylmercury exposure impairs brain development. In adults, low-level exposure to methylmercury appears to primarily harm the cardiovascular system, with higher mercury levels in hair or blood linked to a greater risk of heart attack and progression of atherosclerosis. Therefore, mercury exposure likely counteracts some of the cardiovascular benefit of the omega-3 fatty acids DHA and EPA in fish.[14]

The general public has elevated fish, especially salmon, to health food status because of its omega-3 fatty acid content, but those fatty acids can be easily obtained via algae-based supplements. Eating fish has the negative consequences of excessive exposure to animal protein, methylmercury, and PCBs.

I recommend that you avoid or minimize fish consumption and instead take about 250 milligrams DHA and EPA daily via an algae oil supplement. If you do occasionally eat fish, take care to choose varieties that are low in both methylmercury and PCBs. The US Food and Drug Administration (FDA) and Environmental Protection Agency (EPA) provide a list of low-mercury fish that includes anchovies, cod, flounder, tilapia, salmon, and shrimp.[15] Remember, though, that some of these fish are high in PCBs, especially farmed salmon. Seafood that is lower in both PCBs and mercury include oysters, sardines, rainbow trout, sablefish, and arctic char.

ARSENIC

Arsenic is found naturally in soil, but additional arsenic is present because of human activity. Some areas have increased concentrations of arsenic as a result of industrial pollution, the use of arsenic-based drugs in poultry production,[16] and past use of arsenic-containing pesticides and fertilizers, especially former cotton plantations that are now growing rice. Arsenic-containing pesticides were used until 1970, and trace levels of arsenic remain in those areas.

There are two forms of arsenic: organic and inorganic. Inorganic is the more dangerous of the two. Arsenic exposure leads to the generation of reactive oxygen species, inflammation, and genomic instability and is associated with an elevated risk of several cancers.[17] Common dietary sources of arsenic include brown rice and fruit juice. In 2012 *Consumer Reports* exposed the arsenic levels in brown rice and brown rice products, and since that time I have been recommending the use of other grains instead of rice in the diet.[18] It is important to avoid brown rice because of the risks of arsenic exposure and white rice because of its high glycemic load and lack of nutrient value.

LEAD AND CADMIUM

The heavy metals cadmium and lead also have toxic effects.[19] Negative health effects of lead during pregnancy and childhood include impaired brain development, hearing problems, and increased risk of miscarriage; in adulthood, research has shown an increased occurrence of hypertension and cardiovascular disease.[20] Lead is naturally present in the earth's crust and therefore in soil; however, industrial activities—such as years of using leaded gasoline and lead paints, plus the past use of lead-containing pesticides—have added more risk of exposure.

Like other heavy metals, cadmium is both naturally present in rock and soil and also comes from industrial sources, such as waste incineration and burning of fossil fuels. Cadmium transfers efficiently from soil to plants and is therefore found in most foods. However, human absorption of cadmium is low; we absorb only approximately 6 percent of the cadmium we ingest.[21]

Negative health effects of high cadmium exposure include impaired ability of the kidneys to reabsorb nutrients, leading to excess calcium excretion and negative effects on bone, and increased risk of diabetes, hypertension, periodontal disease, and cancer.[22] According to data published in 2019, the foods causing the most exposure to cadmium are breads and grains (34 percent) followed by leafy vegetables (primarily lettuce, 20 percent), potatoes (11 percent), legumes and nuts (7 percent), and stem and root vegetables (6 percent). The average weekly cadmium consumption was found to be 22 percent of what is considered the tolerable weekly maximum.[23] Shellfish, oysters in particular, are also high in cadmium.[24]

Cocoa products, particularly those originating from Central and South America, have been singled out for high levels of cadmium and lead.[25] Cocoa powder also contains beneficial flavonoids, but because of the lead and cadmium content, it should not be consumed daily, or it should be limited to 1 tablespoon of an African-sourced cocoa per day.

Also, if you grow greens in a home garden, they may be high in lead if the garden is close to a busy street or highway, or if it is adjacent to buildings with flaking paint. It can be a good idea to have your garden soil tested for lead levels. A healthful diet is helpful, as vitamin C, calcium, and iron help to reduce lead absorption.[26] Animal products can also contain lead if animal feed has been tainted with it; and since lead especially concentrates in bones, animal bone broths may be contaminated.

The final point to remember about cadmium and lead is when you eat a plant-centered diet, the phytates in plant foods, particularly in beans, intact grains, and nuts and seeds, bind lead and cadmium and aid in their removal. The organosulfur compounds found in garlic, onions, and ginger also aid in the body's removal of these toxins.[27]

GLYPHOSATE

Glyphosate is the most commonly used herbicide in the world. When it was introduced in the 1970s, it was thought to have no harmful effects on humans because it works on a chemical pathway used mostly by plants and not by humans. However, newer data suggest that this is not the case; both glyphosate and its metabolite aminomethylphosphonic acid (AMPA) may be harmful to humans, and this

Glyphosate Use in US Agriculture, 1974–2014

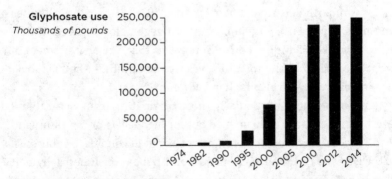

Source: Benbrook CM. Trends in glyphosate herbicide use in the United States and globally. *Environ Sci Eur.* 2016;28:3.

is especially concerning because of the large scope of current glyphosate use.[28]

Occupational exposure to glyphosate by agricultural workers is associated with an elevated risk of non-Hodgkin's lymphoma.[29] The IARC classified glyphosate as "probably" carcinogenic to humans in 2015, citing the limited evidence for its connection to non-Hodgkin's lymphoma as well as sufficient evidence of carcinogenicity in animal studies.[30] The report was met with some controversy, with some in the industry implying that the researchers had excluded data; however, the IARC has a policy of excluding industry-conducted research that is not publicly available, and it stands by its conclusions.[31] The use of glyphosate has increased along with the use of genetically modified crops (especially the corn and soybeans used mostly for animal feed), which are bred to be resistant to it. It is also sometimes sprayed on conventional wheat before harvest.[32] **Buying organic wheat, corn, and soybean products helps to minimize your exposure to glyphosate.**

N-NITROSO COMPOUNDS

N-nitroso compounds (NOCs) are carcinogens formed by a reaction between nitrite and heme iron. Exposure to NOCs occurs mainly from eating processed meats, which contain sodium nitrite as a preservative. Dietary exposure to NOCs has been linked to colon, rectal, and other gastrointestinal cancers, among others. Exposure to NOCs is a major factor in the IARC determination "that processed meat is a human carcinogen."[33]

Note that nitrates naturally present in vegetables do not have these dangerous effects. Formation of NOCs in the mouth and digestive tract is promoted by heme iron, which is present only in animal products, and is inhibited by plant phytochemicals such as vitamin C, vitamin E, and polyphenol antioxidants. The presence of phytochemicals drives the chemical reaction to produce nitric oxide rather than NOCs, leading to a beneficial effect on blood pressure rather than a carcinogenic effect.[34]

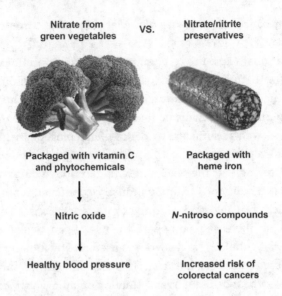

Nitrate from green vegetables   VS.   Nitrate/nitrite preservatives

Packaged with vitamin C and phytochemicals

Packaged with heme iron

↓

↓

Nitric oxide

N-nitroso compounds

↓

↓

Healthy blood pressure

Increased risk of colorectal cancers

HETEROCYCLIC AMINES AND POLYCYCLIC AROMATIC HYDROCARBONS

Heterocyclic amines (HCAs) are formed in all types of meat during cooking at temperatures of about 300 degrees Fahrenheit or higher. Meats that are well-done have the highest levels of HCAs. Intake of more-well-done meat and/or HCAs has been associated with a greater risk of breast, prostate, and pancreatic cancers, and red meat intake is strongly associated with colorectal cancers.[35] The main dietary sources of HCAs are ground beef, steak, beef gravy, chicken, broiled fish, and bacon; grilled chicken is also high in HCAs.[36] Poly-aromatic hydrocarbons (PAHs) are another type of cooking-produced carcinogen found in meat, specifically when the meat is cooked over an open flame.[37]

Avoiding or limiting meat is the most effective way to avoid these carcinogens; however, if you do eat any meat, it is safer to add it to a soup or stew and not to grill it. Make sure to use only a small amount of meat and plenty of G-BOMBS in any meal to limit the damage that HCAs can do to DNA.[38]

ENDOCRINE DISRUPTORS

Endocrine-disrupting chemicals (EDCs), which are found in foods and food packaging, are synthetic chemicals that have the ability to mimic or interfere with the activity of our natural hormones. BPA, probably the most well-known EDC, is present in many plastics and can liners. In an analysis of thirty-three studies, researchers found that higher urinary BPA levels were associated with a greater risk of diabetes, obesity, and/or hypertension.[39]

BPA is not the only concern in packaging. Phthalates are plasticizers, which are additives that make plastics more flexible. In addition to being found in plastic food packaging, phthalates are also found in cleaning, personal care, and cosmetic products. Almost all plastic food packaging likely contains EDCs, even ones that are BPA-free.[40] Most Americans have measurable levels of phthalates in their bodies (measured by urinary excretion of phthalate metabolites); but in a dietary intervention, just three days of eating only fresh foods and nothing canned or packaged in plastic decreased participants' urinary levels of the phthalate DEHP and BPA.[41]

Many POPs, such as PCBs, dioxins and furans, and DDT, also have endocrine-disrupting actions,[42] which is to say they interfere with our endocrine systems, which regulate metabolism, reproduction, and sleep, among other things.

You can do many things to reduce exposure to EDCs, such as limit your consumption of food packaged in plastic, not heat food in plastic containers, use dried beans instead of canned beans, buy tomatoes in glass jars instead of cans, avoid canned coconut milk, minimize your use of animal products, and avoid cleaning and personal care products that have added fragrances.[43]

MICROPLASTIC PARTICLES

Reducing exposure to cans and packaged foods is not enough to be assured of safety. Eight million tons of plastic—especially food packaging, water bottles, and plastic bags—are dumped into the oceans every year, and the tiny microplastic particles they produce can be harmful to marine and freshwater organisms. These microplastic particles smaller than 5 millimeters long are now incorporated into all types of fish and other sea creatures that we eat.[44] It is also important not to eat fish regularly, especially avoiding large fish such as tuna and shark, to be safe from harmful chemicals.

A variety of sources have been cited for microplastic pollution, such as dumping plastic waste into the oceans, where it degrades slowly, and using plastic microbeads as exfoliants in beauty products. Because of their tiny size, these pollutants escape water filtration systems and end up in the oceans or other bodies of water and cause serious environmental and food safety concerns. The risk assessment of microplastic contamination of the oceans is still in its infancy, but at this point we know that microplastic ingestion can cause cancer in rats and has infected the world's seafood supply. These particles are now detected in fish of all sizes, including sardines, wild fish, farmed fish, and particularly shellfish. Countries consuming more shellfish have populations with higher microplastic content in their bodies.

Microplastic particles are carriers of chemicals used in the manufacturing of plastic that can enhance their dangers to wildlife and humans. These plastic by-products are known to cause reproductive problems, endocrine disruption, and altered gene expression.

REGULAR CONSUMPTION OF ORGANIC FOODS IS LINKED TO LOWER CANCER RISK

Eating a nutritious diet that is also "clean"—as free of contamination as possible—clearly pays in health dividends. A study called the NutriNet-Santé Prospective Cohort Study launched in 2009 reported an association between organic food consumption and a 25 percent lower risk of cancer. When the results were divided by cancer site, they showed a reduction in risk for postmenopausal breast cancer, non-Hodgkin's lymphoma, and all lymphomas.[45]

The study was conducted in France and followed 68,946 adults for an average of four and a half years. At the beginning of the study,

participants answered whether they "most of the time," "occasionally," or "never" ate organic versions of sixteen groups of foods,
including vegetables, fruits, grains and legumes, dairy products, eggs,
and meats. Their answers were converted into an overall organic food
score, and participants were divided into four groups of increasing
scores.

The NutriNet-Santé study has reignited the debate over whether
and how organic agriculture affects the health of consumers. The
study had strengths and weaknesses worth noting: the large size of
the population that was studied is a strength, but the follow-up time
was somewhat short (for the outcome of cancer). In addition, the
study didn't confirm that people with higher organic food intake had
lower urinary pesticide levels. A longer study may have revealed a
stronger link with cancer, not a weaker link.

Until now, very little research has been conducted on differences
in health outcomes between consumers of "organic" versus "conventional" agricultural products.[46] Only one previous study, using
data from the Million Women Study based in the United Kingdom,
investigated associations between organic versus conventional foods
and cancer risk. This study found a 21 percent reduction in the risk of
lymphoma, but surprisingly, it also found a small (9 percent) increase
in breast cancer risk.[47]

Separate analyses of the NutriNet-Santé cohort reported that
greater consumption of total organic foods and organic plant foods
was associated with a reduced likelihood of metabolic syndrome
and obesity.[48] Another study in Norway found a link between consumption of organic vegetables during pregnancy and a lower risk of
preeclampsia.[49]

Presumably, differences in the health effects of organic and conventional foods would be due to the lack of synthetic pesticides in
organic agriculture. For example, in 2015, the common synthetic agricultural pesticides glyphosate, malathion, and diazinon were classified as Group 2A carcinogens (probably carcinogenic to humans),
with links to non-Hodgkin's lymphoma—one of the cancers whose

risk was found to be increased in the new study.[50] Most studies on the potential carcinogenic effects of synthetic pesticides have focused on people who work on farms and therefore are exposed to high doses of synthetic pesticides. This occupational exposure has been linked to several cancers.[51] However, exposure from food is still significant. Studies have found that consumers of conventional produce compared with consumers of organic produce have greater urinary concentrations of pesticide breakdown products.[52] Children may be even more vulnerable to pesticides because of their small body size,[53] and prenatal effects are also a concern.

We have been told that pesticide residue on produce is present at levels tens or hundreds of times smaller than the levels generally accepted to cause harm, but obviously, the harm is still present.[54] The NutriNet-Santé scientists propose that there could be a "cocktail" effect—that is, previously unrealized harms caused by the combined effects of chronic low doses of many different pesticides.[55] Clearly, it is wise to avoid exposure to pesticides whenever possible and to buy organic fruits and vegetables.

HEALTHY SOIL MAKES HEALTHY PEOPLE

Modern farming methods have contributed to destroying our health and our planet. According to the World Wildlife Fund, modern farming methods have led to eventual soil erosion and degradation, which has resulted in the loss of half of the topsoil on earth in the past 150 years. This can potentially impact the ability of the earth to feed its growing population. Organic agriculture, on the other hand, has many environmental benefits, such as enhancing soil quality and biodiversity, reducing synthetic pesticide exposure for people who work on farms, and increasing some antioxidant nutrients in produce.[56]

Regenerative organic agriculture takes organic farming a step farther and attempts to protect the value of the soil and restore the natural microbiome, insects, worms, and nutrients as nature intended. This will lead to healthy soil that will also benefit future generations.

The best soil makes the healthiest food, and the beauty and wonder of a plate of real organic food is magical. Think about the people who produced the food, who worked to have it come to life; the protected soil quality and the clean water which produced that food are gifts we are privileged to enjoy. By supporting organic agriculture, we support our planet and our planet's people. We want to leave a planetary legacy to our children and to future generations, and organic and regenerative farming practices and organic eating help make that possible. Try to see food not as an expense, but as an investment you make in your life, your future, and your peace of mind.

Truly organic food may cost more, but think of all the money people spend on expensive clothing, jewelry, cars, dining out; and traveling—while they eat the cheapest food they can find. "Cheap" food is, in fact, not cheap. We just move the costs elsewhere—mostly to increased medical expenses and reduced health and productivity.

To be healthy, you have to eat at least 90 percent of your diet from an assortment of nutrient-rich, colorful plants. The health benefits of regularly eating vegetables and fruits far outweigh the potential negatives of the reality of pesticide residue in many plants. However, in light of what we know about how pesticides can harm us, it is optimal to eat mostly organic foods and foods grown in high-quality soils as much as possible. In doing this we support our own health, our planet, and our planet's people the best way we can.

FIGHTING CANCER AND AGING

This discussion of chemicals and other disease-causing substances helps you see the reasons for the fourth principle of a Nutritarian diet and the science behind various dietary choices. The main reason I recommend a diet of foods rich in phytonutrients from colorful plants is that if you consume sufficient amounts of natural antioxidants to reduce oxidative stress, you resist the leading cause of cancer and the leading cause of aging.

SOME DEFINITIONS

- *Oxidation*—A chemical reaction in which electrons are removed from a molecule

- *Free radical*—A molecule that has an unpaired electron and, as a result, is highly reactive

- *Reactive oxygen species (ROS)*—A common type of free radical made from the reduction of molecular oxygen, producing superoxide, the precursor to other reactive compounds

- *Oxidative stress*—Excess free radicals creating an imbalance favoring pro-oxidants over antioxidants, which can lead to premature aging and disease

The formation of ROS is an inevitable by-product of our metabolic processes as our bodies break down food. As early as the 1950s, it was understood that ROS boost aging.[57] They are produced in the mitochondria of each cell, where a series of reactions transfers hydrogen ions and electrons as part of the process of producing energy. Oxygen is necessary for these reactions; most of it combines with hydrogen ions to form water; however, some oxygen molecules form ROS, which are highly reactive and potentially destructive.

The body has natural antioxidant enzymes to neutralize ROS, and mechanisms to promptly repair oxidative damage. However, if these antioxidant defenses are overwhelmed by ROS, oxidative stress results, in which free radicals are able to damage membrane lipids, cellular proteins, and DNA. The functions of these damaged components are then compromised.[58] This oxidative damage can cause a chain reaction that produces even more harmful free radicals and chronic inflammation.[59] Over many years, the accumulation of oxidative damage contributes to premature aging and diseases, such as cancer, diabetes, dementia, and heart disease.[60]

In addition to endogenous (internal) production of free radicals by our cells' mitochondria, we may also be exposed to free radicals from external sources such as cigarettes, ultraviolet light, pollutants, harmful substances in a poor diet, excessive calories, and alcohol.

Excess calories are the most common cause of inflammation, as overeating (especially of processed foods and animal products) overwhelms the energy production capacity of mitochondria, leading to accelerated production of ROS. In addition, excess calories result in more ROS production by white blood cells, which is exacerbated in obesity. The wave in increased free radical production and oxidative stress after a large meal drives inflammatory effects in adipose (fatty) tissue, which then also compromises the function of pancreatic beta-cells (the cells that produce, release, and store insulin), as these cells are especially sensitive to oxidative stress.[61] In contrast, weight loss and eating slightly less than your caloric needs reduce markers of oxidative stress.[62]

AGEs are also a major dietary source of oxidative stress. Foods high in AGEs include fried foods, broiled meats, high-fat animal foods, and dry-cooked starchy foods such as fried or roasted potato skins and baked goods.[63] Excess calories, especially from high-glycemic foods that lead to sustained and repetitive elevations in blood glucose levels, also promote formation of AGEs in body tissues as sugars react with proteins and lipids, altering their structure and disrupting their normal functions. The production of AGEs contributes to and enhances oxidative stress and inflammation, and conversely, oxidative stress causes production of some AGEs, contributing to atherosclerosis and complications in diabetes.[64]

OXIDATIVE STRESS PROMOTES CANCER, HEART DISEASE, AND DEMENTIA

Oxidative damage to DNA can result in mutations or breaks in DNA strands; failure of the cell to either promptly repair these defects or shut down cell division can then lead to cancer initiation.[65] For example, skin cancer is initiated by oxidative damage that results from exposure to ultraviolet (UV) light. Oxidative stress also contributes to the promotion and progression of other cancers by promoting proinflammatory gene expression patterns, disrupting cell signaling pathways, and encouraging cell proliferation; then, more mutations are created during the progression phase.[66] Cancerous cells also use a high level of ROS to maintain their rapid rate of proliferation.[67]

Oxidative stress is also the major driver behind heart disease. A single meal rich in animal fat and refined carbohydrates results in a measurable reduction in blood vessel function because of the oxidative stress these unhealthful foods cause.[68] It is not your LDL cholesterol level in itself that confers a high risk of heart disease; it is oxidized LDL. In its oxidized form, LDL is more atherogenic (plaque-promoting) than regular LDL. Oxidized LDL is taken up by macrophages and accumulates in the artery wall, accelerating the development of atherosclerosis. It also attracts inflammatory cells and is toxic to endothelial cells. LDL is more susceptible to oxidation under conditions of oxidative stress.[69]

The brain is especially susceptible to oxidative stress because of its high energy and oxygen requirements, weak antioxidant defenses, and polyunsaturated fatty acid–rich cell membranes. The brain's mitochondrial DNA has a limited repair capacity, thus the link to neurodegenerative diseases. Oxidative stress is also directly involved in the deposition of amyloid beta, a characteristic process in the development and progression of Alzheimer's disease, and in dopaminergic neuron loss, characteristic of Parkinson's disease.[70] The brain has a continual need for antioxidants, which are ubiquitously deficient in modern diets.

Because oxidative stress is a natural part of physiology, the human body has a system of antioxidant defenses set up to either prevent or repair oxidative damage to DNA, proteins, and lipids to fend off the potential tissue damage. Dietary antioxidants complement our endogenous antioxidants, building up our defenses against oxidative stress. Conversely, inadequate intake of dietary antioxidants leaves us more vulnerable to oxidative stress. **A Nutritarian diet is carefully designed to optimize the amount and variety of antioxidants to prevent free radical formation, AGEs, and oxidative stress.**

ANTIOXIDANT FUNCTIONS OF CAROTENOIDS AND FLAVONOIDS

CAROTENOIDS

There are more than 750 carotenoids—the yellow, orange, and red pigments found in plants and some algae and photosynthetic bacteria. Carotenoids have considerable capacity to scavenge free radicals and destroy them. The most common carotenoids in our diet are alpha- and beta-carotene, beta-cryptoxanthin, lycopene, lutein, and zeaxanthin. Foods high in carotenoids include kale, carrots, sweet potatoes, spinach, and tomatoes.

Alpha-carotene, beta-carotene, and beta-cryptoxanthin are provitamin A carotenoids—they are our plant sources of vitamin A; the body can convert them to vitamin A, which itself is not found in plant foods. Vitamin A is essential for vision and immune function.[71] All of these carotenoids (provitamin A and non-provitamin A) carry out antioxidant functions. Their importance has been highlighted by a number of studies showing a relationship between higher blood carotenoid levels and longer life.

Carotenoids and Lifespan

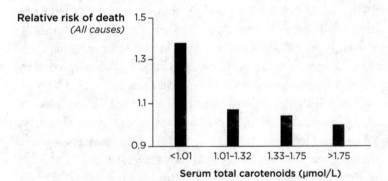

Source: Shardell MD, Alley DE, Hicks GE et al. Low-serum carotenoid concentrations and carotenoid interactions predict mortality in US adults: the Third National Health and Nutrition Examination Survey. *Nutr Res.* 2011;31:178–89.

Population studies show an association between high dietary intake of carotenoid-containing fruits and vegetables and reduced risk of lung, prostate, breast, and head and neck cancers. In a study of more than thirteen thousand American adults, low blood levels of carotenoids were found to be a predictor of earlier death. Lower total carotenoids, alpha-carotene, and lycopene in the blood were all linked to increased risk of death from all causes.[72]

One study on alpha-carotene, which included 15,318 participants over a fourteen-year follow-up period, found a significant trend toward a reduced risk of death from all causes as blood alpha-carotene concentrations increased. In the group with the highest baseline alpha-carotene, risk of death was 39 percent lower.[73] Alpha-carotene is a good marker of vegetable intake, since it is not often found in multivitamins (beta-carotene is present in most conventional multivitamins), and dark green and orange vegetables are rich in alpha-carotene. Another study on several different carotenoids reported that lower total carotenoids, alpha-carotene, and lycopene in the blood were all linked to higher mortality rates and that low lycopene was the strongest predictor of earlier death.[74]

A third study looked at blood carotenoids in relation to leukocyte (white blood cell) telomere length. Telomeres are regions of DNA that shorten with each cell division. Telomere shortening is an indicator of cellular aging, so a longer telomere length is desirable. The highest blood levels of alpha-carotene, beta-carotene, and beta-cryptoxanthin were each linked to a 5–8 percent increase in telomere length. The free radical scavenging action of carotenoids is thought to protect the DNA in the telomere region from oxidative damage.[75]

Carotenoids are also well-known for their dramatic protection against eye disease. Lutein and zeaxanthin selectively accumulate in the center portion of the retina (the macula), where they protect against oxidative damage and maintain good vision. Lutein and zeaxanthin have a special capability to filter blue light. Blue light is a component of sunlight, and although we need blue light exposure during daylight hours to regulate our circadian rhythms, too much blue light can result in oxidative damage to the macula. Lutein and zeaxanthin from the diet, especially leafy green vegetables, accumulate in the macula where they can absorb up to 90 percent of the blue light the macula is exposed to.[76] In addition to filtering blue light, lutein and zeaxanthin reduce glare and enhance contrast and visual acuity.[77] Higher intakes of lutein and zeaxanthin or their food sources (such as spinach and collards) are linked to a lower risk of age-related macular degeneration (AMD), which is the leading cause of blindness in the elderly worldwide. In AMD, photoreceptors in the macula are progressively damaged and/or lost, which impairs vision.[78] This disease is caused by our low intake of protective vegetables in the diet and should not be blamed on aging.

The Age-Related Eye Disease Study 2 trial used a supplement containing 10 milligrams lutein and 2 milligrams zeaxanthin and reported a decrease in the progression of AMD.[79] One cup of cooked spinach contains 20 milligrams of lutein plus zeaxanthin; of kale, 6 milligrams; and of collards, 12 milligrams.[80] A Nutritarian diet, which contains so much more of these compounds but also a much greater symphony of carotenoids and other supportive phytonutrients, is likely to be a hundred times more effective at preventing and treating AMD compared with eating the SAD and taking nutritional supplements that contain just a few of these protective nutrients. Phytonutrients work synergistically and most effectively in their natural packages, which are associated with hundreds of other beneficial compounds. They also work best when the bloodstream is not flooded with empty calories and the toxins and oxidative stress they induce.

In 2009, a group of ophthalmologists, led by Joshua Dunaief and eight other physicians specializing in macular degeneration at the Scheie Eye Institute of the University of Pennsylvania School of Medicine (now the Perelman School of Medicine, or Penn Med), noted that several of my patients had reversed their macular degeneration using the Nutritarian diet. They contacted me because they were interested in researching these remarkable results further. They devised a research project and submitted it to the National Institutes of Health for funding. Unfortunately, the grant was turned down partly because of limits to funding nutritional research and because these physician/

researchers, although they had done much research in ophthalmology, had done none related to nutrition. More research is needed to document the precise effectiveness of this dietary approach, but the lack of more research now should not stop us from reviewing the research and clinical evidence we do have and applying it to save lives.

Carotenoids are also critical for skin health and preventing skin cancer, as they accumulate in the skin and fight oxidative damage from sun exposure. Studies have reported that eating carotenoid-rich foods not only prevents but also can repair skin damage caused by exposure to UV light. Women who were instructed to eat tomato paste (which is rich in lycopene) daily for twelve weeks improved their skin's resistance to UV-induced skin reddening and reduced the amount of DNA damage caused by the same dose of UV light. Supplementation with either beta-carotene or lutein and zeaxanthin produced similar results.[81] The full symphony of nutrients supplied by a Nutritarian diet is the best protection against skin cancer.

PROTECTING YOUR HEART, BLOOD VESSELS, AND BRAIN

Oxidative stress is a major factor in the development of atherosclerosis, and higher dietary intake or higher blood levels of carotenoids have been linked to a reduced risk of cardiovascular disease in many studies. The antioxidant activity of carotenoids is thought to help prevent LDL formation and oxidation.[82] Two weeks of a tomato-rich (lycopene-rich) diet improved endothelial function (the health of the inside lining of the blood vessels), and this was related to antioxidant effects.[83]

Lycopene in particular has heart health–promoting actions apart from its antioxidant activity. It inhibits the enzyme HMG-CoA reductase, which is responsible for cholesterol synthesis and is the same enzyme inhibited by statin drugs.[84] In a meta-analysis of twelve trials, daily supplemental tomato in the diet, using about 1 cup of tomato juice or 3–4 tablespoons of tomato paste, was reported to reduce LDL cholesterol by 10 percent.[85] Lycopene also has anti-inflammatory effects and may have an antiproliferative effect on vascular smooth muscle cells, which help prevent atherosclerosis.[86]

It is important to note here that carotenoid supplements do not provide the same benefits as food-derived carotenoids. Clinical trials on beta-carotene supplementation did not find a reduction in cardio-vascular disease risk, and in some cases risk increased.[87] To get the benefits of carotenoids, you require the entire spectrum of natural compounds found in foods that work synergistically, not just isolated nutrients.

FLAVONOIDS

Flavonoids are a class of phytochemicals that includes green tea cate-chins, berry anthocyanins, soy isoflavones, cocoa flavanols, and citrus flavanones. They also have antioxidant activity and have the critical role of protecting mitochondrial DNA, but they are not dietary anti-oxidants themselves. Although they do have antioxidant activity, they are quickly metabolized once ingested. They do not directly neutralize free radicals but instead enhance cellular defense against damage.

Flavonoid intake is associated with cell stability and resistance to oxidative damage.[88] They improve immune defenses and have direct cancer protective properties, such as enabling the death of damaged cells (apoptosis) before they can become cancerous. Furthermore, when a cancerous cell appears, flavonoids can inhibit its replication. In fact, some studies have demonstrated that freeze-dried berry pow-ders (filled with flavonoids) help precancerous lesions in the digestive tract to regress to normal cells.[89]

Flavonoids are also intimately involved in the cardiovascular sys-tem. They drive a signaling pathway in the endothelium, leading to the production of nitric oxide, which regulates blood flow and blood pressure. Flavonoids counteract oxidative stress by activating the pro-tein Nrf2 and inhibiting enzymes that produce free radicals. They also have other biological effects, such as binding up excess copper and iron before they can cause damage, and inhibiting proinflammatory cytokines and the proinflammatory transcription factor NF-kB.[90] As a result, higher flavonoid intake is linked to a lower risk of hyperten-sion, cardiovascular disease, and overall cardiovascular mortality.[91]

Flavonoids also have anti-diabetic effects. The flavonoids rutin, kaempferol, and quercetin have been shown to inhibit carbohydrate digestion and absorption, either through direct binding to the carbohydrate or by interacting with a transporter involved in absorbing carbohydrate. They also inhibited the carbohydrate digestive enzyme alpha-amylase, further slowing carbohydrate absorption.[92]

In the brain, flavonoids promote the activity of signaling pathways and expression of genes associated with neuron survival. They also limit inflammation and enhance blood flow. Importantly, flavonoids affect synaptic plasticity, that is, the ability of the points of communication between brain cells (synapses) to adapt to changes in the cells' activity. Such adaptation is involved in memory and learning new information. High flavonoid intake is associated with maintenance of cognitive function in older adults.[93]

One study found improvements in learning and memory in older adults with mild memory impairment who were given supplemental wild blueberry juice daily for twelve weeks.[94] Another found an increase in brain activation (based on functional magnetic resonance imaging [MRI]) during a working memory task in a group given freeze-dried blueberry powder (the equivalent of 1 cup of blueberries daily) for sixteen weeks compared with a placebo group.[95] Similarly, multiple interventions with cocoa flavanols in older adults have found improvements in cognitive function.[96]

Flavonoids remain stable during food processing and heating and generally reach the small intestine intact. They are metabolized by intestinal cells, by gut microbiota, or by the liver, and in most cases, it is their metabolites that have the relevant biological activity.[97]

A FULL VARIETY OF VITAMINS, MINERALS, CAROTENOIDS, AND PHYTOCHEMICALS ALL WORK TOGETHER

I classify food into four broad categories: produce, whole grains, processed foods, and animal products. Of the four, only produce and

Foods with Flavonoids

Class	Flavonoid	Food Source
Anthocyanidins/ anthocyanins	Cyanidin, delphinidin, malvidin, pelargonidin, peonidin, petunidin	Berries, grapes, red onions, red cabbage, eggplant, black beans
Flavan-3-ols	Catechins	Green tea, cocoa, grapes, berries, apples, apricots, soybeans
	Proanthocyanidins	Apples, cranberries, cocoa, grapes, pecans, pistachios, stone fruits, cinnamon
Flavanones	Eriodictyol, hesperetin, naringenin, naringin	Citrus fruits
Flavones	Apigenin, luteolin, baicalein, chrysin, vitexin, orientin	Parsley, celery, peppers, thyme, oregano
Flavonols	Quercetin, myricetin, kaempferol, isorhamnetin, rutin, tiliroside, aromadendrin, silymarin, silybin	Onions, scallions, cruciferous vegetables, blueberries, tea, saffron, cranberries, asparagus, coriander, endive, fennel, ginger, okra, peppers, radishes, beans, buckwheat
Isoflavones	Daidzein, genistein, glycitein	Soybeans

Sources: "Flavonoids." Oregon State University, Linus Pauling Institute, Micronutrient Information Center. https://lpi.oregonstate.edu/mic/dietary-factors/phytochemicals /flavonoids. Last updated Nov 2015; Alkhalidy H, Wang Y, Liu D. Dietary flavonoids in the prevention of T2D: an overview. *Nutrients*. 2018 Mar 31;10(4):E438; Panche AN, Diwan AD, Chandra SR. Flavonoids: an overview. *J Nutr Sci*. 2016 Dec 29;5:e47.

whole grains contain antioxidants, phytochemicals, and protective carotenoids and flavonoids. Vegetables, beans, nuts, seeds, and fruits are our main sources of these compounds, and whole grains less so. These important micronutrients are not found in processed foods and animal products. This is a critical reason why diets that are high in animal products and low in natural plants accelerate aging of the body.

Nutrients such as vitamins C and E are also not found in processed foods and animal products. Vitamin C readily donates electrons, meaning it can pair the unpaired electrons of free radicals and thus end their disease-causing potential. This antioxidant action is one of vitamin C's major roles in the body.[98] Vitamin E also functions mainly as an antioxidant. Greens, seeds, and nuts are especially rich in this fat-soluble vitamin. In fact, vitamin E's affinity for fat aids its role in protecting cell membrane lipids and LDL cholesterol from free radicals.[99] Its benefits are realized only when you consume foods rich in vitamin E, which contain several forms of the vitamin and the accessory phytonutrients that maintain its nonoxidized state. Vitamin C and other antioxidants aid the protective effects of vitamin E. These nutrients are most effective and protective when ingested from food and working within the entire food matrix.

THE FOOD MATRIX

The **food matrix** is the complicated construction of whole foods with their unique combination of nutrients, fiber, and biochemical architecture. The food matrix even includes phytonutrients not yet identified and named. Processed foods and nutritional supplements present nutrients with a food matrix that has been altered from that of whole natural plants—especially the removal of fiber and the nutrients bound to fiber.

Often, as in grains like wheat and rice, removing the bran or grinding the grain to a fine powder (making flour) alters the matrix greatly and unfavorably increases the caloric load and the glycemic load of the

A Note on Whole Grains

Wheat flour, semolina, durum wheat, organic flour, stone-ground flours, and enriched wheat flour ARE NOT WHOLE GRAINS. One hundred percent whole wheat flour, made from the whole grain, is certainly much better than refined flours, but it still is not an intact grain when it is ground into a flour and used in a baked good. Grinding the grain increases its glycemic potential and also enables the damage that occurs during baking to more easily accumulate as it cooks. Wheat products are most nutritious when the wheat berry is purchased and then cooked in water. Other grains are best eaten as intact grains and not ground into a flour. They can also be sprouted a bit and then ground coarsely in a blender and soaked and formed to make a grain dish or loaf. When you buy commercial flours, even whole wheat flour, its innards are not kept fresh like those of an intact grain, so it has oxidized. And it has a higher glycemic load compared with the whole intact grain (the wheat berry of wheat grain). Whole wheat pastry flour is even worse because it is ground even more finely and therefore is more glycemic.

food. When food is intact, and especially when it is uncooked, its fiber passes through the digestive tract into the toilet bowl, carrying some calories with it. This reduces the calories absorbed, which enhances weight loss or weight control. This is why fruit juice (which is stripped of fiber) is problematic in the context of weight loss and why raw carrots and raw green peas (which are rich in fiber) are so much lower in caloric bioavailability compared with cooked carrots and peas.

The food matrix also holds together fiber, water, and nutrients to enhance satiety even when the food is chewed. The weight and volume of the water-fiber matrix works to satisfy hunger or appetite with fewer calories. It also leads to better bowel regularity and ease of passage.

Intact whole grains are also rich in vitamin E and contain anti-oxidants. An *intact* grain is one in which the bran (the multilayered outer skin of the grain) and the germ (the embryo for the new plant) are unbroken. The bran and germ contain more of the vitamin E, minerals, and phytochemicals compared with the endosperm, which is the starchy "white" food that is vastly popular and contains almost no micronutrient value. Recommended intact whole grains include wheat berries, steel cut oats, quinoa, buckwheat, barley, millet, teff, and amaranth, but not brown rice, as discussed earlier.

The food matrix may also interfere with the absorption of certain nutrients. Carotenoid absorption from raw vegetables alone is quite low, in part because carotenoids are often tightly bound in the food matrix. Disrupting the vegetable's structure by grating, blending, or juicing it significantly increases the bioavailability of its carotenoids.

Cooking vegetables also improves their carotenoid bioavailability, and the good news is that carotenoids are not destroyed by heat. In fact, when the food matrix is broken down by cooking, the carotenoids become more accessible to the digestive system.[100] Adding some fat to the meal—such as with a nut and seed–based salad dressing—is another method for increasing carotenoid bioavailability. In a study on carotenoid absorption, a fat-containing salad dressing on a salad of romaine lettuce, spinach, carrots, and tomatoes was found to greatly increase absorption of alpha-carotene, beta-carotene, and lycopene compared with a fat-free dressing. Absorption was negligible with a fat-free dressing.[101]

When we cook foods, it is preferable to use a liquid base, such as a stew, or a wokked dish to prevent browning or burning, which creates toxic compounds. We also don't want to overcook vegetables because nutrients are lost through cooking. Woks are primarily used in a number of Asian cuisines and focus on cooking with high heat for a short time. However, the chefs I work with and I have started using woks in a way that brings maximum flavor and health benefit because we wok in a bit of water, not oil, and then turn the flame off and add the flavorful sauce.

KEEP YOUR AGEs LOW AND
YOUR NRF2 ACTIVATION HIGH

The master regulator of the body's response to oxidative stress and toxic compounds is Nrf2 (nuclear factor erythroid 2–related factor 2). It enables the transcription of genes involved in cellular protection. Nrf2 binds to and activates the antioxidant response element (ARE) in genes. These code for the fabrication of detoxification enzymes, antioxidant proteins, and other proteins that help to counteract oxidative stress.[102] Many dietary phytochemicals, including flavonoids, activate Nrf2, which plays a strong role in cancer prevention.[103]

Once activated by phytochemical-rich foods, the ARE builds enzymes such as glutathione peroxidase, catalase, and superoxide dismutase, which scavenge free radicals or convert them into less harmful forms.[104] Nrf2 is both a sensor of dietary phytochemicals and a sensor of oxidative stress. An increase in oxidative stress activates Nrf2, triggering it to activate the production of the antioxidant enzymes necessary to resolve the stress.

Nrf2 is also a regulator of the body's detoxification system, which metabolizes and breaks down drugs, carcinogens, and foreign chemicals. Detoxification occurs in three phases: After an initial chemical alteration in phase I, phase II enzymes metabolize harmful chemicals into less toxic compounds, preventing them from causing damage. Phase III involves transporters that dispose of the substance. The availability of phase II enzymes allows the body to detoxify carcinogens and other harmful compounds; Nrf2 is central to controlling the expression of phase II enzymes.[105]

Nrf2 is especially important in the brain because of the brain's high susceptibility to oxidative stress. Oxidative stress in the brain is linked to neuronal cell death and neurodegenerative diseases.[106] In the cardiovascular system, Nrf2's anti-inflammatory effects protect against atherosclerosis, and Nrf2 signaling also promotes survival and renewal of cardiac muscle cells.[107]

Common Nrf2 Activators

Allicin/diallyl sulfide (garlic)[108]

Anthocyanins (berries)[109]

Ellagic acid (berries and pomegranates)[110]

Goji berries[111]

Flavonoid classes: flavanols, flavones, flavonols, flavanones, stilbenes, isoflavones[112]

Berry anthocyanins[113]

Quercetin[114]

EGCG (epigallocatechin gallate) (green tea)[115]

Sulforaphane (broccoli / broccoli sprouts)[116]

Cruciferous vegetables / isothiocyanates[117]

Resveratrol (grapes)[118]

Acai[119]

Curcumin[120]

Lutein[121]

Cocoa[122]

Lycopene (tomatoes)[123]

Falcarindiol (carrots)[124]

Pomegranates[125]

Hydroxytyrosol (olives)[126]

Carnosic acid (rosemary, sage, other herbs)[127]

Zinc[128]

Apigenin (celery, parsley, chamomile)[129]

Naringenin (citrus fruits)[130]

Luteolin (celery, peppers, oregano, parsley)[131]

DHA and EPA[132]

Soy isoflavones[133]

Black currants[134]

Butyrate (produced by gut bacteria from dietary fiber and resistant starch)[135]

Tocopherols (vitamin E)[136]

Cinnamon[137]

Rosmarinic acid and carnosol (rosemary)[138]

Zerumbone (ginger)[139]

Ursolic acid (apple peel)[140]

Chlorogenic acid (coffee, apples, apricots, chia seeds)[141]

The isothiocyanates in green cruciferous vegetables, such as sulforaphane, are among the most protective phytonutrients—and are particularly effective in activating Nrf2.[142] Once activated by sulforaphane, Nrf2 suppresses the activity of adhesion molecules on the endothelial cell surface to prevent binding of inflammatory cells, therefore halting atherosclerotic plaque development.[143] Multiple studies have shown that sulforaphane or other isothiocyanates, by activating Nrf2, block inflammatory gene expression and oxidative stress in human endothelial cells.[144] Sulforaphane also helps maintain the integrity of the blood-brain barrier, which is crucial for proper brain function, via activation of Nrf2.[145]

Other studies have shown that sulforaphane has protective effects on other types of cells in the cardiovascular system—cardiac muscle cells and vascular smooth muscle cells. There is also evidence that the boost in antioxidant defenses from sulforaphane helps to keep blood pressure down and that sulforaphane inhibits platelet aggregation, which helps prevent both heart attacks and strokes.[146] Isothiocyanates are even more concentrated in green plants when they are young and are exceedingly high in microgreens (immature greens) and sprouts. Broccoli sprouts have a huge amount of these protective compounds.

Every day when we sit down to eat, we have a choice. We can either eat foods that help our bodies and minds, or we can eat foods that harm us and increase dangerous health risk factors. Too many of us continue to make the wrong choice. Most of us simply don't know the implications of this daily decision. There is great pressure to keep eating the same old diet in the same old way. My hope is to change that. I have included a great deal of detail and scientific research in these pages to help emphasize the opportunity we have before us. When we eat a nutrient-dense, plant-rich diet, we fill our bodies with powerful life-giving nutrients and phytochemicals that have the ability to reverse existing disease and protect our health and well-being. The science is clear, and the choice is yours to make.

CHAPTER FOUR: QUICK SUMMARY

THE FOURTH PRINCIPLE OF THE NUTRITARIAN DIET

Synthetic chemicals, toxins, pathogenic bacteria, parasites, and other disease-causing substances should be avoided.

While the damaging effects of environmental toxins are well-known, the SAD exposes people to a wide range of toxins as well. Cooked meats and other cooked animal foods contain dangerous compounds such as heterocyclic amines, polycyclic aromatic hydrocarbons, and lipid peroxides. These compounds are formed during high-temperature cooking and are linked to increased risk of cancer.

Tina Feigley
Lost 70 Pounds and Reversed Type 2 Diabetes

As a busy elementary school teacher, I didn't have time in my schedule for cooking—there were papers to grade, lessons to plan, and the thousand other details that go into creating a positive classroom experience. When I came home exhausted at the end of the day, I invariably reached for fast food, chips, cookies, and candy, and the weight just kept piling on. I was up to 206 pounds. I wasn't surprised when I was diagnosed with type 2 diabetes, but I was shocked to learn that my HbA1c number was 11—meaning, totally out-of-control diabetes. I was determined to take massive action. I did not want a lifetime of medication and the problems that diabetes brings; my dad had diabetes and my brother currently has it, and I did not want to go down that road myself.

I was determined to learn what I could do to get well, so I read Dr. Fuhrman's book *The End of Diabetes* and immediately started following a Nutritarian diet. I told my doctor that I wanted to follow this nutritional plan rather than start taking medication. I asked him to give me twelve weeks to change my numbers. He was very reluctant but agreed to give me six weeks to get my HbA1c to below 7 percent. It came in at 6.9 percent, so he said if I kept doing what I was doing, then he wouldn't put me on diabetes medication. I learned from Dr. Fuhrman that six weeks is not enough time to record the benefits because HbA1c registers three months of glucose readings. He was right; my next reading in six months was 5.1—totally normal. I had lost 70 pounds and was no longer diabetic.

What I learned from Dr. Fuhrman's book saved my life. Before that, I honestly did not know that you could reverse diabetes. I thought once you were diagnosed, you had to go on medication and just live with this disease. When I read *The End of Diabetes* and realized that I could do something to change my situation, I knew I had to do it. It was easier than I thought, and it gave me amazing energy.

I now enjoy cooking, and I love what I eat. I was delighted to discover I lost the cravings and desire for junk food. I don't miss any of it. I love my new eating style, with all of the greens, beans, veggies, and fruit. I have lost a ton of weight and I feel so good! Within twelve months, I dropped from a size 22 to a size 8. I have tons of energy now,

I'm more confident, and I feel good about my appearance. When I was overweight, I felt very self-conscious all the time; I hated going out to eat because I could feel people staring at me. And in the summer, I wouldn't go to the beach because I was too fat.

The first two to three weeks were quite hard for me, but when I saw my weight dropping and my glucose levels dropping, I thought either I can continue what I'm doing or I can give up and face a lifetime of medications and health problems. I chose not to give up, and boy, was it worth it! I love eating this way now and encourage other people to do the same. This diet style will help you to take back your life and enjoy it to the fullest!

In addition, we are exposed to packaging chemicals such as BPA, and other toxins such as dioxin, food additives, arsenic, toxic metals, and even the weed killer glyphosate and other chemicals used on commercial crops.

Acrylamide and AGEs

Acrylamide is formed when foods are fried, baked, roasted, or otherwise cooked with high, dry heat; it is not produced during boiling. It has been classified as a Group 2A carcinogen (probably carcinogenic to humans) by the IARC, and numerous studies document significant risk.

AGEs (also known as glycotoxins) are created when acrylamide and similar chemicals bind to proteins in foods or proteins in the body. AGEs are a diverse group of highly oxidant compounds that accumulate in the body and cause cumulative pathogenic damage leading to disease, premature aging, and death. Though these toxins are associated mostly with baked, high-glycemic carbohydrates, the highest levels of AGEs are found in cooked animal products.

Persistent Organic Pollutants

Some pesticides and other industrial chemicals remain in the environment despite not having been produced or used in decades. A number of these now-banned substances are POPs. These substances are released into the environment from waste incineration, the burning of fuels, and forest fires. Because they are fat-soluble, POPs accumulate in the fatty tissues of animals; therefore, our main route of exposure is via fatty animal–based foods such as fish, butter, and ground beef.

N-Nitroso Compounds

NOCs are carcinogens formed by a reaction between nitrite and heme iron. Exposure to NOCs occurs mainly from ingestion of processed meats, which contain sodium nitrite as a preservative. Dietary exposure to NOCs has been linked to colon, rectal, and other gastrointestinal cancers.

Heterocyclic Amines and Polycyclic Aromatic Hydrocarbons

HCAs are formed in all types of meat during cooking, at temperatures of about 300 degrees Fahrenheit or higher. Meats that are well-done have the highest levels of HCAs. Intake of more-well-done meat and/or HCAs has been associated with a greater risk of breast, prostate, and pancreatic cancers, and red meat intake is strongly associated with colorectal cancers. PAHs are another type of carcinogen found in cooked meat, specifically when meat is cooked over an open flame.

Endocrine Disruptors

EDCs, which are found in foods and food packaging, are synthetic chemicals with the ability to mimic or interfere with the activity of our natural hormones. BPA, probably the most well-known EDC, is present in many plastics and can liners. High BPA levels have been associated with a greater risk of diabetes, obesity, and/or hypertension.

HEALTHY SOIL MAKES HEALTHY PEOPLE

Many environmental benefits are associated with organic agriculture, such as enhancing soil quality and biodiversity, reducing exposure to synthetic pesticides for farmworkers, and increasing some antioxidant nutrients in produce. Regenerative organic agriculture takes organic farming a step farther—it attempts to protect the value of the soil and restore the natural microbiome of insects, worms, and nutrients. This will lead to healthy soil that will also benefit future generations.

PREVENTING OXIDATIVE STRESS

Consuming a sufficient amount of natural antioxidants reduces oxidative stress, which is a leading cause of aging. Oxidative stress is a condition in which free radicals are able to damage membrane lipids, cellular proteins, and DNA.

No magic pill or potion will ensure your longevity. The key to living healthfully to 100 years of age is to eat a nutrient-dense, plant-based diet that is rich in lifespan-enhancing, anticancer superfoods. The power over your health outcome is in your hands.

The accumulation of oxidative damage over many years contributes to premature aging and diseases such as cancer, diabetes, dementia, and heart disease. Foods high in AGEs are a major dietary source of oxidative stress. The production of AGEs enhances oxidative stress and inflammation. In turn, oxidative stress causes production of yet more AGEs, which contributes to atherosclerosis and complications associated with diabetes.

CATALYSTS FOR OXIDATIVE STRESS

In addition to the production of free radicals by our cells' mitochondria, we may also be exposed to free radicals from external sources such as cigarettes, UV light, pollutants, poor diet, excessive calories, and alcohol. Being overweight and/or eating large meals results in accelerated production of ROS and an increase in oxidative stress, which drives inflammatory effects in fat cells. In contrast, maintaining a favorable weight and eating slightly under your caloric needs reduces markers of oxidative stress.

DNA Damage and Cancer

Oxidative damage to DNA can result in mutations or breaks in DNA strands. The failure of the cell to either repair these defects or shut down cell division can lead to cancer initiation. Oxidative stress also contributes to the progression of cancers by promoting proinflammatory gene expression patterns, disrupting cell signaling pathways, and encouraging cell proliferation; then, more mutations are created during the progression phase.

Oxidative Stress and Heart Disease

Oxidative stress is the major driver behind heart disease. A single meal rich in animal fat and refined carbohydrates results in a measurable reduction in vascular function because of the oxidative stress these unhealthful foods cause. The presence of oxidized LDL accelerates the development of atherosclerosis, attracts inflammatory cells, and is toxic to endothelial cells.

Oxidative Stress and Brain Health

The brain is especially susceptible to oxidative stress because of its high energy and oxygen requirements, weak antioxidant defenses, and polyunsaturated fatty acid–rich cell membranes. The brain's mitochondrial DNA has a limited repair capacity, thus the link to neurodegenerative diseases. Oxidative stress is also directly involved in the development and progression of Alzheimer's disease and Parkinson's disease.

THE "A" TEAM: ANTIOXIDANTS

An antioxidant is a substance, either produced by the body or derived from food sources, that inhibits oxidation. It does this by binding to free radicals, which neutralizes them and thus prevents or repairs oxidative damage to DNA, proteins, and lipids. The antioxidants we get from a healthful diet complement the antioxidants our bodies makes naturally, building up our defenses against oxidative stress. Conversely, inadequate intake of dietary antioxidants leaves us more vulnerable to oxidative stress. A Nutritarian diet is carefully designed with the precise amount and variety of antioxidants to prevent free radical formation, AGEs, and oxidative stress.

Carotenoids

Carotenoids are yellow, orange, and red pigments that have anti-oxidant activity and a considerable capacity to scavenge free radicals. More than 750 varieties are found in plants and some algae and photosynthetic bacteria. The body converts alpha-carotene, beta-carotene, and beta-cryptoxanthin into vitamin A, which is essential for vision and immune system function. Higher blood carotenoid levels have been linked to

- Increased telomere length and longer life

- Skin health and the prevention of skin cancer

- Reduced risk of cardiovascular disease

- Reduced risk of LDL formation and oxidation

- Improved endothelial function

- Eye health

It is important to note that carotenoid supplements do not provide the same benefits as food-derived carotenoids. To get the benefits of carotenoids, you require the entire spectrum of natural compounds found in foods that work synergistically, not just as isolated nutrients.

Flavonoids

Flavonoids are a class of phytochemicals that includes green tea cate-
chins, berry anthocyanins, soy isoflavones, cocoa flavanols, and citrus
flavanones. Though they have antioxidant activity and protect mitochondrial DNA, they are not dietary antioxidants themselves. They
do not directly neutralize free radicals but instead enhance cellular
defense against damage. Their intake is associated with

- Encouraging cell stability and resistance to oxidative damage

- Improving immune defenses

- Enabling the death of damaged cells before they can become
  cancerous

- Inhibiting the replication of damaged/cancerous cells

- Promoting the production of nitric oxide, which regulates blood
  flow and blood pressure

- Counteracting oxidative stress by inhibiting enzymes that
  produce free radicals

- Lowering the risk of hypertension, cardiovascular disease, and
  total and cardiovascular mortality

- Inhibiting carbohydrate digestion and absorption

In the brain, flavonoids promote neuron survival, limit inflammation, and enhance blood flow. High flavonoid intake is associated with
the maintenance of cognitive function in older adults.

THE FOOD MATRIX

The food matrix refers to the unique combination of nutrients,
fiber, and biochemical architecture of whole foods. Processed foods
and nutritional supplements present nutrients with a food matrix
that is altered from that of whole natural plants. When the bran
is removed from grains and the grains are ground into flour, their

matrix is unfavorably altered, and their caloric and glycemic loads are increased. When food is intact, the food matrix holds together fiber, water, and nutrients to enhance satiety, and the fiber passes through the digestive tract, carrying some calories with it, thus reducing the number of available calories.

NRF2: THE BODY'S "MASTER REGULATOR"

Nrf2 is a protein that enables the transcription of genes involved in cellular protection. It is the master regulator of the body's response to oxidative stress and toxic compounds. Nrf2 binds to and activates the antioxidant response element (ARE) in genes that code for detoxification enzymes, antioxidant proteins, and other proteins that help to counteract oxidative stress. Many dietary phytochemicals activate Nrf2, which plays a strong role in cancer prevention. The isothiocyanates in green cruciferous vegetables such as sulforaphane are particularly effective in activating Nrf2.

Felecia Suber
Her Blood Disease Is Normal for the First Time in Thirty Years

For more than thirty years, I suffered from a rare blood disorder that caused me to have dangerously high platelets. I finally got a diagnosis in 2016 that it was JAK2-induced thrombocytosis—a more serious illness, with splenomegaly and other complications. Doctors gave me a long, scary synopsis of what my future looked like: bone marrow biopsies, bone marrow transfusions, powerful and toxic drugs, chemotherapy. I thought I was going to pass out when my doctor began explaining what we needed to do. Out of all this information, I heard him say, "Depending on how your immune system responds." So I refused all the treatments and began my quest to find out how I could boost my immune system naturally.

I felt like it was divine intervention when I came across Dr. Fuhrman's book *Super Immunity*. I read it and reread it. With the knowledge

Continued on next page

that I gained, I began to adopt a 100 percent Nutritarian diet. Within one year, I had recovered! My platelets are now within normal range for the first time in more than thirty years!

I used to see my oncologist every three months, then every six months. Now I go just once a year. I told my doctor exactly what I did with Dr. Fuhrman's Nutritarian diet to heal my body. I thought he'd be excited, but he was like, "Okay," and seemed a bit confused—not excited—that I had made a remarkable recovery. He still tries to push prescriptions on me "just in case." I refuse and tell him, "I got this." I sure wish excellent nutrition was the number one prescription for all patients.

More of my results in one year: my blood pressure was 152/74 mm Hg and is now 106/68; overall total cholesterol is now 110; HbA1c is normal at 5.1 (it was borderline before); I've lost 88 pounds so far. I was borderline anemic, but now I am no longer anemic. I take zero medications. I walk more than five miles every day. I ride my bike thirty miles a week. Now I consider myself 58 years young!

I read Dr. Fuhrman's book and listen to the audiobook every six months to keep my knowledge fresh. I've given away more than twenty copies of it and keep several in my car trunk at all times to share with others. Sorry my story is so long—I'm just so darn happy and wanted to share!

CHAPTER FIVE

We Can Prevent Cancer

Unfortunately, most of us have no idea that a great number of diseases, including heart disease, stroke, diabetes, and many cancers, are the direct result of a poor diet. We assume that genetics, age, or other factors beyond our control cause these diseases. The reality is that our bodies are powerfully resistant to disease when provided with an optimal diet and micronutrient excellence. And that is why we have the ability to reverse course and heal ourselves. I have shared a great deal of research thus far about a variety of topics, and now I want to delve into the topic of nutritional defense against cancer.

We have become the victims of the high-tech, mass-produced food culture fueling a cancer epidemic unrivaled in human history.[1] The high rates of cancer we see today in the United States are a recent historical phenomenon; in past civilizations of populations eating natural foods, cancer was almost unheard of. Studied populations that eat mostly natural, unprocessed plant foods have longer life expectancies and low rates of cancer.[2] For example, if we review the Blue Zones around the world, areas where people eat more gathered and farmed plant foods and live longer as a result, we simply do not see cancer as a common cause of death.

Today we have at our disposal thousands of scientific studies that document the power of many plant foods to protect against cancer,

Thomas Johnson
Recovered from Ulcerative Colitis and Testicular Cancer

I was raised on the SAD. My mother, being a true Italian, indulged every holiday and tradition with endless amounts of food that I loved to eat. If I wasn't eating lasagna or pasta with meatballs covered in cheese, I was eating chicken or fish in some rich and creamy sauce. Junk food like hotdogs, hamburgers, and pizza as well as cold cuts, milk, and cheeses filled our refrigerator; and cakes, cookies, and donuts filled our pantry. Such food and eating habits were passed along to me with great love and the best of intentions. Although fortunate to have had such a wonderful family, little did I know the drastic effects it was having on my body.

For as long as I can remember, I was always sick in some way. Growing up, I was constantly on medications for colds and flu, and I suffered daily from allergies and asthma. My nose was always running, my throat was always sore, and my stomach was often upset. At age 5, I was hospitalized with fever of more than 105, and my elementary school attendance was filled with long illnesses that often kept me out of school for weeks at a time. At age 15, I was diagnosed with ulcerative colitis shortly after a long course of the antibiotic Biaxin was given to me to treat mononucleosis. I was told at the time that ulcerative colitis was "incurable" and that I would be sick for the rest of my life. My doctor also told me that there was no link between diet and the disease and that I would be on medication for the rest of my life.

For the next twenty-five years, I struggled with this horrible condition. Even on a good day, I still felt daily cramping and pain. During

a flare-up, I would need to go on more serious medication—high-dose steroids—and then deal with all the side effects. Along with this, starting at age 24 when I became an elementary school teacher, I was constantly exposed to illness and was regularly taking antibiotics to deal with what seemed like an endless case of strep throat and bronchitis. I would easily take six to ten different courses of antibiotics during any given school year. I was also still struggling with constant allergies and taking nasal decongestants like they were candy.

In the late summer of 2006, at age 35, a flare-up of the ulcerative colitis was so violent that I was hospitalized. It was like nothing I had ever experienced, and in just over a week, I dropped 45 pounds. I went from being 6 feet, 1 inch tall and 198 pounds to around 150 pounds—at which point I stopped weighing myself. My doctors tried everything, but nothing worked. Ulcerations had spread throughout the entire colon, and they were bleeding uncontrollably. The amount of blood loss and crippling pain were beyond description. My life was at risk.

Desperate to save my life, the doctors did a procedure that inserted a tube into my jugular vein. I was given a powerful combination of medications that eventually worked. After almost a month in the hospital, I began to recover and was sent home, but this time on an even more powerful and dangerous medicine—a strong immunosuppressive agent that had negative side effects that included putting me at high risk of cancer.

In the spring of 2012, at age 41, I was diagnosed with a form of testicular cancer called a classic seminoma. Tumor markers and tissue samples confirmed this diagnosis. The doctors wanted to remove the tumor and follow up with chemo and radiation, but given my history of ulcerative colitis, they—and I—were very concerned that my body was not going to react well to this treatment. Also, the immune-suppressing drugs I was taking for the colitis would be problematic with such treatment. I was left in a quandary: Do I do just the surgery and not the follow-up chemo and radiation, which would put me at risk of the cancer spreading to other locations? Or do I have the surgery, chemo, and radiation and risk the ulcerative colitis? I didn't like either option.

Incredibly, during that spring I saw Dr. Fuhrman's PBS special on healthy eating and was greatly inspired by his message. The doctors were talking to me about five-year and ten-year survival rates, but I had a 2-year-old boy at the time—so I wanted the fifty-year plan! Dr. Fuhrman's special made me think that there might be another way—a way

Continued on next page

that also addressed the cause of the disease in the first place. Against my doctors' wishes and advice, I held off on all conventional treatments and instead went to a healing retreat to follow Dr. Fuhrman's nutritional approach. When I told one of my doctors that I was going to hold off on his recommended treatment, he said, "You are going to die and I am not going to be part of it!" and then kicked me out of his office.

I had learned from Dr. Fuhrman that if you give your body the right foods, and care for your body through exercise and rest, your body will be able to heal itself and transform itself in ways you never imagined. Even things like cancer and ulcerative colitis can disappear. Within only a few weeks of my adopting these dietary changes, my body began to transform. My energy level, clarity of mind, and overall physique made dramatic changes for the better.

Sonograms of the tumor taken that summer showed that it was beginning to shrink in size, and I began pulling myself very slowly and gradually off the medication for the ulcerative colitis because I was, for the first time since I was 15, feeling no symptoms in my gut. Fortunately, I found an amazing immunologist who monitored my progress and supported my efforts. Eventually I was able to completely remove myself from medicine for the colitis, and the tumor shrank to nothing. I was able to rid myself of cancer and ulcerative colitis without any of the conventional treatments and without any of the side effects associated with them. Along with that, my overall health has so greatly improved it is hard to fully quantify the before and after results.

Now, at age 47, I can report with amazement that the cancer remains undetectable, the ulcerative colitis is gone, the allergies that plagued me my whole life are also gone, and I have not needed to take an antibiotic or medicine of any kind since 2012. I continue to live the lifestyle Dr. Fuhrman prescribes in his books. Every day over the past six years, I have followed Dr. Fuhrman's dietary guidelines and have not deviated from the course once. I have never felt better, stronger, healthier, and more emotionally balanced. I thank him for saving my life and allowing me to thrive physically for the first time ever. This is the gift he has given me, and I am forever indebted and grateful.

P.S. I am raising my 8-year-old son following the Nutritarian diet, and he is so incredibly healthy, so incredibly fast, and strong as an ox for his size! He loves eating this way for his health and for our planet. Dr. Fuhrman's gift to my life, and the generations to follow, is priceless!

and yet our population is heavier and more cancer-prone than ever before. While we are pouring billions of dollars into drug companies and cancer centers, millions of people continue to die from what should be largely preventable diseases. Almost nothing has been done to teach people about the power of nutritional excellence to protect against cancer, and the public remains confused.

A Nutritarian diet is designed to be the optimal diet not just for slowing aging and preventing cancer, but also for potentially reversing early-stage cancer in many cases. Research has shown that the same foods and phytochemicals that are most effective at preventing cancer also work to reverse early-stage cancers and prevent recurrence of cancer in people who already have received a diagnosis of cancer.[3] Whether you are looking to prevent cancer or want to attack cancer with the most powerful nutritional protocol, this program is for you.

MOST CANCERS CAN BE ATTRIBUTED TO OUR TOXIC FOOD ENVIRONMENT

A cancer cell is essentially a normal cell whose DNA has been damaged to the point that the cell can no longer control its replication. When our needs for phytonutrients are met, human cells have all the functional features necessary to protect themselves from chemical damage to their DNA. But our nutrient-poor and toxic diet enables inflammatory elements and defects to eventually accumulate to the point where carcinogenic changes occur.

The process that is creating our modern epidemic of cancer is twofold. One aspect involves the exposure of our cells to damaging stresses such as:

CHEMICAL CARCINOGENS, ASBESTOS, MICROPLASTIC PARTICLES, INSECTICIDES, RADON, ACRYLAMIDE, TOBACCO, FRIED FOODS, EXCESSIVE ANIMAL PROTEIN, HETEROCYCLIC AMINES, NITROSAMINES, ALCOHOL, AND SWEETENERS.

The second aspect is our woefully insufficient dietary intake of plant-derived phytonutrients. This deficiency renders our cells incapable of functioning to their fullest potential for repair and maintenance. Our cells have powerful built-in mechanisms to remove or destroy toxic substances, inhibit DNA damage, repair broken DNA cross-links, and remove cells that are injured or abnormal before they become cancerous. These robust cancer-protective mechanisms are activated only when we consume lots of vegetables.

According to the American Institute for Cancer Research, about 50 percent of common cancers are preventable by not smoking, limiting sun exposure, maintaining a healthy weight, exercising regularly, and following a healthful diet.[4] I propose that the percentage is much higher and that **at least 90 percent of the cancers now seen in modern societies could be avoided if we optimized our dietary intake and engaged in a reasonable anticancer lifestyle**.

Many earlier societies and people who live in isolated regions of our planet have been free of the common cancers seen today. The reason is that those people were still eating mostly natural plant foods.[5] Today, modern nutritional science has identified the most powerful and effective plant foods that protect against cancer. These are the foods used in the design of the Nutritarian diet. Of course, the earlier in life one begins to eat this healthful diet, the more powerful and complete the protection.

The National Cancer Institute recommends eating five servings of fruits and vegetables each day; however, scientific studies suggest that more is better and that much, much more is much, much better at reducing cancer risk.[6] Unfortunately, very few Americans follow this very minimal recommendation.

Despite more than a hundred billion dollars in cancer research, invested largely in the development of drug chemotherapy and screening and detection techniques, we are still not winning the war on cancer. Cancer has surpassed heart disease and has become the leading cause of premature mortality for Americans between the ages of 45 and 64.[7] From 1999 to 2015, overall rates of death from cancer in the

United States fell by about 1.4 percent per year in women and 1.8 percent per year in men. At the same time, new cancer diagnoses have remained stable in women and declined 2.2 percent per year in men.

The two major reasons for these tiny improvements are a continuing decline in cigarette smoking, which affects more cancers than just lung cancer, and the significant reduction in physicians' prescribing estrogen and progesterone for postmenopausal women, leading to lower breast cancer deaths. However, there have been significant increases in the incidence of some cancers, such as malignant melanoma, myeloma, thyroid and liver cancers, and lately even breast cancer. The recent uptick in breast cancer incidence in particular is thought to be partly due to the rise in obesity.[8] Eating more processed foods and gaining more body weight induces more cancers, and the trend of cancers occurring earlier in life is alarming.

Survival rates for advanced cancers are still quite low, despite the availability of the most updated and advanced medical care. It is worth noting that almost 75 percent of advanced cancers are

Colorectal Cancers Increasing in Young Adults (age <50)

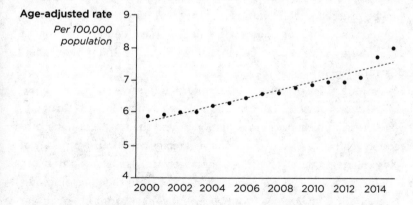

Sources: Siegel RL, Fedewa SA, Anderson WF et al. Colorectal cancer incidence patterns in the United States, 1974–2013. *J Natl Cancer Inst*. 2017 Aug 1;109(8); National Cancer Institute. Surveillance, Epidemiology, and End Results (SEER) Program. https://www.seer.cancer.gov.

recurrences, originally diagnosed at earlier stages.[9] Chemotherapy
for cancer does not typically kill all the cancer cells in the body; some
residual cells survive. Years later, those few surviving cells can multiply aggressively, and the cancer can be deadlier, with the cancerous
cells having developed resistance to chemotherapy. When chemotherapy is used without optimal nutritional therapy, the chances of
the body recognizing, attacking, and removing these residual abnormal cells is slim.

Anticancer Solution: A Dietary Protocol Rich in Greens

G-BOMBS

Greens, **B**eans, **O**nions, **M**ushrooms, **B**erries, and **S**eeds

Hundreds of scientific studies document that these foods prevent cancer and also prolong lifespan in people who have cancer.

All fruits and vegetables are excellent sources of nutrients, but
colorful vegetables, particularly green ones, are more helpful in
reducing cancer. This is because they contain higher amounts of
cancer-protective compounds.[10] Even chlorophyll, the green plant
pigment essential for photosynthesis, has huge benefits because it
binds tightly with several carcinogenic chemicals and inhibits their
digestion.[11]

Among all vegetables, the cruciferous family—including broccoli, Brussels sprouts, cauliflower, kale, bok choy, collards, arugula,
watercress, and cabbage—has demonstrated the most dramatic protection against cancer. Cruciferous vegetables contain a symphony of
phytonutrients with potent anticancer effects. Isothiocyanates and
indoles, which are phytochemicals derived from the glucosinolates
in cruciferous vegetables, inhibit several cancer-promoting cellular
processes, including cancer cell growth and proliferation and angiogenesis.[12] Indole-3-carbinol (I3C) is especially protective against

What Is Angiogenesis?

It means the growth of new blood vessels, particularly when blood vessels sprout new branches to feed the nutritional needs of growing tissues.

hormone-associated cancers because it helps the body excrete estrogen and other hormones. Both I3C and sulforaphane have been found to blunt the growth-promoting effects of estrogen on breast and cervical cancer cells.[13]

Tumors produce chemical signals to encourage angiogenesis. Cancer is like an alien growing inside you that is hijacking your biology to aid its own growth. Plant-derived phytochemicals defend your "earth" against alien invaders. They do not allow cells to become dysplastic (abnormal), to proliferate, and to become cancerous (multiply beyond control); and these phytochemicals inhibit angiogenesis, which is needed to support abnormal growth and metastasis.

What makes this research even more fascinating is the gene-diet interaction, in which certain foods activate a battery of genes that initiate DNA repair and other protective mechanisms. These phytochemicals, such as isothiocyanates and indoles from green vegetables, are not only antiangiogenic, but they also have been shown to provide protection against exposure to environmental carcinogens by inducing detoxification pathways to neutralize potential carcinogens. These compounds also fight cancer by inhibiting cancer cell replication and promoting cancer cell death. These compounds from greens also effectively curb inflammation and have antiviral and antibacterial effects. Their anti-estrogenic effects help to prevent hormone-related cancers.

Eating cruciferous vegetables regularly is associated with a decreased risk of breast, prostate, and colorectal cancers.[14] Consumption of these vegetables is also associated with reduced cardiovascular

Cruciferous Vegetables and Longevity:
Study of 134,796 Chinese Adults

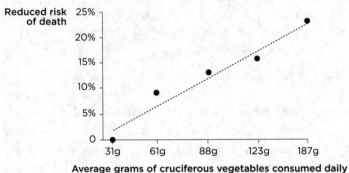

Source: Zhang X, Shu XO, Xiang YB et al. Cruciferous vegetable consumption is associated with a reduced risk of total and cardiovascular disease mortality. *Am J Clin Nutr.* 2011;94(1):240–46.

mortality and all causes of death, with no threshold where benefits diminish.[15] This means that the more of them you eat, the greater protection you get.

As we have seen, the antioxidant response element (ARE) is a segment in our genes that code for detoxifying enzymes, antioxidant proteins, and other proteins that protect against oxidative stress. The Nrf2 transcription factor is activated by cruciferous-derived isothiocyanates to bind to the ARE and drive expression of these genes, which then enables the cell to neutralize oxidative molecules and resolve and repair oxidative damage.[16]

Sulforaphane (found in broccoli, Brussels sprouts, and cabbage) and other isothiocyanates also activate Nrf2 to suppress adhesion molecules on the endothelial cell surface, thereby preventing binding of inflammatory cells and thus retarding atherosclerotic plaque development.[17] The activation of Nrf2 in endothelial cells maintains vascular elasticity and inhibits oxidative aging of the blood vessels.[18]

One interesting gene containing an ARE segment that is activated by isothiocyanates codes for glutathione-S-transferase (GST), an enzyme involved in the biotransformation, detoxification, and removal of harmful compounds. Certain common alterations in the gene for GST are thought to increase breast cancer risk by slowing and reducing the detoxification of carcinogens. Interestingly, higher cruciferous vegetable intake restores downregulated detoxification and reduces the risk associated with these gene defects.[19] This means that eating cruciferous vegetables can lessen and remove the risk associated with genetic predispositions for cancer.

Myrosinase is an enzyme found in the cell wall of cruciferous vegetables that catalyzes the chemical reaction that forms these beneficial compounds. These invaluable isothiocyanates are not in the vegetable; rather, they form in the mouth as you chew the vegetable. When the cell wall is crushed, the glucosinolates within the cell vacuoles mix with the myrosinase that was compartmentalized in the plant cell wall, so the isothiocyanates and indoles can be formed. This means that the better you chew your vegetables and salads, the more beneficial nutrients will be formed.

If you have high blood pressure or heart disease and you are looking to reverse it, eating lots of green cruciferous vegetables—both raw and cooked—will aid your recovery. Because myrosinase can be deactivated by heat, it is important to eat some raw cruciferous vegetables every day and chew them well. Shred some red cabbage or Chinese

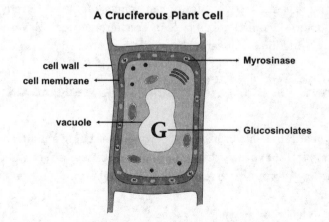

A Cruciferous Plant Cell

cell wall

cell membrane

Myrosinase

vacuole

G

Glucosinolates

cabbage onto your salad, add a little arugula or watercress, and use a bit of ground mustard seed for flavor.

If we blend or juice these greens when raw, or chew them well when they are raw, the enzyme gets released from its membrane packet, mixes with the glucosinolate, and then these anticancer compounds are formed. If you cook the vegetable too much before you chew or blend it, however, the myrosinase will be deactivated, which will prevent the formation of the protective isothiocyanates.

To add cruciferous vegetables to stews or soups without inactivating myrosinase, blend them first while raw, using a bit of liquid, and then add them to the pot to cook. The isothiocyanates formed in the blender will remain stable through cooking. Add some raw cruciferous vegetables to your salads. Arugula, watercress, cabbage, or bok choy can supply myrosinase that can activate some of the glucosinolates from cooked greens that traveled into the stomach without conversion, thus enhancing isothiocyanate production. Cooking vegetables in a wok can also be helpful, as they are typically only partially cooked, so they retain more myrosinase activity compared with steamed or boiled vegetables.

Studies on cruciferous vegetables in people who already have cancer also show promising results. Studies performed on men with prostate cancer involving broccoli and sulforaphane-rich broccoli sprouts showed profound benefits inhibiting the growth of cancer.[20] Even a study using frozen broccoli showed beneficial effects.[21] Cruciferous vegetables work for breast cancer too. In one study, postmenopausal women ate 14 cups of cruciferous vegetables per week, and a reduction in cancer markers was observed within six weeks.[22]

BEANS, ESPECIALLY SOYBEANS, PROTECT AGAINST CANCER TOO

The health benefits of beans go beyond their being a main source of dietary carbohydrate; they also have anticancer effects. Certainly, their low-glycemic properties, their high fiber and resistant starch content, and their phytate content all contribute to their anticarcinogenic properties.[23] The fermentation of beans in the gut also creates compounds with anticancer effects. One of them is the short-chain fatty acid butyrate, which inhibits cell growth and proliferation and induces programmed cell death in tumor cells; it also has antioxidant activity, which protects normal cells from oxidative DNA damage, contributing to colorectal cancer prevention.[24]

Soybeans are especially protective against cancer, even compared with other beans. In a meta-analysis of fourteen prospective studies including more than 1.9 million participants, greater legume intake was associated with a 9 percent reduction in colorectal cancer risk, but soybean intake resulted in an even greater 15 percent reduction.[25] Similarly, in a meta-analysis of ten prospective studies, bean intake was associated with a dose-dependent reduction in prostate cancer of 3.7 percent per 20 grams of legumes eaten per day, with soybeans showing the most benefits.[26] Soy consumption is also linked to a reduced risk of breast, endometrial, and prostate cancers in large meta-analyses.[27]

Soybeans may have particular anticancer effects against hormone-related cancers because of their phytoestrogens called *isoflavones*. Isoflavones have different effects depending on different types of estrogen receptors, allowing for anti-estrogenic effects in breast tissue that contribute to cancer prevention and also estrogen-mimicking effects in the skeletal system that help maintain bone mass.[28] However, the anticancer effects of soybeans are not limited to those involving estrogen. Isoflavones also inhibit the proliferation of cancerous cells and angiogenesis.[29] Higher soy intake is also linked to a reduction in the risk of stomach, lung, and colorectal cancers, in addition to cancers related to the reproductive system.[30]

You should be aware, however, that soy protein powder and isolated or concentrated soy protein–based processed foods do not promote health. Because the amino acid distribution of soy protein powders is very similar to that of animal protein and is highly concentrated, such powders can increase IGF-1 levels too much, which is not helpful in preventing or treating cancer.[31] Eating edamame or tempeh and using dried soybeans in soups and stews are the most favorable and protective ways to utilize soy. Tofu and soymilk have the fiber of the bean removed, and though they may still have benefits, they do not have as much anticancer potential. Using the whole bean, fiber and all, along with other beans, lentils, and peas, offers more potent protection.

EAT MORE SCALLIONS, ONIONS, AND GARLIC

The *Allium* family of vegetables, which includes onions, garlic, shallots, leeks, and scallions, has strong anticancer properties. Meta-analyses have found that garlic consumption is associated with a reduction in the risk of prostate cancer and gastric cancer, and epidemiological studies show inverse correlations between onion and garlic intake and the incidence of many cancers.[32] A large European study found a 56 percent reduction in colon cancer risk, a 73 percent reduction in ovarian cancer, an 88 percent reduction in esophageal cancer, and a 71 percent reduction in prostate cancer in participants who consumed the greatest quantities of onions and garlic.[33]

The organosulfur compounds in allium vegetables are responsible for both their flavor and their health effects—and are the source of the eye irritation that often accompanies chopping onions. In addition to the sulfur compounds, the anticancer effects of onions and garlic are likely due to their rich flavonoid content, especially quercetin. Allium-derived compounds affect a number of cell-signaling pathways, leading to anti-inflammatory effects and inhibition of cancer-related processes such as the proliferation and migration of cancer cells.[34]

Similar to the isothiocyanates in cruciferous vegetables, organosulfur compounds in allium vegetables are produced when the raw vegetables are chopped, crushed, or chewed. For example, within ten to sixty seconds of crushing garlic, a compound called allicin is formed that then breaks down into several different organosulfur compounds.[35] These compounds have a wide range of powerful anti-cancer effects, including stopping the growth of cancer cells, activating Nrf2, and inhibiting angiogenesis.[36] Garlic also enhances the immune system.

Onion and garlic phytochemicals have beneficial effects in the cardiovascular system, too, such as inhibiting cholesterol synthesis, platelet aggregation, inflammation, and vascular smooth muscle cell proliferation.[37] Even garlic extract supplements have been found to reduce blood pressure and total and LDL cholesterol.[38]

MUSHROOMS ARE ANTICANCER SUPERSTARS

Mushrooms have a distinguishing characteristic among plant foods—namely, they are not exactly plant foods. They are fungi, and their cell walls contain bioactive polysaccharides, called beta-glucans, that can interact with receptors on immune cells, including macrophages, dendritic cells, T cells, and natural killer cells.[39]

Mushroom polysaccharides have both direct and indirect anticancer effects. Directly, these polysaccharides have shown antiangiogenic and antiproliferative effects on cancerous cells.[40] Mushrooms powerfully affect angiogenesis.[41] Fat cells produce multiple angiogenic factors (including leptin, angiopoietin, and other growth factors) that stimulate new blood vessel growth (neovascularization) to facilitate fat cell expansion and growth. Mushrooms inhibit this process, thus deterring the storage of fat in the body. Simultaneously, these antiangiogenic properties of mushrooms inhibit the growth of abnormal cells and cancer, because abnormal cells cannot replicate rapidly and become dangerous without developing their own robust blood supply. So mushrooms prevent abnormal cells from replicating and help the immune system seek out and destroy such cells at the same time.

Indirectly, mushroom polysaccharides enhance immune surveillance by increasing the activity of natural killer (NK) cells, which are immune cells that detect and destroy cancerous and virus-infected cells.[42] Clinical trials using high-dose mushroom extracts support the potential of mushrooms for improving immune function, as much of this research suggested that mushroom extracts counteracted immune suppression in patients undergoing chemotherapy.[43]

Mushrooms also contain specialized lectins that adhere to cancer cells and inhibit their growth.[44] The antiangiogenic, antiproliferative, and other anticancer effects of many mushroom varieties—including white, cremini, portabella, oyster, maitake, and reishi—have been studied in stomach, colorectal, breast, and prostate cancer cells.[45]

Mushrooms are particularly relevant to the prevention of breast cancer. Mushroom phytochemicals inhibit aromatase, an estrogen-producing enzyme. Most common mushrooms, including white button, cremini, and portabella, display strong aromatase inhibition.[46] In one study, frequent consumption of mushrooms (approximately one button mushroom per day) was associated with a 64 percent decrease in the risk of breast cancer.[47]

A dose-response effect was found in a meta-analysis of ten studies on mushroom intake and breast cancer; this means that the more mushrooms eaten daily, the greater the reduction in risk. At 20 grams of mushrooms a day, the risk reduction was 60 percent, and results were similar in premenopausal and postmenopausal women.[48]

In clinical trials on cancer patients, mushrooms are generally administered as concentrated supplements; however, one study on men with prostate cancer and biochemical recurrence found that white button mushroom powder decreased PSA (prostate-specific antigen) levels in 36 percent of patients.[49] These immune-modulating effects facilitate the body's attacks on microbial invaders and developing tumors.[50]

The immune-boosting effects of mushroom polysaccharides may also protect against respiratory infections. One study in healthy volunteers found that eating white button mushrooms daily for one week enhanced salivary antibody secretion, suggesting that regular intake of mushrooms improves immunity.[51]

I recommend eating only cooked or dehydrated mushrooms, because several raw mushroom varieties contain a potentially harmful substance called agaritine, and cooking significantly reduces agaritine content.[52]

BERRIES AND POMEGRANATES (AND FLAVONOIDS) ARE REMARKABLE

Berries and pomegranates (as well as passion fruit, plums, citrus fruits, cranberries, and cherries) are rich in flavonoids, which are concentrated in their skins and give them their deep blue, red, and purple colors.[53] Although flavonoids are abundantly present, they are not the only phytochemicals in berries; ellagic acid (also found in pomegranates), resveratrol, and other polyphenols also contribute to the cancer-preventing effects of berries, which include inhibiting inflammation, modulating gene expression and DNA methylation, promoting cell cycle arrest in cancerous cells, and inhibiting angiogenesis.[54] Even dehydrated berry gels and powders have shown positive results in cancer prevention trials in people with both precancerous and cancerous conditions.[55]

Pomegranates, like berries and mushrooms, have antiangiogenic properties, but they also contain natural aromatase inhibitors. Pomegranates and mushrooms are two of the few foods that contain these compounds, which limit the amount of estrogen available to foster growth of breast and prostate cancers.[56] When 8 ounces of pomegranate juice was given daily to men with prostate cancer, the time it took for their average PSA to double increased from 15 months to 54 months.[57] A similar study looked at men with a rising PSA who were given either 1 or 3 grams of pomegranate extract daily. After 18 months median PSA doubling time increased from 11.9 months to 18.5 months, with no difference between the two doses.[58]

Many other low-sugar fruits not regularly found in supermarkets, such as passion fruit, guava, kumquats, and sour cherries, also supply an abundance of flavonoids yet are low-glycemic. Kumquats and mandarinquats (a hybrid of kumquats and mandarin oranges) are unique in that the entire small fruit with the skin is tasty and packed with chemoprotective compounds, such as limonene, which has been demonstrated to have antiproliferative effects on human prostate cancer cells and also to protect against skin cancer.[59] That's right— eating the skins of citrus fruit protects our skin. That's pretty cool! So don't underestimate the anticancer potential of including many of these powerful fruits in your diet. Plus, they taste great.

NUTS AND SEEDS: THE ONLY FATTY FOODS THAT FIGHT CANCER

One of the most interesting nutritional findings from hundreds of research studies over the past two decades has been the note-worthy effects of eating nuts and seeds on enhancing healthspan and lifespan. Daily consumption of nuts and seeds has been associated with a reduced risk of death from cancer.[60]

In women, nut consumption during adolescence (one or more servings of nuts daily compared with less than one serving per month) was found to be associated with a 24 percent lower risk of develop-ing breast cancer later in life.[61] Walnut consumption has also been linked to suppression of breast cancer.[62] Researchers reported that eating 2 ounces of walnuts daily for as little as two weeks significantly changed gene expression in confirmed breast cancers. In a clinical trial, women who had undergone needle biopsies to evaluate breast lumps were asked to eat 2 ounces of walnuts daily until follow-up surgery two weeks after the initial biopsies. Tissue evaluation showed that expression of 456 identified genes had changed for the better. Also in women, daily nut consumption has been associated with a reduced risk of colorectal and pancreatic cancers in large prospective studies.[63] In the largest epidemiological studies in nutritional history, such as the Physicians' Health Study, the Nurses' Health Study, and the Adventist Health Studies, daily nut and seed consumption con-sistently has demonstrated protection against all causes of death.[64]

Seeds in particular have demonstrated marked protection against common cancers, undoubtedly because of their rich lignan content.

Flaxseeds, chia seeds, and sesame seeds are rich in phytoestrogens known as lignans, and these compounds have anti-estrogenic effects that have been studied for their potential to protect against breast and prostate cancers.[65] Enterolignans, metabolites of lignans that are formed in the digestive tract, blunt estrogen production and activity. Plant lignans also increase concentration of sex hormone–binding globulin, which blunts the effects of estrogens.[66]

Flaxseeds are the richest in lignans (85.5 milligrams per ounce), followed by chia seeds (32 milligrams per ounce) and sesame seeds (11.2 milligrams per ounce).[67]

Observational studies have shown a decrease in breast cancer risk with greater consumption of flaxseed, and randomized controlled trials also have had promising results.[68] Though the effects are more pronounced when started earlier in life, even trials on those who already have cancer show benefit. In one such trial, women recently diagnosed with breast cancer were assigned to eating either a placebo muffin or a muffin containing 25 grams of flaxseed each day for about thirty-five days until they underwent surgery. Tissue from their tumors analyzed before the intervention and at the time of surgery demonstrated significant tumor cell death and reduced tumor cell proliferation in the flaxseed group.[69] Even more impressive was a study that followed women with breast cancer for ten years and found a 71 percent decrease in mortality in women with greater lignan consumption.[70]

Lignans also counteract prostate cancer progression. A study on men with prostate cancer provided with thirty days of flax supplementation (30 grams per day) demonstrated reduced cancer proliferation. Another study assigned men with prostate cancer to flaxseed supplementation, a low-fat diet, or both for thirty days until prostatectomy. The low-fat diet had no effect, but both flaxseed groups showed reduced tumor cell proliferation.[71]

The anti-inflammatory effects of lignans, evidenced by reductions in C-reactive protein levels, contribute to their anticancer and cardiovascular benefits. Reductions in blood pressure have been reported in studies of whole flaxseed or lignan supplementation. Flaxseed has established blood pressure–lowering abilities, which are likely a combined effect of both its fiber content and its lignan content. The combined anti-inflammatory effects of lignans and fiber make flaxseeds and chia seeds important foods to consume daily.[72]

TOMATOES AND OTHER CAROTENOID-CONTAINING FRUITS AND VEGETABLES

Population studies show an association between high dietary intake of carotenoid-containing fruits and vegetables and reduced risk of lung, prostate, breast, and head and neck cancers.[73] Low blood carotenoid levels are associated with a greater risk of premature death.[74] Circulating carotenoids also travel to the skin, where they help to prevent oxidative damage from sun exposure, protecting against skin cancer.[75]

A high dietary intake of fruits and vegetables provides a wide spectrum of carotenoids. Vegetable juices (carrot, tomato, and spinach and other greens) represent a particularly potent form of carotenoids. Cooked vegetables provide more absorbable carotenoids compared with raw vegetables. Carotenoids are embedded in the matrix of the vegetables, and some of the cellular structure must be disrupted (for example, by blending, heating, or running through a juicer) to make the carotenoids more accessible to the digestive system.[76] Eating raw vegetables with fat-rich nuts, seeds, or avocado also improves absorption.[77]

Cooked tomatoes are rich in lycopene and other potent antioxidants; tomatoes are the primary dietary source of lycopene, providing about 85 percent of the lycopene in the diet.[78] Lycopene has anticancer activities,[79] and greater consumption of lycopene and higher blood lycopene levels are both associated with a lower risk of prostate cancer.[80] Two clinical trials on tomato supplementation in men with prostate cancer found reductions in PSA, and one of the trials noted reductions in a marker of DNA damage in prostate tissue as well as

increased apoptosis in tumor cells.[81] Carotenoid-rich foods offer protection against breast cancer, and high lycopene intake is most powerfully associated with lower rates of breast cancer in multiple studies.[82]

MORE BODY FAT = MORE CANCER

Regular exercise is linked to a reduced risk of breast, gastric, esophageal, endometrial, and colorectal cancers.[83] Exercise has also been
linked to better quality of life and better survival in patients with cancer.[84] More than just burning calories and reducing body fat, exercise
increases cellular efficiency and waste removal. Cells age faster with
the accumulation of cellular and chemical debris, but the body has
the means to engulf this debris with membranes and carry it to the
lysosomes, where they are burned for energy. In other words, exercise
augments the self-cleaning activity of cells and the destruction of
aged and worn-out parts (called autophagy), which slows aging and
reduces the risk of cancer.[85]

Extra fat on the body is a detriment to our survival, as fat cells
secrete angiogenesis promoters, tumor necrosis factor, and interleukin 6. These compounds create a state of chronic inflammation
and raise the inflammatory marker CRP (C-reactive protein) produced
by the liver, increasing the risk of cancer.[86] Fat accumulation can even
spill over into organs, such as the liver and heart, leading to cellular
dysfunction.

Swollen fat cells are "lipotoxic," meaning they are a source of
inflammatory, disease-promoting compounds, increasing the risk for
many cancers and worse outcomes in patients who have been diagnosed with cancer, such as recurrence of breast cancer, lower survival
rates in colorectal cancer and in prostate cancer, and more aggressive
disease progress.[87]

Most people have been brainwashed to think that oil, such as
olive oil, is a healthy food, but I am saying the opposite here: Olive oil
can be a major contributor to causing cancer if the oil significantly
increases your body fat. All oil has 120 calories per tablespoon, and

Increase in Olive Oil Consumption, 1995–2018

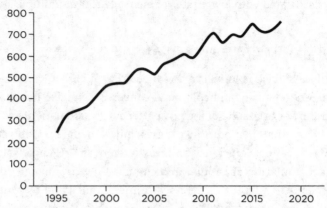

Sources: US Department of Agriculture, Economic Research Service. Oil Crops Yearbook 2019. https://www.ers.usda.gov/data-products/oil-crops-yearbook/. Last updated 20 August 2019; Ash M, Dohlman E. "Oil Crops Year in Review: US Soybean Demand Powered by Record 2006/07 Supply." June 2008. https://downloads.usda.library.cornell.edu/usda-esmis /files/jw827b648/hm50ts447/v405sb103/OCS-yearbook-06-18-2008_Special_Report.pdf.

those calories are rapidly absorbed. More oil equals more body fat, and more body fat means more cytokine and lipokine secretion, which then activates aromatase activity. This creates estrogen stimulation of breast and prostate cell growth, with accompanying angiogenesis promotion to permit cancer and metastasis. Oil calories add up quickly, with most Americans consuming more than 500 calories a day from added oils and fats, making it impossible for them to lose their dangerous body fat.

Nuts and seeds suppress appetite, while oil stimulates appetite and does not satiate, most often resulting in excess calories and an increase in abnormal and dangerous levels of body fat. Cutting out the oil in your diet and using a less caloric portion of whole nuts and seeds for the fat in your salad dressings and sauces are effective ways of reducing body fat, and proven methods for preventing cancer and heart disease.[88]

WHAT ELSE YOU CAN DO
TO AVOID AND PREVENT CANCER

ALCOHOL: NOT SO FUN ANYMORE

Since alcohol is metabolized into a carcinogenic compound (acetaldehyde) by the body, excessive alcohol use has been considered a risk factor for cancer; however, smaller amounts are also dangerous. The 2014 *World Cancer Report* issued by the IARC concluded that no amount of alcohol intake is safe when considering cancer risk.[89]

Alcohol use is considered a causative factor for several cancers: mouth, pharynx, larynx, esophageal, colorectal, breast, and liver cancers.[90] In women, alcohol may also increase estrogen levels, further contributing to greater breast cancer risk.[91] More than one hundred studies have looked at the association between alcohol intake and breast cancer, and pooled analysis indicates that even light to moderate drinking is associated with a moderate (more than 10 percent) increase in breast cancer that increases with higher intakes of alcohol.[92]

YOU NEED FOLATE, NOT FOLIC ACID

Folic acid is not folate. Folic acid is a synthetic imitation of the real thing—*folate*—from vegetables, which offers protection against much more than just birth defects. This broad protection cannot be accomplished by taking folic acid supplements. Key to the avoidance of cancer is to steer clear of supplements that contain folic acid and foods fortified with folic acid. It is too easy to overdose with this synthetic "nutrient" that promotes cellular replication and cancer.

Folate is abundant in green vegetables, beans, and other plant foods. In fact, people following a Nutritarian diet typically have blood levels of folate that are above the normal range. That extra amount of folate in the bloodstream is not hurtful; instead, it shows that a person is eating healthfully and has been exposed to hundreds of other protective compounds found in vegetables. Only people who don't eat sufficient amounts of vegetables could be low in folate.

Folate plays a critical role in supporting our health as adults, beyond its known positive effects in pregnancy and childhood.[93] It is one of the many nutrients necessary for red blood cells to transport oxygen properly, and it also supports the functioning of our nervous and cardiovascular systems. Adequate amounts of B vitamins, including folate, are important for maintaining good cognitive function throughout life.[94]

Folate deficiency is harmful, but excess, in the form of folic acid, can be dangerous. Synthetic folic acid is found in supplements and fortified foods and is twice as absorbable by the human body compared with natural folate.[95] Foods fortified with folic acid include nutritional yeast, packaged cereals, and white bread. While the body converts some synthetic folic acid to folate, it has a limited capacity to do so. Much of the remaining folic acid that is not converted circulates in the blood and tissues unmodified.

It is unknown exactly what unmodified folic acid does in the human body, but it has the potential to disrupt normal folate metabolism, and there is substantial evidence that it can reduce the protective functions of immune cells and promote cancer.[96] Cancer cells have more folate receptors on their surfaces and produce more folate-dependent enzymes than normal cells; excess circulating folic acid feeds into this process and allows cancer cells to proliferate. Excess folic acid may also lead to cancer development by producing changes in gene expression.

The thinking that folic acid could promote cancer is not new. In the 1940s, high doses of folic acid were given as an experimental treatment to leukemia patients, and the rate of cancer cell proliferation increased. This finding was the origin of the use of antifolate drugs for cancer chemotherapy today.[97] Numerous studies have found folic acid supplementation to be linked to increased cancer risk, and serious concerns have been raised over the potential cancer-promoting effects of foods being fortified with folic acid.[98] A 2011 meta-analysis of six folic acid supplementation trials found that the incidence of cancers was 21 percent higher in the folic acid supplementation

Folic Acid Versus Folate During Pregnancy

Health authorities universally recommend folic acid supplementation to women of childbearing years and especially during early pregnancy to prevent neural tube defects because our population eats so unhealthfully, without sufficient intake of vegetables and beans. This recommendation is likely responsible for a climbing rate of childhood cancers, particularly acute lymphocytic leukemia and possibly brain tumors in childhood as well. Acute lymphocytic leukemia and other children's cancers have been linked to women's poor diets and lack of green vegetables before pregnancy.[99]

Folic acid supplementation is not only cancer-promoting; it also allows a false sense of security because it takes the focus off of having to eat greens and make other healthful food choices before and during pregnancy. No sense of urgency or importance is placed on eating right. This synthetic and simplistic solution to a complicated problem may be increasing cardiac birth defects, childhood respiratory illnesses, and childhood cancers.[100]

If women instead were encouraged to eat green vegetables and a healthful diet to meet their folate needs, a broad spectrum of benefits would result. When we get folate from a healthful diet, it comes naturally packaged in balance with hundreds of other cancer-protective micronutrients for both the mother and her child.

groups than in the control groups.[101] Scientific studies also support the conclusion that synthetic folic acid from multivitamins is cancer-promoting, whereas folate from food is protective.[102]

Studies on multivitamin use and breast cancer have produced inconsistent findings because of confounding variables. Some ingredients in multivitamins are helpful, and others are hurtful; however, several studies that have demonstrated increased breast cancer risk indicate folic acid as the major culprit.[103] The bottom line is that

Kelly Molino
Completely Recovered from Autoimmune Hepatitis

In April 2016, at age 50, I had my blood tested because I had dark urine and was unexpectedly losing weight and feeling fatigued. The results were frightening: My liver enzymes were twenty times higher than normal, and my doctor said, "Go right to the hospital." I found out I had autoimmune hepatitis. Thinking and hoping that these initial results might have been a fluke, I opted to retake the blood test. To my dismay, I discovered that my condition was getting worse. Fortunately, at around the same time, my husband happened to see Dr. Fuhrman on one of his PBS shows, and by coincidence, I had a friend who had been cured of another autoimmune disease by following Dr. Fuhrman's Nutritarian diet.

Thank God I began Dr. Fuhrman's autoimmune protocol diet the very next day. This involved eliminating the two most frequent triggers of autoimmune disease: animal products and gluten. In addition, I began eating a high-micronutrient diet based on Dr. Fuhrman's autoimmune protocol.

I also saw a liver specialist and asked his opinion about nutritional treatment. He told me that I needed to start taking this seriously, because I was "about a month or so away from being hospitalized due to liver failure." He also told me I needed to begin taking steroids right away and that I would need to take them for the rest of my life. I was firm in my decision to give Dr. Fuhrman's protocol for autoimmune disease a chance. I became a member of DrFuhrman.com and posted a question in the online Ask the Doctor forum: "Given my ALT (alanine transaminase) liver enzyme levels, which are around 500 units per liter (normal is around 7–50) do I have any time to try your diet before beginning the steroids?"

The response from Dr. Fuhrman filled me with hope and confidence that his was the right approach. He said I could give the diet a try for two weeks, and if the enzymes didn't decrease by then, I could go on a low-dose steroid until the diet kicked in. Then he would help wean me off the steroids, because long-term steroid use can increase risk of cancer and other serious health problems. This was all the encouragement I needed to give this diet my all. I am eternally grateful to him for his quick and encouraging answer.

My husband and I charted a graph of my blood work results, which showed just how quickly the diet improved my health. In just two weeks, my liver enzyme numbers not only stopped rising, but they started going down. And by four weeks, my enzymes dropped by more than half. I was elated, because I never had to take any steroids!

I'm amazed by the power of food. All in all, it took about a year on Dr. Fuhrman's diet plan for me to be completely cured. I will be forever in debt to God, Dr. Fuhrman, and his associate Dr. Benson, for helping to heal me and teach me the truth about healthy food. Not only am I healed, but I am also healthier than I have ever been in my life. I wish everyone could know about the power of superior nutrition.

---

avoiding both supplements with folic acid and foods fortified with folic acid is important for the prevention of cancer. Luckily, we can get the right amount of folate from our diet.

THROUGH NUTRITIONAL EXCELLENCE

Clearly, we should try to avoid consuming drugs, pesticides, fungicides, and other man-made chemicals and synthetic nutrients. Data even suggest that synthetic nutrients in supplements, such as beta-carotene, vitamin A, and vitamin E, have negative effects that also increase cancer rates.[104]

When we live in a healthful manner, we can avoid needless medical care, dangerous drugs, unnecessary medical tests, invasive medical procedures, and exposure to the potential harm that can come from modern medical care. To live long and healthfully, we want to live in a way that allows us to avoid the need for medicinal substances and medical interventions as much as possible.

This eye-opening nutritional science is powerful—and perhaps shocking to many. But given the research, it is hard to deny that we have the ability to protect ourselves from the vast majority of cancers through nutritional excellence. These life-saving and life-giving tools are well within our grasp. In the coming chapters, I ask you to trust me to take you on a delicious journey that will transform your life.

CHAPTER FIVE: QUICK SUMMARY

A cancer cell is essentially a normal cell whose DNA has been damaged to the point that the cell can no longer control its replication. Most diseases, including most cancers, are the result of a poor diet. Dietary influences contribute more to causing cancer than other environmental influences, such as exposure to pesticides, chemicals, pollution, and infectious agents—though all may contribute. A Nutritarian diet allows micronutrient excellence to fuel the body's powerful disease defenses. The same foods and phytochemicals shown to be most effective at preventing cancer also work to reverse early-stage cancers and prevent recurrence of cancer in people who already have received a cancer diagnosis.

CANCER: TOXIC ENVIRONMENT AND POOR DIETARY CHOICES

Cancer has surpassed heart disease to become the leading cause of premature death for Americans between the ages of 45 and 64. Although from 1999 to 2015 overall rates of death from cancer and new cancer diagnoses in the United States fell slightly, there has also been a significant increase in the incidence of some cancers, notably breast cancer—thought to be secondary to the growing girth of Americans' waistlines.

Two main factors drive our modern epidemic of cancer: our exposure to chemical carcinogens and to toxic foods. The second includes our insufficient intake of plant-derived phytonutrients—a deficiency that renders our cells incapable of functioning to their fullest potential for repair and maintenance.

A DIETARY PROTOCOL RICH IN VARIOUS FOODS
Greens

Colorful vegetables are more helpful in reducing cancer because they contain higher amounts of cancer-protective compounds.

Green vegetables are the most powerful cancer fighters because they contain numerous plant pigments and phytochemicals that bind carcinogenic chemicals and enhance defenses. Cruciferous vegetables supply isothiocyanates, such as sulforaphane and indole-3-carbinol, that

- Inhibit cancer cell growth and proliferation
- Help suppress angiogenesis
- Neutralize potential carcinogens
- Promote death of cancer cells (apoptosis)
- Curb inflammation
- Have antiviral and antibacterial effects
- Have anti-estrogenic effects that help to prevent hormone-related cancers

Beans, Particularly Soybeans

Beans, lentils, and peas have strong anticancer benefits because of their rich flavonoid content as well as other highly protective phytonutrients. Soybeans are particularly protective against stomach, lung, and colorectal cancers because they contain isoflavones, a type of phytoestrogen. Avoid soy protein powder and isolated or concentrated soy protein–based foods, however, because their amino acid profile is similar to that of animal protein and can raise IGF-1 levels too much.

Scallions, Onions, and Garlic

Vegetables in the *Allium* family, such as onions, garlic, shallots, leeks, and scallions, contain organosulfur compounds that have a wide variety of powerful anticancer effects, including stopping the growth of cancer cells, activating Nrf2, and inhibiting angiogenesis. They also contain quercetin, a flavonoid, which has anti-inflammatory effects that help inhibit the proliferation and migration of cancer cells.

Mushrooms

The cell walls of mushrooms contain beta-glucans, which are bio-active polysaccharides. Beta-glucans enhance immune surveillance by increasing the activity of natural killer cells that detect and destroy cancerous and virus-infected cells. In addition, mushrooms contain specialized lectins that adhere to cancer cells and inhibit their growth, and phytochemicals that inhibit aromatase, an enzyme that increases the risk of breast cancer.

Berries and Pomegranates

Berries and pomegranates (as well as passion fruit, plums, citrus fruits, cranberries, and cherries) are low in sugar and abundant in flavonoids, which are concentrated in their skins. Berries contain ellagic acid (also found in pomegranates), resveratrol, and other polyphenols that inhibit inflammation, modulate gene expression and DNA methylation, promote cell cycle arrest in cancerous cells, and inhibit angiogenesis.

Nuts and Seeds

Phytochemicals from nuts and seeds have antioxidant, anti-inflammatory, and antiproliferative effects. Seeds—particularly flax-seeds, chia seeds, and sesame seeds—are rich in phytoestrogens known as lignans. These compounds have anti-estrogenic effects that have been studied for their potential to protect against breast and prostate cancers. Enterolignans, metabolites of lignans formed in the digestive tract, blunt estrogen production and activity. Lignans also have anti-inflammatory effects, which contribute to their anticancer and cardiovascular benefits.

Carotenoids: Tomatoes and Other Vegetables

Fruits and vegetables—particularly tomatoes, carrots, and greens—provide a wide spectrum of carotenoids. High levels of carotenoids in the blood are associated with reduced risk of many cancers; in contrast, low levels of carotenoids are linked to a greater risk of premature death. Tomatoes—specifically cooked tomatoes—are rich in lycopene, which is associated with a reduced risk of prostate and breast cancers. The most efficient way to make carotenoids more bioavailable is by blending, heating, or juicing them in order to break down the cell structure. When eating salads of raw vegetables, include some fat-rich nuts and seeds to improve the absorption of carotenoids.

AVOID EXCESS BODY FAT, ALCOHOL, AND OTHER TOXINS AND SYNTHETIC NUTRIENTS

Extra fat on the body is dangerous because fat cells secrete an array of harmful substances, including angiogenesis promoters, tumor necrosis factor, and interleukin 6. Swollen fat cells are "lipotoxic," increasing the risk for many cancers and also poorer outcomes in people who have cancer. Excess body fat also promotes the production of estrogen, which is an additional factor connecting obesity with breast and prostate cancers.

Alcohol use is considered a causative factor for several cancers: mouth, pharynx, larynx, esophageal, colorectal, breast, and liver. In women, alcohol may also increase estrogen levels, increasing the risk of breast cancer.

Besides avoiding drugs, pesticides, fungicides, and other man-made chemicals, common ingredients in supplements increase cancer risk. Folic acid, a synthetic form of folate (which is abundant in green vegetables and beans), is found in supplements and fortified foods. There is substantial evidence that excess folic acid can promote cancer by inhibiting the functions of immune cells and producing changes in gene expression. Synthetic nutrients in supplements, such as beta-carotene, vitamin A, and vitamin E, have negative effects that may increase cancer risks.

CHAPTER SIX

The Struggle to Lose Weight

People overeat because of many different social, emotional, recreational, and addictive influences. Practically our entire population is sickly and overweight today. Although health authorities announced in 2016 that 70 percent of Americans were overweight or obese, I disagree with that too-low percentage. That 70 percent is based on a BMI of 25 as separating the "overweight" category from the "normal weight" category. I would argue that the percentage of overweight Americans is much higher because a BMI of 24 and above should be considered overweight.

If we look at the healthiest, most long-lived individuals in the Blue Zones around the world, we find that males have BMIs of less than 24 and females of less than 23. Though there are always some exceptions, given differences in human shapes and skeletal mass, according to data demonstrating the relationship between lifespan and BMI,

An average-height (5 feet, 5 inches) female should not weigh more than 126 pounds (57 kilograms), which is a BMI of 21.0.

An average-height (5 feet, 10 inches) male should not weigh more than 165 pounds (75 kilograms), which is a BMI of 23.7.

Daniel Shuman
Lost 152 Pounds and Transformed His Life

The opportunity to take what I've learned and share that knowledge is priceless. At 29 years old, I weighed 355 pounds. I was a miserable food addict and feared for my life. After the birth of my second son, I decided that I wanted to live a long life; but I feared that given my habits, inability to control my eating, and family history, that was unlikely.

I began by eliminating all processed foods and limiting refined sugars. After a few months, and about 20 pounds lost, I started to learn more and more about nutrition. My father had read *Eat to Live*, which was recommended to him by his general practitioner. I decided to give the Nutritarian diet a shot, and that kicked my weight loss into high gear and made this so easy.

Now I have lost 150 pounds! There is not one area of my life that hasn't been improved by my weight loss and lifestyle change. My social, professional, and even intimate life all changed for the better. I can now play with my children without getting winded. I can now run four miles at a pace of under ten minutes a mile, when in the past, going up the one flight of stairs to my apartment left me winded. I can bike forty to fifty miles as if it's nothing.

The world has opened up to me. I live in a diverse neighborhood in Queens, New York. Strangers of all cultures who've witnessed my progress stop me constantly with words of encouragement, and requests for advice, which I'm elated to share. I feel as if my entire identity has changed for the better.

The Blue Zones

Blue Zones are demographic regions of the world that author and educator Dan Buettner wrote about in 2005, saying that people in those areas commonly live much longer, active, and healthier lives than people in other areas of the world. Blue Zones generally have the highest percentage of centenarians, that is, people who live past the age of 100 years. Centenarians are not overweight; there are no obese centenarians, and most of them have a comparatively lower BMI, less than 21.[1]

The five Blue Zones Buettner identified are

- Sardinia (an Italian island)
- Okinawa, Japan
- Seventh-Day Adventists living in Loma Linda, California
- Nicoya Peninsula (Costa Rica)
- Ikaria (a Greek island)

For most people with normal musculature, even a BMI right below 23 is likely too much weight. For example, at 5 feet, 10 inches tall, my athletic weight throughout my entire life has been between 145 and 150 pounds. My body fat is around 10 percent and has remained there my whole life; my BMI is between 21 and 22. I have done weight training regularly the entire time, and no matter how much I do, my weight does not increase, given my Nutritarian diet. In order to get bigger than that, I would have to eat a diet with more animal products, which would reduce my healthspan and lifespan.

In other words, increasing one's BMI, even when it is mostly muscle, is still lifespan-unfavorable. The shortest-lived people in the United States are linemen on football teams who have fed themselves to become excessively large. The National Institute for Occupational Safety and Health's NFL Mortality Study in 1994 disclosed that football players as a whole had a normal lifespan; however, the linemen had a 50 percent increased risk of early cardiovascular death that was directly related to body size. Those with the largest body size had a six-fold increased risk of early heart disease death, compared with other positions occupied by leaner and smaller athletes.[2] Eating to bulk up for football (or for any reason) is not a wise move for later-life health.

Now let's take a closer look at those people in the United States whose BMI is favorable (less than 23), which is less than 15 percent of the population. We find that the majority of those in this favorable BMI category either smoke cigarettes, are addicted to drugs (legal or illegal), are alcoholics, or have some medical condition—such as depression, anxiety, digestive disorders, autoimmune diseases, asthma, or an occult cancer (one that cannot be detected yet)—that is contributing to or responsible for their lower BMI. Most Americans with a favorable BMI have a significant health issue, so that group is not representative of healthy people who have achieved a favorable weight through eating right and exercising. Only 2.7 percent of Americans report they eat healthfully, have a favorable BMI, don't smoke, and exercise regularly.[3] These data from the National Health and Nutrition Examination Survey (NHANES) recorded people who followed these four basic principles of healthy living: a "good" diet, moderate exercise, not smoking, and keeping body fat under control—not super high standards of excellence.

Researchers also studied the NHANES BMI data and determined that higher intake of fruits and nonstarchy vegetables was associated with good health and favorable weight. The combined cup equivalents of fruit and nonstarchy vegetable intake compared with the total gram weight of all food consumed daily served as the primary variable determining BMI. Waist circumference, LDL cholesterol, fasting glucose level, and insulin and triglyceride levels were all determined by this simple ratio of healthy foods to total calories. The results showed that the more low-calorie vegetables a person eats, the better that person's health and longevity.[4]

97.5 percent of American adults get a failing grade on healthy lifestyle habits.

Some interest groups try to persuade us that we can be healthy and overweight at the same time. They often present data which suggest that people who are mildly overweight have better health statistics than people of normal or lower body weight. But this is simply a deceptive statistic; the truth is that being sick makes you lose weight, and most normal-weight people in the United States are sickly or smoke, so they are not representative of the very small percentage of people in this country who are slim and eat well. Even being mildly overweight significantly shortens lifespan, when compared with that 2.7 percent of normal-weight individuals who earned that favorable weight by living healthfully, rather than smoking and drinking. Even being slightly overweight has an unfavorable effect on health and lifespan.[5]

I am always amazed to see so many popular and well-known health gurus and health leaders who are overweight and unhealthy themselves. If their diet advice were sound, they would not be overweight!

WHY DIETS FAIL

It is very difficult to control your appetite when your diet is not nutritionally sound. The more unhealthfully you eat, the stronger the internal signals are that command you to eat more food.

Why do humans overeat themselves into health difficulties? Other animals become overweight only when offered unhealthy foods (in captivity), which then derails their natural appetite controller, or *appestat*.

Eating calorically dense, low-nutrient processed foods develops an increased desire for excessive calories and affects your brain. You become increasingly dependent on overeating to feel well. The food addicts who need a massive diet change the most are the ones who have the strongest food addictions. Unfortunately, too many people have thrown in the towel, having failed so many times at trying to lose weight that they have become resigned to staying overweight and sick the rest of their lives.

However, food addicts need to know that their addiction is not their fault. They have not been taught how "fake" foods have been designed to hook them, and they have not been taught how to undo the damage. Once they know how this works and how easy it is to lose weight once their intake of nonstarchy vegetables is very high, they can succeed. When people, even those with food addiction and emotional overeating habits, receive the proper guidance and support, they find that changing to a healthful diet is both comfortable and pleasurable, rather than a constant struggle.

The most common barrier to weight loss is a heightened and uncomfortable sensation of hunger that drives overeating and makes dieting fail; it is simply too physically uncomfortable to eat less. In my thirty years of experience in this field of study, I have seen (and published my observations) that enhancing the micronutrient quality of the diet significantly lessens feelings of hunger.[6] My fellow researchers and I recorded and stratified the results from 768 individuals who had chosen the Nutritarian diet approach, and we found that the better they adhered to the nutrient-dense, plant-rich diet style, the more likely their hunger sensations were transformed. Those who adhered at least 90 percent to the prescribed food recommendations for months recorded a decreased and changed perception of hunger—they simply desired fewer calories. They found that their feelings of hunger were less intense; were felt in the throat and upper chest, rather than in the stomach; and were more manageable.

RESOLVING TOXIC HUNGER AND FOOD ADDICTIONS

Fatigue, weakness, stomach cramps, tremors, irritability, and headaches are commonly interpreted as hunger, but they are what I call *toxic hunger*. These sensations resolve gradually for the majority of people who adopt a high-nutrient diet and are replaced with a less distressing sensation, which I have labeled "true" or "throat" hunger.

It is well-documented that a diet low in antioxidant and phytochemical micronutrients leads to heightened oxidative stress and a buildup of toxic metabolites, promoting disease and cancers.[7] It has also been shown that a higher intake of nutrient-rich plant foods decreases measurable inflammatory by-products, lowering risk of disease.[8] When a diet is low in the micronutrients and phytochemicals found in plants, inflammatory by-products build up in the body and not only create disease, but also make you eat more. The increased toxins create uncomfortable sensations that direct you to eat too much and too frequently.

When your body is more "toxic," you experience discomfort when no longer digesting food, most commonly felt as fatigue. This is withdrawal from a poor diet. Fatigue is a common symptom of withdrawal and is not a symptom of hunger. Just as smoking cigarettes more frequently staves off the pain of withdrawal from nicotine, overeating and eating more frequently to feel more "energy" mitigates the withdrawal fatigue that occurs after digestion ceases, for people eating nutritionally deficient diets. This is the reason so many people remain overweight. The worse the nutritional quality of your diet, the more calories you need to consume to feel "normal."

The digestive cycle has two phases: the *anabolic phase*, or the eating and digestive phase, and the *catabolic phase*, or energy utilization phase, which begins when digestive activities cease. During the catabolic phase, when the body is utilizing stored energy reserves, the body can most efficiently heal, repair, and self-clean. Digestion inhibits the waste-processing activity of the liver. During the catabolic phase, when toxins are mobilized and processed for removal, people experience uncomfortable symptoms that they perceive as hunger. One can halt these uncomfortable withdrawal symptoms by eating again, which ends the catabolic phase and reduces detoxification. Such withdrawal symptoms drive overeating behavior and are a major factor leading to obesity.

Anabolic Phase	Catabolic Phase
Absorptive phase	Utilization of stored energy

Healthful eating is more effective for long-term weight control because it modifies and diminishes the sensations of withdrawal-related hunger. If you are struggling with excess weight, the eating approach I recommend helps to diminish that sensation of discomfort, and the subsequent desire to continue to eat, and therefore leads to your consuming fewer calories—the amount you really need. The reality is great news for anyone struggling to lose weight: Those intense and often irresistible urges to eat lessen with time as you eat healthier and get healthier.

Fake food, sweets, and excess animal products all exacerbate toxic hunger's effectiveness in driving up appetite. Imagine that you drink six cups of coffee a day or smoke a pack of cigarettes: You can recognize the headache and shakiness that appear as your caffeine or nicotine levels get low, and you are relieved and feel better again when you have that next cup of coffee or that next cigarette. Those uncomfortable feelings are really your body striving to fix itself as it engages and intensifies the removal of noxious waste. You can quell your caffeine withdrawal headache by more frequent intake of caffeine, and you can quell your fatigue and other food-withdrawal symptoms by eating again.

But make no mistake about this: Fatigue is not a symptom of hunger; it is a symptom of withdrawal from bad food choices.

As soon as the body's intake, digestion, and assimilation of food are complete, the catabolic utilization of glycogen reserves and fatty acid stores begins to meet the body's energy needs. Hunger normally increases in intensity as glycogen stores are diminishing toward the end of glycolysis and should not normally begin at the start of the catabolic phase when glycolysis begins.

However, the switch from the anabolic phase (when calories are entering the body) to the catabolic phase (when stored calories are being burned for energy) is an important physiological change. The liver goes from storing nutrients during the anabolic phase to becoming engaged in utilizing stored reserves and breaking down and removing waste during catabolism. This heighted mobilization and elimination of cellular waste products brings about the symptoms of toxic hunger.

In short, the more closely you follow the Nutritarian eating approach I recommend, the more quickly you move beyond the physical and emotional hunger pains that are all too common. This means that over time, when you follow the Nutritarian diet, the withdrawal symptoms such as fatigue, headaches, tremors, stomach cramps, and mood changes become resolved. It's important to note that increased anger, hostility, anxiety, and depression can also be symptoms of food addiction and withdrawal, but these symptoms will also resolve with time.[9]

In contrast, true hunger occurs hours later when glycogen stores are nearing completion, so you can replenish your glycogen stores, preventing gluconeogenesis. Gluconeogenesis is the utilization of muscle tissue for needed glucose once glycogen stores have been depleted. True hunger does not fuel fat deposition; it exists to protect lean body mass from being used as an energy source.

When a diet is low in antioxidants, phytochemicals, and other micronutrients, intracellular waste products such as free radicals, advanced glycation end products (AGEs), lipofuscin, lipid A2E, aldehydes, and others accumulate.[10] It is well-established in the scientific literature that these waste products and toxic substances contribute

Withdrawal from Eating Highly Processed Foods

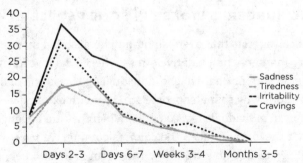

Number reporting withdrawal symptoms

Source: Adapted from Schulte EM, Smeal JK, Lewis J, Gearhardt AN. Development of the highly processed food withdrawal scale. *Appetite*. 2018 Dec 1;131:148–54.

to disease and are associated with typical food-withdrawal symptoms, which can feel similar to those experienced when someone goes through withdrawal from an addictive drug.[11] These uncomfortable symptoms, widely misperceived as hunger, are relieved by eating, which halts catabolism and halts the detoxification process.

We live in a society where the default eating style includes a low intake of colorful plant foods and an abundance of processed foods and commercially baked goods. I want to make this point again because it is vital: For most people who do not eat a healthy diet composed of mostly vegetables, fruits, nuts, and berries, the commonly experienced sensations of hunger are instead the symptoms of withdrawal from a diet that is inadequate in micronutrients. Such a diet makes withdrawal symptoms from proinflammatory metabolic wastes almost ubiquitous.

There is growing evidence that food addiction is the most prevalent clinical pathological condition.[12] A full understanding of the nature of hunger is essential for people who aim to lose weight and keep it off permanently. If we don't manage the heightened (withdrawal) hunger, we are doomed to fail. And those symptoms cannot be managed unless we improve the micronutrient quality of what we eat. What we eat is the major factor that influences how much we desire to eat.

TOXIC HUNGER IS WORSE IN SOME PEOPLE

Evidence suggests that overweight individuals build up more inflammatory markers and oxidative stress when fed a low-nutrient meal compared with normal-weight individuals.[13] This is one of the important reasons why so many people struggle with extreme weight.

Certain individuals feel so poorly when they are no longer digesting that they have to keep the calories coming. We used to call this *hypoglycemia*, but now we know this term is invalid because most people who feel poorly when their glucose is low are just undergoing withdrawal from their low-nutrient, high-protein diet; all the nitrogenous by-products such as ammonia, urea, and uric acid are making them feel sick. These individuals are not suffering from low glucose; they are suffering from the detox that accompanies the low glucose. Of course, these detox symptoms are relieved when they eat more protein to stop the withdrawal. So-called hypoglycemic symptoms are almost always resolved by a Nutritarian diet within four weeks, except when these symptoms are truly due to excessively low glucose from an insulin-secreting tumor or from a rare defect in glucagon production. Then, it is not a dietary problem, but a medical one.

The heightened inflammatory potential in people with a tendency for obesity is marked by increasing levels of irritants noted in blood and urine such as lipid peroxidase and malondialdehyde and reduced activation of hepatic detoxification enzymes.[14] This supports my experience that people who are prone to obesity get more uncomfortable withdrawal/hunger symptoms during the nondigestive (catabolic) stage when breakdown and mobilization of toxins are enhanced. The result is the need to eat again and overconsume calories. This is a vicious cycle promoting continuous (anabolic) digestion, frequent feedings, and increased intake of calories.

Chronically overweight people in the typical American food environment feel "normal" only by eating too frequently or by eating a heavy meal, so that the anabolic process of digestion and assimilation continues right up to the beginning of the next meal. They need excess calories in order to feel normal. Such people are almost always eating or digesting what they have eaten. Some people even wake in the middle of the night with a need to eat something.

Adequate and comprehensive exposure to phytonutrients in conjunction with removal of proinflammatory foods can reduce inflammation and inflammatory markers and lower the levels of metabolic waste products that are waiting to be released. It has been easily demonstrated that people on the high-nutrient diet are able to go for longer periods without feeling "hunger" symptoms.

Other studies have documented a decrease in appetite with ingestion of greater amounts of fiber and/or micronutrients.[15] A

Canadian study published in 2008 found that hunger before meals and appetite after meals were reduced in women who took multi-vitamin and mineral supplements.[16] My Nutritarian approach, however, is much more comprehensive and effective since supplemental ingredients cannot duplicate the anti-inflammatory and broad biochemical benefits of whole natural foods. The Nutritarian protocol is unique and noteworthy, given the highly significant reductions in appetite, blood pressure, LDL cholesterol, fasting glucose levels, and body weight in people who have made the change to a high-micronutrient diet.[17]

The micronutrient density of a diet has a strong effect on whether an overweight person can reset his or her appetite and maintain a normal weight. After an initial phase of adjustment during which the person experiences toxic hunger due to withdrawal from proinflammatory foods, a high-nutrient-density diet can relieve the intense symptoms that spur a heightened desire for excess calories.

Furthermore, a vast body of research documents the protective benefit of a micronutrient-rich diet against cancer and cardiovascular disease, as we have seen.[18] If clinicians can with confidence assure their patients that they will not experience uncomfortable sensations of hunger after the "detoxification" stage is over, they can keep those patients motivated to withstand the withdrawal symptoms they experience early in the dietary transition. Those patients will then experience substantial and sustainable weight loss, as well as the prevention of many major chronic diseases.

In caring for thousands of individuals following a nutrient-dense, plant-rich Nutritarian diet, I have observed that for the vast majority, the withdrawal symptoms usually come under control in less than one week. Withdrawal can be intense for some people, with occasional flu-like symptoms of low-grade fever, headaches, body aches, sore throat, irritability, and depression. But during live-in programs or immersions that I offer, the discomforts people feel coming off the SAD resolve usually within four to five days of clean, nutrient-rich eating and they feel much better.

HOW MUCH SALT IS ACCEPTABLE IN THE DIET?

I write about salt in more detail in my book *The End of Heart Disease*.

If you live in the United States, your lifetime probability of developing high blood pressure is around 90 percent.[19] This is a sobering statistic, since high blood pressure increases a person's risk of developing heart disease (heart failure, heart attack, or sudden cardiac death), kidney failure, and stroke. Treatment that lowers high blood pressure decreases the risk of stroke and heart failure.[20]

After many years of high salt exposure, blood pressure starts to rise. By the time this occurs, cutting down on salt does not so easily resolve the problem. This is because the blood vessels have been damaged, and the nervous system's sympathetic tone has been unfavorably modified. High salt intake also promotes cardiac fibrosis, increasing risk of arrhythmias or an irregular heartbeat.[21] Similarly, the kidneys contain around one million tiny, delicate filters composed of blood vessels, which lose their function from the excessive pressure, leading to a disorder known as "hypertensive nephrosclerosis," a major cause of kidney disease.

Numerous observational studies and randomized controlled trials document that high sodium intake increases blood pressure.[22] The evidence implicating excess sodium intake as a major cause of high blood pressure levels has been called "overwhelming."[23] A large, long-term lifestyle intervention study published in 2007 showed that a 25–35 percent reduction in dietary sodium over ten to fifteen years resulted in a 25–30 percent lower risk of negative cardiovascular outcomes.[24] It is estimated that a 50 percent decrease in sodium consumption in the United States could prevent at least 150,000 deaths annually.[25] According to a meta-analysis of sixty-one studies, the lower an individual's blood pressure, down to at least 115/75 mm Hg, the lower the risk of stroke and heart attack.[26] There is no "threshold" below which the risk does not decrease, assuming the lower blood pressure is "earned" through healthy eating, exercise, and salt avoidance and is not just medicated downward.

But it is not all about blood pressure. The interesting finding from many different studies is that high salt intake is linked to increases in all-cause mortality and that its death-hastening effects occur in people who are not "salt sensitive" to its blood pressure effects. In other words, even if your blood pressure is not high, high sodium intake increases death from all causes.[27]

TOO MUCH SALT

If we ate just natural foods without added salt, we would most likely consume about 500 to 750 milligrams of sodium a day. Real food supplies the perfect amount of minerals people need to maximize their health. The human body was designed to function on food, and early humans did not consume salt. Our Stone Age ancestors consumed a diet consisting of mainly fruit, vegetables, nuts, seeds, fish, insects, and wild game. All the sodium that humans require, as well as the other minerals we need, is present in those natural foods.

This eat-what-you-can-find diet continued for approximately one hundred thousand generations, during which time salt was not added to food. Humans developed agriculture around three hundred generations ago, and the Industrial Age again changed our diet over a span of five to ten generations until now. The "Processed Food Era" really started only after World War II, which was two to three generations ago. This means we live with the "thrifty genes" of a hunter-gatherer diet that were selected over a long period to conserve sodium, not get rid of it because our ancestors had to deal with low salt intake, periods of starvation, and caloric inadequacy.[28] The problem is that today, most areas of the world consume five to ten times as much sodium as would be found in a natural "unsalted" diet.

Since almost all Americans and modern industrialized societies consume so much salt, we have to look at isolated or primitive populations to really see the long-term result of low salt intake. It is still possible to find pockets of people living on mostly natural food diets, without added salt. Indigenous tribes in New Guinea, the Amazon Basin, the highlands of Malaysia, and rural Uganda all eat very little salt. Hypertension is unheard of in these regions, and blood pressure does not rise steadily with age as it does in the United States and other countries with high salt intakes. The most elderly members of these populations have blood pressure readings similar to those we see in children. When salt is introduced into these salt-free cultures, however, blood pressure climbs.[29] In all human populations studied by medical anthropologists, salt-free cultures (that is, those that do not use salt as a condiment) show almost no increase in blood pressure even into old age. By contrast, blood pressure rises significantly over many years in all human populations in which salt is added to food in significant quantities. The result is that most people, in all those salt-added societies, sooner or later end up with high blood pressure.

Despite some people claiming that they are not salt sensitive because they have low blood pressure on a high-salt diet, over the years, high salt intake takes its toll, and almost everyone develops high blood pressure. The majority of Americans develop hypertension by the time they are in their 60s. Of those who are lucky enough to escape it up to age 65, 90 percent will still develop high blood pressure if they live past age 80.[30] By this time, it would not be so easy to cut out the salt and fix all the damage.

For maximum disease prevention, sodium levels should likely be held to the levels that are normal to our biological needs, which is likely fewer than 1,000 milligrams per day. Average adult sodium intake in the United States is around 4,000 milligrams for every 2,000 calories consumed. Natural foods contain about half a milligram of sodium per calorie or less. Except for some sodium-wasting medical conditions or unusual individuals, it is rare for a person to require more sodium than is present in real food in its natural state.

Even the Centers for Disease Control reports that salt kills far more Americans than tobacco (or anything else), and almost 70 percent of all Americans, including everyone older than 40, should cut their salt intake by nearly two-thirds, to 1,500 milligrams per day. Medications cannot do nearly what diet improvement and salt reduction can do, and more and more physicians and scientists recognize this. Just cutting out salt can return blood pressure to normal, which can cut the risk of heart disease by nearly 70 percent.

We know high blood pressure is one of the most powerful predictors of earlier mortality. A comprehensive meta-analysis published in 2014 that evaluated data from sixty-six countries and 107 randomized intervention trials evaluated the potential of limiting dietary sodium to 2,000 milligrams per day (2 grams, or roughly 1 teaspoon), as recommended by the World Health Organization.[31] This study was robust, using twenty-four-hour urine tests and dietary records, and also took the duration of the intervention into account. Researchers determined that if this salt reduction were implemented, approximately 1.65 million cardiovascular deaths would be prevented every year.

Even more dramatic were the study's calculations that showed cardiovascular deaths would be reduced 40 percent more if the salt intake were lowered from 2 grams to 1 gram of sodium a day. If the study's population made that change, 2.3 million deaths would be prevented, 40 percent of which would have occurred before the age of 70. Clearly, the lower the sodium intake, the better.

The researchers found that the average intake of sodium worldwide in 2010 was 3,950 milligrams per day, nearly double the 2,000 milligrams recommended by WHO and close to triple the ideal limit of less than 1,500 milligrams recommended by the American Heart Association. In their meta-analysis of controlled intervention studies, the researchers also found that reducing dietary sodium lowered blood pressure in all adults, with the largest effects seen in older individuals, blacks, and those with preexisting high blood pressure. The researchers acknowledged that these data may underestimate the full health impact of excessive sodium intake, which is also linked to a much higher risk of nonfatal heart disease, kidney disease, and stomach cancer (the second most deadly cancer worldwide).

- Average sodium consumed daily worldwide: **3,950 milligrams per day**

- Percentage of world population consuming more than the WHO-recommended 2,000 milligrams of sodium per day: **99.2**

- Number of worldwide cardiovascular deaths per year attributed to excess sodium intake: **1.65 million**

NATURAL SALTS AND SEA SALTS

Salt, or sodium chloride—NaCl—can be mined from the ground or harvested from the sea. All salts, whether from the sea, salt mine, or salt marshes, can trace their origins to the ocean. These days, various types of salt are being marketed—from "Celtic salt," an expensive version of sea salt, to the rare *fleur de sel* from France and Peruvian Pink and Hawaiian Black Lava salts. But the truth is, all of these products are just NaCl in different sizes, shapes, colors, textures, and prices. The different flavors that people perceive are mostly related to texture. Sea salts, for instance, are composed of larger, flakier crystals, and when you bite into a larger crystal of salt, the flavor is different. However, when you cook with sea salt, any differences disappear because the salt dissolves into the liquid ingredients.

Devotees of sea salts and specialty salts claim they taste better and are nutritionally superior because of the trace amounts of minerals they contain. But the truth is: salt is salt. The presence of a tiny amount of minerals (such as a hundredth of a milligram) in a salt does not make all the excess sodium less damaging.

Even if the salt did contain a larger amount of minerals, would its excess consumption no longer be harmful? Of course not. It is the tremendous magnitude of sodium exposure that makes it a risky food, and that exposure is not made harmless by a bit of extra mineral content. It's like saying that because green vegetables are so high in minerals, if you consume them with large amounts of sodium, the minerals will make the excess sodium safe.

No type of salt provides significant nutritional benefits. All of these salts, even the so-called mineral-rich ones, are more than 98 percent NaCl, regardless of where they originally came from. The amounts of trace minerals they contain are negligible and have no significant effect on human health. Your best source of minerals is food; and yes, vegetables contain all the trace minerals humans need, in amounts that are meaningful to human health.

HOW MUCH SODIUM DOES ONE CONSUME?

If a serving of a food contains 100 calories, it would not naturally have more than 50 milligrams of sodium. So if you note that 100 calories of a packaged food contains 200 milligrams of sodium per serving, you know that 150 milligrams of sodium were added to what was naturally in the food. **I suggest that you not add more than 300–400 milligrams of extra sodium to your day's dietary intake over and above what is natural in foods.** This allows you to have one serving of something each day that has some sodium added to it, such as a low-salt tomato sauce, but all other foods should have only the sodium that Mother Nature put there.

Remember, processed foods can contain 1,000 milligrams or more of sodium per serving, and many typical restaurant meals contain 2,300 to 4,600 milligrams.[32] And it's not just the usual fast food villains adding up the sodium. Seemingly innocent, healthy foods can be part of the problem too. One cup of vegetable broth can provide 940 milligrams of sodium, and 1 cup of canned beans can rack up 770 milligrams. Two tablespoons of Italian dressing on your salad could add 486 milligrams, and 1 cup of regular pasta sauce could contain 1,100 milligrams.

Why are processed foods so loaded with sodium? Because salt heightens flavors, reduces bitterness, and enhances sweetness. It is perfect for processed foods. It is cheap, it keeps foods from becoming discolored, and it extends shelf life. It also binds water and makes foods weigh more, so you pay more for a heavier package.[33] Consumer research has revealed that unless food products are salty enough, people do not like the way they taste. Consumers have gotten used to higher and higher levels of salt; after years of exposure to high-salt foods, their taste buds lose their sensitivity to sodium, and they also lose the ability to taste other subtle flavors in real foods. At this point, all food tastes bland unless it is oversalted. A population that heavily salts their food demands packaged foods and restaurant meals that also contain lots of salt, because they have become accustomed to this flavor profile. I promise that when you stop eating foods loaded with salt for a few months and then eat a processed food item that you would have enjoyed previously, you will find it to be almost inedible as a result of the amount of salt you now taste.

ADD FLAVOR—BY REDUCING SALT

The natural flavor of food—without added salt—is an acquired taste. Gradually, your taste preferences should change and you will learn to prefer food without salt. Be creative and use other flavoring agents, such as herbs, spices, onion, roasted and raw garlic, lemon or lime juice, vinegar, or lemon pepper. Experiment with fresh herbs instead of the dried versions. Fresh mint, cilantro, and dill add interesting flavors. I use many types of salt-free herbal seasoning blends.

Condiments such as ketchup, mustard, soy sauce, teriyaki sauce, and relish are high in sodium, so read labels, choose low-sodium versions, and use them sparingly to keep salt intake to the approved level. You can make seasoning blends to your own liking. Here is an example:

1 teaspoon ground celery seed
2½ teaspoons crushed marjoram
2½ teaspoons crushed summer savory
1½ teaspoons crushed thyme
1½ teaspoons crushed dried basil
1 teaspoon crushed garlic

You can also add dehydrated onion, oregano, chili powder, cumin, or any other favorite flavor.

THE FOOD-INJURED BRAIN

Many mechanisms are at work to induce overeating and make dieting difficult for so many of us. It is well-accepted these days that fast food, junk food, and concentrated calories from commercially baked goods stimulate dopamine in the brain's reward system. Over time, we become more insensitive to dopamine and require more and more food to get the same level of stimulatory pleasure.

Animal studies have shed light on the mechanism of food addiction, as the structure of the brain of rats changes when the rat is fed junk food. The number of dopamine neuroreceptors in the brains plummets, and some areas of the brain shrink, decreasing learning capacity.[34]

Researchers in one study found that feeding rats junk food progressively degraded their brain's reward system.[35] They tested two groups of rats: One was fed a diet of high-fat, high-calorie foods, and the other received a normal diet. The rats fed the unhealthy diet quickly became obese and less active, and developed a preference for unhealthy foods. At mealtimes, rats received mild electric shocks. The rats fed the normal diet immediately stopped eating, but the rats on the junk food diet continued eating even while they were being shocked. They became desensitized and compulsive. This relates to

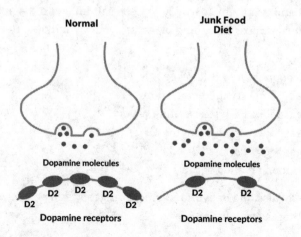

Normal

Junk Food Diet

Dopamine molecules

Dopamine molecules

D2 D2 D2 D2 D2

D2 D2

Dopamine receptors

Dopamine receptors

how fast food induces compulsive eating, and the diminished brain function, in turn, affects how poor eaters behave socially.

Like the rats fed fast food, the vast majority of Americans have diminished dopamine function as a result of overconsumption of junk food, and they also have diminished chemosensory perception, such that they are less able to smell and taste. This is especially evident in the elderly, who have a reduced ability to enjoy the more subtle flavors of real foods, so they become more reliant on using salt and sugar to help them experience flavor.[36] People with Alzheimer's disease show even greater olfactory deficits than healthy elderly, an effect related to the degree of dementia.

Our inability to enjoy healthy foods is linked to alterations in the brain that ultimately become permanent as we progress toward dementia. It is ironic that eating unhealthy foods damages us in a way that makes healthy eating unappealing. The good news is that we can recover; however, the recovery process requires conscious effort because unhealthy diets can alter personality as well as taste preferences.

After we are chronically overstimulated with concentrated and rapidly absorbed calories, our cravings intensify. Eventually, after chronically eating fast food, the enjoyment we once got from eating high-fat, low-nutrient foods turns into the intense desire of food cravings, which are difficult to satisfy. Interestingly, fast food cravings and the loss of control in eating have similar causes to the loss of self-control that impulsively violent people experience. Rich, unhealthy diets simultaneously impair the function of dopamine, serotonin, and other neurotransmitters in the brain, and these changes correlate with anger and violence in both animal studies and human studies.[37] People who consume an unhealthy fast food diet have less self-control and a reduced ability to control food intake; they are also more hostile and easily angered. Unnatural foods destroy us from the inside out and affect how we treat each other. In contrast, a healthy diet with adequate phytochemicals, nutritional diversity, and omega-3 fatty acids elevates serotonin levels and normalizes receptor sensitivity, which can improve mood and eliminate the need to eat for emotional reasons.[38]

Along with other researchers, I have noted depression in people withdrawing from processed food, sweets, and fried foods.[39] A study published in 2011 evaluated the consumption of fast food (hamburgers, sausages, pizza) and processed pastries (muffins, doughnuts, croissants) with a median follow-up of 6.2 years. These researchers found that fast food and commercially baked goods are linked to depression in a dose-dependent manner.[40] The results revealed that people who ate even just two servings a week of fast food or commercially baked goods, compared with those who ate little or none, were 51 percent more likely to develop depression, and the risk was even greater for people who consumed more. Up to this point, the relationship between depression and some components of our diet, such as omega-3 fatty acids and B vitamins, has been extensively studied and confirmed;[41] but the role of fast food and white bread products has received little attention.

What is not recognized as easily is how much a diet of fast food and white flour–containing commercially baked goods contributes to dysthymia, that is, a minor and chronic flattening of emotion and loss of happiness and excitement about life. In other words, not everyone who consumes processed food becomes depressed, but the majority of people experience mood alterations, their happiness and satisfaction with life are diminished, and their concentration and creativity are weakened. The resolve to improve one's life and health is weakened by the exposure to unhealthy food.

RECOVERY FROM FOOD ADDICTION, BINGE EATING DISORDER, AND EMOTIONAL OVEREATING

Abstinence from food triggers is the secret to recovery from food addiction and related disorders. Abstinence is radical, but it produces the best and most consistent results with people who suffer from food addiction. It is easier to do this all the way because you are not going to be in continuous turmoil trying to make decisions over what to eat and whether you should cheat and how much you can get away with. When you have no decisions to make, and you have made the firm commitment to *just do* it, no matter what, a big emotional weight is taken off your shoulders. That's when the magic happens and you can really see how great this Nutritarian program works, and how easy it is to lose the excess weight.

Whether we are talking about addiction to cocaine, nicotine, alcohol, or junk food, people need a period of abstinence to recover from their addictive triggers. This often means enlisting other people to help you remove temptation and easy access to processed foods in your immediate environment, especially when you are in the initial stages of recovery. Saying "I'll give it a try" does not cut it. What that really means is you'll be back to an addictive pattern of behavior as soon as things get difficult. To change bad habits, you need to be 100 percent committed and to plan ahead to prevent obstacles from sabotaging your results.

You can start with semiabstinence and then get ready for the real deal soon after. Start by "crowding out" the unhealthy food from your diet. Begin with a lunch overhaul: Eat a large salad or a bowl of vegetable-bean soup and a piece of fruit every day. Prepare healthful salad dressings in advance. Next, add a simple breakfast of oatmeal (with flaxseeds, chia seeds, or hemp seeds) and fruit; then soon after that, add a large plate of wokked vegetables for dinner. The following chapters provide you with all the menus and recipes you will need.

Eating so much healthy food will help you begin the process of retraining your taste buds. Then, within a few weeks, you have to make a commitment to dietary excellence—by eliminating sweeteners, flours, oils, and salt. That is the only way to really change your taste preferences and learn how to prefer natural foods. In other words, without a long enough period abstaining completely from sweeteners, salts, and oils, you are going to have weakened taste, and you will never learn how to prefer the taste of more mildly flavored natural foods.

The period of lunch overhaul can get you to set your date, do your learning, and prepare for your adoption of a program with complete abstinence from junk food, fast food, oils, sweeteners, salt, and commercially baked goods—the most addicting foods. After a few months of being 100 percent compliant with healthful Nutritarian eating, you simply will no longer desire these foods anymore.

WHY LUNCH IS THE MOST IMPORTANT MEAL:

- You are probably out of the house.

- Temptation is around in the workplace.

- It is best not to eat a heavy meal before bedtime.

- It is best to eat soon after (not before) exercising.

- A hearty lunch with beans and nuts keeps you from wanting to eat before dinner.

TRANSFORMING YOUR FOOD PREFERENCES

When people come to my office overweight and with medical problems, I ask them, "What kind of results do you want? What type of diet do you want me to design for you? Do you want me to devise a diet program so you drop 15 pounds this month and 10 pounds the next and get rid of your diabetes, high blood pressure, high cholesterol, and headaches and get off all drugs in the next few months? Or do you just want to make some moderate improvements, drop 5 pounds, and still require medications?"

They routinely tell me that they want the most powerful program—the one that will get them completely well and off all medications. Then I instruct them that I will be making all the decisions about what they will be eating. I offer you the same advice:

- Don't think about what you "think" you should eat.

- Don't eat what you "feel like" eating or what you "like."

- Don't make any decisions about what to eat.

The decisions you have been making based on what you think you should eat and what you like to eat have gotten you into this health difficulty. Now all those decisions need to be mine, not yours, and *you must eat what I want you to eat even if you don't like it.* I will offer plenty of good-tasting food choices, and you can mostly eat abundantly, but you can't expect to like the flavors as much as those of your old diet—at the beginning, that is.

I promise you that if you follow my eating instructions for six months, your diabetes, high blood pressure, and excess weight will disappear. Your taste buds will have a chance to become more sensitized and healthier, and you will love eating the Nutritarian way. But in the beginning, when you don't love it, you just have to do it.

Most diets have poor results because most people don't follow them and can't stay on the restrictions imposed.[42] This is not surprising. If people are still partaking in their addictive triggers all the time and don't eat healthfully enough, they are going to find it nearly impossible to restrict calories or continue eating healthfully. And if people are not properly counseled about what to expect, how to beat food addiction, and how to feel calorically satisfied, it is unlikely they will stick with any plan in the long term.

What makes the Nutritarian eating style so unique is that it is the most powerful program to reverse disease, prevent cancer, and extend lifespan; and Nutritarians eat this way to achieve lasting good health, no matter how much they weigh. We are not "on a diet"—instead, we have chosen to eat this way because the benefits are profound. And as time goes on, we prefer Nutritarian foods and recipes. However, many people will never get to the point where they prefer eating this way forever; if they hope to do so, they must stick strictly to the diet style for about six months.

Many people smoke cigarettes, drink alcohol, and take opioids even though they know that these choices are bad for their health and are ruining their lives. Obviously, the same is true for diet; many people will not choose or be able to commit to the healthful eating needed to earn back good health. However, just as with drug addiction, the more people accept the support, education, and services offered to them in their efforts to beat their demons of addiction, the better the outcome.

GETTING WELL FROM DEPRESSION

Depression doesn't have one specific cause, but unhealthy dietary factors contribute to depression. This is almost never mentioned during discussions about mental health difficulties. A dietary pattern that includes fried food, sweetened desserts, processed meats, and refined grains has been associated with depression, and the consumption of whole natural foods has been shown to be strongly protective against this mood disorder.[43] Research published in 2016 also documented that increasing consumption of vegetables and fruits significantly elevated psychological well-being and people's "happiness levels."[44] This study was one of the first major scientific attempts to explore psychological

Getting Well from Depression

There are natural approaches to aid people who suffer from depression that may enable recovery without their resorting to taking medications for the rest of their lives. One of the problems with medications for depression is that people can develop dependence on the medications over time, such that depression is exacerbated when they stop taking the medication. In addition to eating right and exercising, the following might be helpful in combating depression:

Morning light therapy—A 2005 meta-analysis showed morning light therapy to be as effective as antidepressants.[45]

EPA/DHA supplementation—Multiple studies indicate that EPA is more effective than DHA for helping depression.[46]

Saffron—This culinary spice has been investigated in several studies for its potential to ameliorate mild to moderate depression.[47]

SAMe (S-adenosyl methionine)—This methyl donor involved in neurotransmitter synthesis demonstrates effectiveness for battling depression.[48]

well-being beyond the traditional findings that eating more fruits and vegetables can reduce the risks of cancer and heart attacks.

The researchers followed more than twelve thousand randomly selected individuals for more than two years, in conjunction with the Australian Go for 2&5 campaign, which promoted the consumption of two portions of fruit and five portions of vegetables every day. Happiness benefits were detected for each extra portion of fruit and vegetables consumed, up to eight portions a day. Researchers found that subjects who changed from consuming almost no fruits and vegetables to consuming eight portions a day experienced an increase in life satisfaction that was equivalent to moving from unemployment to employment. This well-being improvement occurred within twenty-four months.

One of the study's authors, Redzo Mujcic, from the University of Queensland, said, "Perhaps our results will be more effective than traditional messages in convincing people to have a healthy diet. There

---

St. John's Wort—A 2008 Cochrane analysis documented improvement in depression similar to that attained with antidepressant drugs.[49]

5-HTP (5-hydroxytryptophan) or L-tryptophan—A few studies suggest positive effects over placebo with these precursors to serotonin.[50]

Coconut oil—Low cholesterol may exacerbate depression in some susceptible individuals. In these cases, coconut oil may be helpful in raising cholesterol to increase neurotransmitter production. One of the most intriguing issues about diet and health is the inconsistent data suggesting increased risk of depression and suicide in people with the lowest cholesterol levels.[51] Lots of people offer their opinion regarding these inconsistent findings, but the reality is that we just don't know for sure if this is causal or secondary to the psychopathologies, and we are also not certain whether raising cholesterol levels through diet will help improve depression. Nevertheless, with major depression and accompanying very low cholesterol, we leave no stone unturned that might be contributory and helpful.

is psychological payoff now from fruits and vegetables—not just a lower health risk decades later."[52]

A depression-inducing dietary pattern is not solely due to sugar and white flour; however, the dangerous effects on the brain of these high-glycemic carbohydrates are now well-established. A 2015 study showed a dose-dependent effect of high-glycemic-load foods (white flour and sweetening agents) on levels of depression.[53] Many people have noted a link between eating sugary foods and feeling down the next day, but now we know the effects are cumulative, are long-lasting, and can be severe.

The data were collected from roughly seventy thousand women in the Women's Health Initiative Observational Study (none of whom suffered from depression at the study's start) who had baseline measurements taken between 1994 and 1998 and then again three years later. Diets with a higher glycemic index, including foods rich in refined grains and added sugar, were associated with greater odds of developing depression. Researchers also found that eating high-fiber foods such as whole grains, whole fruits, and vegetables lowered the odds.

The fast food–consuming public not only experiences depressed moods and clinical depression; this same dietary pattern results in obesity, metabolic syndrome, prediabetes and diabetes, and concentration deficits and learning difficulties. Throughout the body, excess sugar is harmful; even a single instance of elevated glucose in the bloodstream can be harmful to the brain, resulting in slowed cognitive function and deficits in memory and attention.[54] In healthy young people, brain imaging demonstrated that the ability to process emotion is compromised with elevated blood glucose levels.[55] Sadness and anxiety can occur with the increasing stimulation of the brain caused by calorically concentrated processed foods. And this result is not restricted to people with diabetes.

In the brain, excess sugar impairs both our cognitive skills and our self-control, because having a little sugar stimulates a craving for more. The mixture of sugar, salt, and oil derails the ability of the body to control calories or be satisfied with normal amounts of food. Fast food creates human eating machines—individuals with no caloric "off" switch. The regular consumption of fast food creates a lack of self-control that is like turning on an obesity-driving switch that leads to diabetes and other life-threatening diseases. Having these health problems then magnifies any emotional problems.

Too often, physicians turn only to drugs to treat depression. Without the awareness and implementation of nutritional improvements, it is difficult for them to achieve satisfactory results on behalf of their patients.

PUTTING FIBER TO THE TEST

High-fiber plant foods contain more than just fiber; they are rich in life-extending phytonutrients. It's no wonder that when we measure fiber consumption in a population, we see profound health and lifespan benefits for those populations that consume the most fiber. When we track the amount of food-derived fiber eaten, we can ascertain the amounts of plant foods eaten in a population and therefore measure their health. In the United States, the average American eats fewer than 20 grams of fiber a day. In other areas around the globe, for instance, in certain regions of rural China and rural Africa where heart attacks and cancers are practically unknown, the populations consume up to 100 grams or more of fiber a day.[56]

On the basis of analyzing the diets of modern-day primitive hunter-gatherer tribes and human fossilized feces, scientists estimate that our Paleolithic ancestors ate this high amount of fiber. It appears that both our nonhuman primate relatives and our ancient human ancestors lived mostly on high-fiber plant food. This is not surprising, considering that 99 percent of the time that humans have existed on this planet as a distinct species, they did not have the tools needed to hunt and kill big animals for food—they were eating mostly fiber-rich plants.

As I have said, eating foods high in fiber suppresses the appetite and makes it almost impossible to be overweight. When we get enough fiber, our desire to eat is shut down. Our gut flora metabolizes fiber into short-chain fatty acids, which bind to and activate receptors on the surface of our cells and in our brain, which shuts off our appetite. When we don't eat enough high-fiber foods, we find it easy to overeat because the body is not fully satisfied.

More fiber means more production of the short-chain fatty acid butyrate and more downregulation of the appestat in the hypothalamus. Fiber also holds water in the gut, signaling volume receptors that induce the feeling of satiation. Therefore, it becomes easy to maintain this type of diet without wanting to overeat and without having to eat thimble-size portions of food.

Overall, eating less and eating less frequently extend the lifespan of most species of living creatures, including primates and humans. But people have been brainwashed to believe that eating more, having larger children, and feeding to maximize growth are healthy and that snacking all the time enhances healthspan and lifespan; in fact, however, the opposite is true.

"INTERMITTENT FASTING" EXTENDS LIFESPAN

Intermittent fasting means interspersing reduced caloric intake in your weekly dietary pattern. You might do this by drinking just juice or water for a day or part of a day, or by skipping dinner. Being in a fasting state directs the body's cells to shift toward maintenance and repair processes rather than growth.[57] Episodic restriction of calories enhances healing and stimulates stem cells to repair tissue. Both repeated water fasting and time-restricted eating have been found to help prevent age-related decline in stem cells and age-related telomere erosion.[58]

Human studies on intermittent fasting suggest it improves insulin sensitivity and promotes weight loss, too.[59] This repair mode that the body undergoes while fasting is associated with a reduction in insulin and IGF-1 signaling, leading to reduced inflammation, improved insulin sensitivity and stress resistance, slower cell growth, improved immune function, and reduced oxidative stress.[60] Importantly, periodic fasting also stimulates the removal of damaged cells and molecules, essentially cleansing cells and tissues of these damaged components.[61]

A nutrient-dense, plant-rich diet full of healthful phytochemicals drives the production of enzymes that enable the excretion of carcinogens and other potentially harmful compounds. But adding periodic calorie reduction to an already nutrient-rich diet intensifies the body's natural ability to heal and repair by extending the amount of time the body is in a heightened repair mode. Even as a healthful eater, occasionally skipping dinner or substituting a vegetable juice for a meal can help optimize your health and slow aging even more.

EXTEND THE OVERNIGHT FAST PRIMARILY IN THE EVENING

The most important thing to remember about intermittent fasting or eating for health and healthspan is to not go to bed at night with a full stomach. Eat dinner earlier so that you finish eating at least three hours before bedtime; this way, food can be digested and the stomach can empty before you go to sleep. If it is impossible for you to eat that early, then eat very lightly in the evening, such as having only salad and fruit. Evidence suggests that eating an early dinner, skipping dinner, or eating a lighter dinner offers more health benefits than eating breakfast later in the morning.[62] In weight loss–intervention trials, restricting nighttime eating and shifting the main meal or majority of calories to earlier in the day increase weight loss, improve blood glucose levels, and reduce inflammation.[63]

Women who had a longer nightly fast also slept a greater number of hours.[64] Because insulin sensitivity is higher in the morning and lower in the evening, consuming the majority of calories earlier in the day aligns better with the circadian rhythm of insulin sensitivity and is beneficial for keeping glucose levels stable.[65] In women with breast cancer, an overnight fast of fewer than thirteen hours compared with one of thirteen hours or more was associated with a 36 percent increase in the risk of breast cancer recurrence over seven years.

If you sometimes get hungry late at night or right before bedtime, just go to sleep; try not to eat more food late at night.

Scott McPeak

Lost 60 Pounds and Got Rid of Diabetes, High Blood Pressure, Fatty Liver, and More

I grew up in South Dakota with a steady diet of meat and dairy. By age 50, I was just over 6 feet tall, weighed 270 pounds, and had type 2 diabetes, high blood pressure, fatty liver disease, and a host of other life-threatening conditions.

I had been feeling very tired and run-down, so I went to the doctor. My blood pressure was 201/110, my HbA1c was 11, my blood glucose was 385, my cholesterol was 270, and blood vessels were growing into my corneas. I also had fatty liver disease, melanoma, a tumor in my jaw, and neuropathy in my hands and feet. It was an overall breakdown of my immune system, and I felt close to death. My doctor wanted me to take five different medications to treat my symptoms and basically just give up. I agreed to take medication for diabetes but was determined to get better some other way.

My wife researched ways to reverse type 2 diabetes and found Dr. Fuhrman's book *Eat to Live*. I began my journey to health that day—literally—because I switched to a Nutritarian diet that very day. I also started walking regularly and never looked back. Believe it or not, after a short time I learned to love the Nutritarian eating style. My taste buds changed, and I fell in love with all kinds of new foods. Many vegetables that I never would have eaten before now tasted great.

The transformation in my health was dramatic. Two months after starting this healthy eating style, I was able to stop taking the drug for

diabetes. My last doctor visit told the whole story. My HbA1c level had dropped to 5, my blood glucose was 83, my total cholesterol was 120, and my blood pressure was 107/78. Plus the nerve pain in my hands and feet had vanished.

The blood vessels receded from my corneas, and today my liver is back to normal functioning. I had surgery to remove the melanoma and the tumor in my jaw, and I lost 60 pounds, which brought me down to 210 pounds. Now I have fully recovered from all my medical issues. This has inspired my wife and family to adopt this way of eating too. They saw how it saved my life. Dr. Fuhrman's books will change your life, too; they certainly changed mine. I feel more clarity in life and have renewed hope. I am now able to enjoy life and live it to its fullest.

Dr. Fuhrman's help led me to a healthy lifestyle that is so wonderful for me and my family. Thanks, Dr. Fuhrman!

CHAPTER SIX: QUICK SUMMARY

People overeat because of many different social, emotional, recreational, and addictive influences. The unhealthier your diet, the stronger the internal signals that urge you to gorge on unhealthy food, so it is imperative that you follow a healthful, Nutritarian eating style that is rich in micronutrients.

THE FALLACY OF BEING MILDLY OVERWEIGHT AND HEALTHY

Many "experts" claim that people who are mildly overweight have better health statistics than people at or below normal weight. This is deceptive, because most Americans who are at a normal weight are sickly or smoke, so they do not represent the small percentage of people who are slim and eat right.

WHAT IS FOOD ADDICTION?

People who eat a calorically dense diet that is high in processed foods invariably develop an increased desire for excessive calories and become increasingly dependent on overeating to feel well. This leads to food addiction—a chronic, compulsive overconsumption of highly palatable foods despite negative health consequences. Addictive foods excessively stimulate the reward centers of the brain, creating an almost irresistible craving for them.

DEALING WITH HUNGER

I have documented that enhancing the micronutrient quality of the diet dramatically changes and lessens the experience of hunger. A Nutritarian diet led to less intense feelings of hunger felt primarily in the throat and upper chest, rather than in the stomach. Subjects also found that they desired fewer calories.

The digestive cycle has two phases: the anabolic, or eating and digestive phase, and the catabolic, or energy utilization phase. During the catabolic phase the body can most efficiently heal, repair, and self-clean. Also during this phase, the liver's glycogen reserves and fatty acid stores are used to meet the body's energy needs, and toxins are mobilized and processed for removal.

"TRUE" VS. "TOXIC" HUNGER

When we eat an unhealthy diet that is high in processed foods, oils, sweeteners, and animal products, we experience "toxic hunger"—a combination of uncomfortable sensations caused by the withdrawal (or detoxification) from inflammation, oxidative stress, and a buildup of toxic metabolites. Withdrawal symptoms drive overeating behavior because we eat frequently and overeat to prevent discomfort.

When we eat a healthful diet that is rich in micronutrients, we experience "true hunger"—a mild sensation that occurs as the body's store of glycogen is almost depleted. Healthful eating is more effective for long-term weight control because it modifies and diminishes the sensations of withdrawal-related hunger. This allows people to be comfortable while consuming substantially fewer calories.

Toxic hunger fuels food addiction because the quickest way to feel better and relieve hunger symptoms is to eat more of the unhealthy foods that caused the symptoms. This leads to an endless cycle of misery. A diet that only restricts portion sizes of unhealthy foods cannot be sustained because of the unrelenting desire to consume more calories.

THE FOOD-INJURED BRAIN

Fast food, junk food and commercial baked goods stimulate dopamine in the brain's reward system. Over time, we become less sensitive to dopamine, and require increasing amounts of processed food to get the same level of pleasure. This also causes diminished chemosensory (taste and smell) perception, which is especially evident in the elderly. Unhealthy diets impair the function of dopamine, serotonin, and other neurotransmitters in the brain; these changes can increase propensity for anger and violence. In addition, the consumption of fast food and commercial baked goods is linked to a higher risk of depression in a dose-dependent manner.

ABSTINENCE IS THE KEY TO SUCCESS

The key to overcoming food addiction is to abstain from your food triggers, especially commercial baked goods and fast foods. Abstinence frees you from the emotional turmoil surrounding what to eat, or when you can cheat. To change bad habits, you need to be 100 percent in, and to plan ahead to prevent obstacles sabotaging your results.

INTERMITTENT FASTING EXTENDS LIFESPAN

Studies suggest that intermittent fasting can improve insulin sensitivity, promote weight loss, reduce inflammation and oxidative stress, slow cell growth, and boost immune function. Adding periodic calorie reduction to an already nutrient-rich diet only intensifies the body's natural ability to heal and repair itself.

CHAPTER SEVEN

We Can Reverse Disease

If you are currently on any medications, it is crucial to consult with your physician when you embark on a Nutritarian program. For instance, if you take medication for high blood pressure, it is important that you and your doctor have a plan to gradually lower that medication as your blood pressure improves so your blood pressure does not become too low and you do not become dangerously overmedicated. The Nutritarian diet will lower your blood pressure quickly, and that can place you at risk of fainting or damaging your kidneys if your medication is not adjusted accordingly. In most cases, you eventually will be able to discontinue your medication as you lose weight and improve your health.

Let me be crystal clear: Food is the cause of, *and should be the solution for*, high blood pressure and high cholesterol. Everything else is just window dressing. It is a fact that heart disease is a food-created issue and that a superior diet can restore your health relatively quickly. If you strictly follow the meal plans in this book, you can significantly lower your blood pressure and cholesterol in fewer than three weeks. In the vast majority of cases, you will not need medication anymore, ever. It works.

Martin Becker
Lost 120 Pounds and Recovered from Cardiovascular Disease and Other Illnesses

In 2014, I was 66 years old, and suffered from obesity, atrial fibrillation, an enlarged heart, high blood pressure, swollen legs, chronic back pain, and a host of other maladies. I was taking 11 different medications to manage my symptoms. My longtime cardiologist was taking a "wait and see" attitude toward my care—but I knew I didn't have any time to wait. I had a congenital prolapsed mitral valve heart murmur that was getting larger. My health was deteriorating rapidly—my weight had ballooned up to 266 pounds. I had a 51-inch waistline and a BMI of 40.6. I was desperate for a change. So in 2014, I went for a second opinion. It was a decision that likely saved my life.

The new doctor conducted the examination and reviewed my blood tests, and then got straight to the point: If I didn't change my way of eating, I'd be dead in ten years. Those words hit me like a two-by-four. I didn't want to die.

That day, I received the most important prescription of my life. He wrote down the title of Dr. Fuhrman's book *The End of Dieting* and told me to read it if I was really serious about changing my life and becoming healthy. So I did just that, and it was the beginning of my journey to wellness.

Raised on the standard American diet, with plenty of eggs, meat, cheese, white bread, and milk, I was obese as a child. I lost weight in high school and college, but after graduation, the pounds started to pile on again. In my job as an organization management consultant, I traveled and entertained clients regularly—and all of that high-end dining and drinking made it hard to maintain a reasonable weight, even though I kept going on diets. I started suffering from shortness of breath, and had trouble walking through airports on my business trips.

I read *The End of Dieting* and immersed myself in the Nutritarian way of eating. I was making progress, when my journey was interrupted by emergency cardiac surgery. The leak in my heart valve had gotten so much worse, that the only alternative was open-heart surgery to repair the mitral valve. As I went through the long recovery and cardiac rehabilitation, I felt like I was given a second chance at life. But because

I had failed so many times, on so many diets, I wondered: Could I take this chance? But I knew my life literally depended on it.

I re-read *The End of Dieting* and totally committed myself to the Nutritarian lifestyle. I learned diets don't work; the goal is to be healthy—and the weight loss will follow. I got the *Eat to Live Cookbook* and enjoyed many of the delicious recipes. For my wife and me, having a huge salad with Dr. Fuhrman's wonderful salad dressings is our key meal of the day. We enjoy cooking the many wonderful dishes that are available on the website.

Four years have passed since my surgery, and I don't take any medications at all. I have a normal heart rhythm and blood pressure. Also no edema, no breathing issues, no back pain, no fatigue, lots of energy, and I sleep well.

I have lost 120 pounds and now weigh 146, with a BMI of 22.2 and a 33-inch waist. Regaining my health inspired me to begin a new career as a certified health and wellness coach. Dr. Fuhrman's teachings have given me a sense of hope and purpose, and now, at 70 years old, I have the time to pay it forward.

A Nutritarian diet style does more than address one or two heart disease risk factors. Unlike taking a cholesterol-lowering medication or one or more drugs to lower blood pressure, this eating style addresses and repairs scores of elements that govern your future cardiovascular health. For example, the inner lining of the blood vessels will become smoother and less inflamed and the vessel walls will become more elastic; thus, more oxygenated blood will be able to fill the coronary arteries. And so importantly, your LDL cholesterol no longer will be oxidized. Remember, oxidized LDL is the bad actor. A Nutritarian diet floods the body with antioxidants that radically lower oxidized LDL. This means you become resistant to heart disease, diabetes, and cancer and improve your immune function, which protects you against virus-induced inflammation and dangerous infections too.

Is it possible to predict whether a heart attack is likely to happen in the near future? Yes, because when the lining of the blood vessels (the endothelium) becomes ragged and inflamed and the LDL molecule becomes heavily oxidized, the risk of having a heart attack is very high.[1] These markers measure endothelial inflammation and cholesterol oxidation, so when they are elevated, a person is in trouble. The good news is that a Nutritarian diet quickly normalizes these values, preventing danger and death within weeks—not years—of starting the program.

Why is a Nutritarian diet the most effective program to eliminate vascular inflammation, lower blood pressure, and reverse heart disease? Because it checks every box to slow aging and maximize health, including body fat reduction, ideal nutrient bioavailability, antioxidant capacity to reduce inflammation, and nutritional completeness. I have treated tens of thousands of patients using this diet style, with incredible results; the complete reversal of high blood pressure, angina, and high cholesterol has been routine and predictable.

A Nutritarian diet is a lower-glycemic approach compared with a low-fat vegan diet and contains more micronutrients, antioxidants, and favorable exposure to fatty acids for protection of cells and protection against cardiac arrhythmias. A Nutritarian diet has all the benefits that are gained from restricting animal products, without the deficiencies that doing so can impose.

Low-carbohydrate ketogenic and paleo diets are very popular and may have some effectiveness for weight control, but they fail miserably in terms of long-term cardiovascular risks. We give studies the most credence when they meet three criteria: First, they examine many thousands of people; second, they go on for many years, even decades; and third, they look at "hard" endpoints, such as heart attack or cardiovascular death. Proponents of low-carbohydrate diets, such as keto and paleo, point to short-term studies on weight loss and cardiovascular risk markers ("soft" endpoints) to back up their claims. Yes, it is true that when people cut out foods such as white flour, sugar, white rice, and white potato, they lose weight; and that initial weight loss leads to improvements in indicators of cardiovascular and metabolic health. However, it is important to remember that these are short-term studies. Diets that produce weight loss in the short term do not necessarily promote health and longevity in the long term.

High-credence studies, some of which I note below, have established that diets such as paleo, Atkins, and ketogenic, which restrict carbohydrates and include much more fats (and are therefore low in fiber and phytonutrients), are proven formulas for a shortened lifespan. Plus, no long-lived population, such as those in the Blue Zones, ever ate a diet that even remotely resembles these high-protein approaches.

It is important to note that improvements in "soft" endpoints, indicators such as higher high-density lipoprotein (HDL, the "good" cholesterol) levels or lower triglycerides, don't always translate into longer lifespan. Even a cookie diet can look good in the short run if people get sick of eating just cookies and therefore eat fewer calories. The main problem is that little side effect called "early death."

At the European Society of Cardiology's 2018 meeting, the results of a large new study were presented that were based on an analysis of data from the National Health and Nutrition Examination Survey (NHANES) in the United States.[2] Researchers looked at data from more than twenty-four thousand participants, with an average follow-up time of 6.4 years. Participants were divided into groups on the basis of their percentage of daily calories derived from carbohydrates. When presenting the major findings, the lead author stated that the findings suggested that low-carbohydrate diets are "unsafe" and "should not be recommended."

Other study findings included the following:

- Compared to the highest-carbohydrate group, the lowest-carbohydrate group showed an increased rate of total deaths (32 percent increase) and of cancer (35 percent), coronary heart disease (51 percent), and cerebrovascular disease (51 percent) deaths.

- Interestingly, the relationship between a low-carbohydrate diet and total mortality was more pronounced in nonobese people (48 percent increase) vs. obese people (19 percent increase).

The researchers went on to strengthen this data by performing a meta-analysis of nine previous studies, adding up to more than four hundred thousand participants and an average follow-up time of 16 years. These data showed significantly higher risks for total mortality (15 percent higher), cardiovascular death (13 percent), and cancer mortality (8 percent) as a result of restricting intake of carbohydrate-rich plant foods such as fruit.

Another study analyzed data from the Nurses' Health Study and Health Professionals' Follow-up Study—a total of 85,168 women and 44,548 men who were followed for twenty-six years and twenty years, respectively.[3] The researchers used multiple dietary surveys to assign participants a total low-carbohydrate diet score, plus an animal low-carbohydrate score and a vegetable low-carbohydrate score. The study findings included the following:

- As the animal low-carbohydrate score increased, all-cause mortality rates increased by 23 percent in both men and women. Cancer mortality was increased by 28 percent over the baseline diet.

- In contrast, as the vegetable low-carbohydrate score increased, all-cause mortality decreased by 20 percent.

- More than 12,500 deaths were recorded. Combining the decreased deaths that occurred from a high-vegetable diet with the increased deaths that occurred from more animal products in the diet gave a 43 percent increased death from all causes comparing the high-animal product, low-carb diet with the high–plant food, low-carb diet.

We have seen many studies comparing animal protein with plant protein, and the results consistently show that as high-protein plant foods increase in the diet and high-protein animal products decrease, lifespan is enhanced.[4]

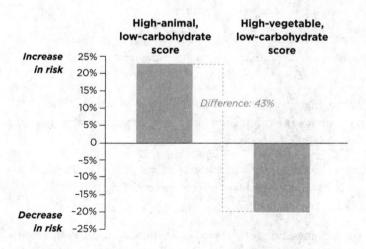

Source: Fung TT, van Dam RM, Hankinson SE et al. Low-carbohydrate diets and all-cause and cause-specific mortality: two cohort studies. *Ann Intern Med.* 2010 Sep 7;153(5):289–98.

Even a small increase in plant protein to replace animal protein gives significant benefit, demonstrating a huge potential to influence disease occurrence on a global scale. This is just from a 3 percent substitution of plant food for animal food.[5] This means that if your daily diet contains 1,500 calories and you remove just 45 calories (3 percent of the total) from eggs and replace them with 45 calories from beans or hemp seeds, your risk of death will go down by 19 percent. Amazing!

3% of plant protein to replace protein from	Change in risk of death from all causes
Processed red meat	-34%
Unprocessed red meat	-12%
Poultry	-6%
Eggs	-19%
Dairy	-8%

A diet heavy in meat and other animal products, with low levels of fiber and phytochemicals, will eventually take its toll:

- Animal protein raises IGF-1 levels in the blood, which is linked to increased risk of several cancers.[6]

- Carnitine and choline from meat and eggs are converted by gut bacteria to the proinflammatory compound trimethylamine oxide (TMAO), which promotes atherosclerosis.[7]

- Meat is high in advanced glycation end products (AGEs), which can damage the cardiovascular system, especially in people who have diabetes.[8]

- Excess heme iron (the type of iron found only in animal products) has pro-oxidant effects, contributing to cardiovascular disease and dementia.[9]

- Arachidonic acid promotes inflammation, which may increase cancer risk.[10]

- N-nitroso compounds are carcinogenic compounds are found mostly in processed meats.

- Heterocyclic amines and polycyclic aromatic hydrocarbons are carcinogenic compounds formed during high heat or flame cooking of any meat.[11]

I have shown you only a small part of an overwhelming amount of evidence that reveals the benefits of eliminating animal protein from our diets and replacing it with phytonutrient-dense plants and legumes. **There is no longer a question about whether we can reverse disease and live longer on a diet of vegetables, legumes, fruits, nuts, and seeds; the question is only how quickly and completely we can turn our health destiny around.**

BUTTER IS BACK—REALLY?

Some people will argue that fat is not the villain—rather, it is white flour and sugar. This faction insists that you can eat all the oil, butter, eggs, and fatty meats you want and still not get heart disease. They use the faulty argument that since baked goods and sweets cause heart disease more than animal products, and hemorrhagic stroke incidence appears to be higher as saturated fat intake and cholesterol go down, it's okay to eat more cholesterol and saturated fat. All you have to do is avoid processed grains and sweets to protect yourself. Views like this were popular years ago with the Atkins diet, but surprisingly many people still adhere to these dangerous views today despite these conclusions being opposed by the overwhelming majority of the world community of nutritional scientists who study health and diet.

Almost all the review studies and conferences of scientists meeting to discuss this issue come to the conclusion that when fruits, vegetables, beans, intact whole grains, nuts, or seeds replace animal products, oils, and saturated fats in the diet, we always see lifespan benefits. Yes, some studies demonstrate that substituting white pasta and white rice for butter or animal products does not show benefit, but this doesn't mean much because such foods do not promote health.

The final conclusion of a 2017 "science update" on saturated fatty acid (SAFA) consumption and the risk of heart disease and stroke included this statement: "In conclusion, strong evidence supports the partial replacement of SAFA-rich foods with those rich in [polyunsaturated fatty acids] to lower LDL [cholesterol] and reduce [coronary heart disease] risk."[12] Saturated fat has not been exonerated at all; too much saturated fat is still dangerous. These research scientists are advising people to eat less butter and coconut oil and more olive oil and nuts and seeds.

Sugar and white flour products and other refined carbohydrates may be worse than saturated fats, but foods high in saturated fats such as meats and butter are still worse than olive oil. This does not mean that olive oil is a superfood, because when the oil is replaced with nuts or seeds, we continue to see dramatic health benefits.[13] I do not recommend modulating your risk by changing the type of fat you use; instead, I recommend that you eat whole plant foods with lots of vegetables, using seeds and nuts as your main source of fat, not oils.

The other reason higher intake of saturated fat in the diet has not shown substantial increased risk of cardiovascular disease is because the diets with lower amounts of saturated fats contained more low-fat processed foods (such as white bread) and lower-fat animal products, such as egg whites and white chicken meat, which promote heart disease almost as much as higher-fat animal products. Both approaches cause disease, and should not be recommended.

In other words, replacing butter with white bread and pasta may not show any benefits, and may even worsen health outcomes; and cutting more fat off meat or eating more egg whites also does not contribute to health. However, replacing butter with walnuts or beans will show a radical benefit. *Butter is not back*: the proper message is that even cutting out butter and red meat is not good enough to protect against heart attacks; you have to eat real whole plant foods, meaning fruits, vegetables, beans, nuts, and seeds.

All animal products and processed foods increase the risk of cardiovascular disease, even if they are low in saturated fat.

There will always be people who want to embrace eating butter, oil, and whatever else they desire. Lots of people look to "science" or "experts" to find support for their heavy consumption of animal products, so parties with vested interests are able to prey on the nutritional ignorance of the general population. There is even a growing "carnivore diet" craze whose followers have been persuaded that eating a diet of mostly meat is healthy. But thousands of studies show otherwise.

THE ATTACK AGAINST NUTS AND SEEDS— GETTING NUTTIER ALL THE TIME

There are also popular low-fat vegan gurus who are completely fat-phobic and believe that all fatty foods (including longevity-promoting foods like walnuts) should be avoided. They advocate that people who are overweight or who have heart disease should avoid not just oils, but also avocados and all nuts and seeds.

These radical fat-avoiding vegans also often advocate against taking supplemental DHA and EPA, stating it is harmful or not needed with a vegan diet. Their rationale is that any fat intake will interfere with weight loss and visceral fat reduction. This extremely low-fat dietary protocol has been shown to improve heart disease in the short term; but unfortunately, those adopting this plan are at higher risk of dementia, depression, and even increased death from heart disease in the long run compared with those using a reasonable amount of seeds and nuts and including a small amount of DHA supplement in their diet. Such low-fat diets have helped many people, given the fact that they are a huge improvement over the dangerous SAD, but they still cannot be recommended because they place people at serious and needless long-term risk.

I publicly advocate against following such (anti-fat) recommendations and consider it critical to protect against DHA deficiency (as discussed in previous chapters), which is especially rampant in elderly people following extremely low-fat vegan programs.

Adequate amounts of omega-3 fatty acids help prevent loss of brain volume, a sign of brain aging. In a study of older women, a 3.2 percent higher omega-3 index was associated with 100 cubic millimeters greater hippocampus volume measured eight years later. Total brain volume was also greater in women with a higher omega-3 index.[14]

A consistent and striking finding from scores of studies is the increase in cardiovascular deaths when nuts and seeds in the diet are avoided. However, these data are ignored or denied by those advocating an extremely low-fat vegan diet. It is exactly like meat-promoting gurus who deny the validity of every study showing that more meat is dangerous. Unable to shut down the avalanche of evidence, they try to discredit the studies or the messenger. But when we look at large studies that involve thousands of participants studied over a long time and that have hard endpoints, they all show the same thing: Eating nuts extends lifespan. Or to put it another way: The lack of sufficient nuts or seeds in the diet increases premature death, particularly cardiovascular death.

This is demonstrated in every long-term study examining this issue (regardless of whether studies were funded by the nut industry), such as

- The Adventist Health Studies[15]

- The Iowa Women's Health Study[16]

- The Nurses' Health Study[17]

- The Physicians' Health Study[18]

The pooled results of these four large, prospective studies reported a 37 percent reduction in heart disease risk in participants eating nuts more than four times per week compared with those who never or rarely ate nuts.[19]

The enormous Adventist Health Study-2 is one of the most important investigations in the world today because it studies a healthy population that includes a high percentage of vegans and vegetarians. Its dramatic results were consistent with those of prior studies: a 39 percent decreased occurrence of cardiovascular mortality in people who eat nuts and seeds at least five times a week compared with fewer than once a week.[20] The population being studied ate mostly a plant-centered diet. The researchers stated that "protein from nuts and seeds was linked to a 40% decrease in the risk of death from [cardiovascular disease], and adjustment for vegetarian diet patterns did not change the results." Besides dietary habits, the researchers looked for a variety of ways to disprove the findings, adjusting the data for differences in age, sex, smoking habits, exercise, weight, and hypertension. But the protective qualities of nuts remained unchanged.

You can't just ignore all these findings by claiming that the results have been tainted by nut industry funding or lack of adequate controls. Only some of the studies were partially industry funded, and all the long-term studies corroborate each other. True, an industry is more likely to help fund research if it thinks the results will turn out in its favor. But it's ridiculous to think that some of the world's leading nutritional researchers heading these studies are falsifying data to make nuts look good because they are getting rich from the nut industry.

The results of the Adventist Health Studies have been corroborated in a meta-analysis of seventeen carefully vetted studies and every large-scale epidemiological trial; they essentially show the same thing—avoiding nuts and seeds in the diet increases the risk of premature death.[21] Multiple other meta-analyses also performed dose-response analysis and reported dose-response associations between nut intake and a lower risk of all-cause mortality.[22] This means that the more nuts someone ate, the less likely that person

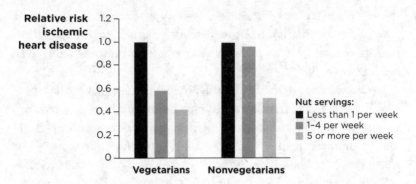

Source: Fraser GE, Sabate J, Beeson WL, Strahan TM. A possible protective effect of nut consumption on risk of coronary heart disease. The Adventist Health Study. *Arch Int Med*. 1992;152(7):1416–24

was to die during the follow-up periods of the studies. The same lifespan benefits have been demonstrated when diabetics eat more nuts and seeds.[23]

The Adventist Health Studies have shown cumulative benefits from the combination of several health behaviors. In vegetarian participants who exercised regularly, expected age at death was 7.8 years older in the group who ate nuts regularly versus the group who did not.[24] The Adventists are the most carefully studied of all the Blue Zone populations and supply critical findings as the healthiest population in the United States.

This reduced risk of coronary heart disease seen in epidemiological studies is supported by studies investigating potential cardioprotective mechanisms of nuts and seeds. These studies suggest that nuts and seeds reduce oxidative stress and inflammation, promote insulin sensitivity, and improve endothelial function, in addition to reducing cholesterol; plus, there is strong evidence from human trials that seeds, flaxseed in particular, reduce systolic and diastolic blood pressures.[25]

A few nutritional gurus advise radically excluding nuts, seeds, avocado, and all oils from patients with heart disease because of evidence that patients with advanced heart disease showed short-term improvements after a very low-fat vegan diet. But to assure the long-term safety of such exclusion, a much larger population and a longer follow-up period would be needed, just like the epidemiological studies mentioned above that have shown excluding nuts and seeds to be dangerous. Radically excluding fat (and DHA supplementation) from the diet of sick people, with so much evidence demonstrating danger, is an irresponsible default position.

Many mechanisms are responsible for the lifespan-promoting effects of nuts and seeds. They are rich in beneficial compounds such as phytosterols, polyphenols, and tocopherols, and they also enhance the absorption of nutrients from other foods—not merely six hundred different carotenoids, but an entire class of lifespan-lengthening (fat-soluble) phytonutrients. This absorption—of terpenes, diterpenes, saponins, and numerous vitamin E fragments (tocopherols and tocotrienols)—is enhanced with increased fat in the meal.[26] There is large genetic variation in absorption and bioavailability from person to person, as well as a variety of testing methods used in such studies, so the precise effects on each person may vary.

Some drugs look favorable and pass FDA scrutiny, but then over the years in general use, when more side effects come to light, a drug may be taken off the market or given a "black box warning"—that is, the strictest warning of possible side effects and hazards associated with the drug. If an independent body of scientists reviewed all the evidence, they would be forced to issue a black box warning for vegan diets that exclude nuts and seeds and are not supplemented with DHA.

> **WARNING:** FOLLOWING A LOW-FAT VEGAN DIET THAT EXCLUDES NUTS, SEEDS, AND ALL DHA-CONTAINING FOODS AND SUPPLEMENTS CAN BE HAZARDOUS TO YOUR HEALTH, INCREASING RISK OF DEPRESSION, DEMENTIA, AND CARDIAC DEATH.

I wrote two books on this subject that go into these issues in more depth—*The End of Diabetes* and *The End of Heart Disease*. The additional information in those books is worth reading if you have diabetes or heart disease. My dietary approach has been shown to work and has been documented to be the most effective way to lower blood pressure and cholesterol.[27]

Simply put: A healthier diet is safer and more effective than one that is less healthy. Almost all diets other than the Nutritarian diet have some flaw or flaws that make them less protective. Some doctors and those in the health community may try to persuade you that you can eat plenty of oil, meat, or butter or all the high-glycemic white potato or white rice you want without compromising the effectiveness of their dietary program. Those are compromises, however, and you don't want to compromise when your life is on the line.

I have used the Nutritarian approach with thousands of patients throughout my thirty-plus years of practicing nutritional medicine. My clinical experience and research have shown that in most cases your cholesterol and blood pressure numbers will be lower when you follow this plan than they would be if you take medication, and the high levels of antioxidants and phytochemicals almost wipe out oxidized LDL. I have seen the most serious and advanced cases of heart disease, including severe obstructive coronary artery disease, cardiomyopathy, and heart failure, be reversed when patients follow this program.

My father-in-law, who did not follow my advice until after he had a heart attack, had a low ejection fraction of 25 percent after that heart attack, which indicated significant damage to the myocardium, with some heart failure. He finally made significant improvements in his diet, and a few years later when he repeated the echocardiogram his ejection fraction had completely returned to normal. His cardiologist was amazed, stating he never sees this kind of reversal. Of course, you can't bring dead heart tissue back to life; but there is usually a significant amount of "hibernating myocardium" (tissue that is injured and not functioning) that can be turned around with increased oxygenation, optimal nutrition, and the enhancement of stem cell activity from nutritional excellence.

Normalizing your blood pressure, cholesterol, and inflammation through dietary excellence and losing your excess fat simultaneously are one hundred times more protective than merely covering up some blood test marker with drugs. This reflects true healing, which can save your life.

ADDRESSING TYPE 2 DIABETES WITH A NUTRITARIAN DIET

The nutritional plan that I originally outlined in *The End of Diabetes* is one that I have used successfully with thousands of my patients. This program is high in micronutrients and fiber and is caloric and glycemic favorable. It is designed to address the specific and urgent health challenges of people with type 2 diabetes and to resolve them quickly. You will swiftly see radical improvement to your health while achieving safe weight loss. All you need is the commitment to see it through. These meal plans and recipes are part of the next chapters in this book.

This targeted program allows most people with type 2 diabetes to eliminate all their diabetes medications within ninety days. Thousands of individuals across the country have lost weight and normalized their cholesterol levels and blood pressure and ended their diabetes with this approach. It is designed for people who want to take aggressive action in their battle against diabetes. By strictly following the meal plans, you will avoid needless years of suffering and extend your healthspan and lifespan.

That's right. It is possible to gradually reduce and eventually free yourself from the medications, glucose monitoring, HbA1c measurements, medical appointments, and devastating complications tied to this life-threatening disease. All of this is possible if you are committed to addressing the underlying cause of the disease: your diet. To put it simply: Food got you into this mess, and food can get you out of it.

Managing diabetes with medication is just that—managing. There is no chance of reversing it, and the possibility of eventual complications associated with the disease remains. In contrast, through nutritional excellence you will be able to significantly reduce your risk of cardiovascular disease and other complications of diabetes. Medication alone cannot provide the same degree of protection. Medication may keep blood glucose levels under control, but it can't restore your health the way healthy living can. Losing weight, flooding your body with the micronutrients it has been missing, and lowering inflammation are all key to your future health.

MANY MEDICATIONS MAKE DIABETES WORSE

Treating type 2 diabetes with medication without changing your diet is an approach doomed to failure. The majority of medications used to lower blood sugar place additional stress on your already-failing pancreas. In addition, these medications have serious side effects, and some increase the risk of heart failure and cancer.[28] Many also cause weight gain, which over time worsens diabetes.[29] It's a vicious cycle. As you gain weight, you become more insulin resistant and you need more medication. Your pancreas pushes to meet these higher insulin demands and eventually can no longer do so. At that point, you may need to take insulin, too, which causes further weight gain, and the pathology advances.

The Action to Control Cardiovascular Risk in Diabetes (ACCORD) trial made headlines in 2008 and 2009 when the study was halted by a safety review board of the National Institutes of Health (NIH) and the results were published in the *New England Journal of Medicine*.[30] More than ten thousand people with type 2 diabetes, with a mean age of 62.2 years who either had cardiovascular disease or were at high risk for developing the condition, were enrolled in the trial. Shockingly, more medications and better glucose control were found to result in more deaths—fifty-four, to be precise—before the NIH stopped the study after 3.5 years. The drugs used by diabetics in the study included insulin, sulfonylureas, and a class of drugs known as thiazolidinediones. That class includes rosiglitazone, which in a study published in 2007 was linked to an increased risk of cardiovascular problems.[31] My opinion is that sulfonylureas should be taken off the market and stopped in every patient who is on them; they are just too dangerous and accelerate the burnout of the insulin-producing beta-cells in the pancreas.[32]

The five-year results of the ACCORD trial, when subjects in the aggressive treatment arm were transitioned to the routine care group, confirmed that more intensive lowering of blood glucose, more intensive lowering of blood pressure, and treatment of blood lipids with a fibrate and a statin drug do not reduce cardiovascular risk in people with established type 2 diabetes.[33] In fact, statin drugs, the standard of care in type 2 diabetes, have been shown in many studies to raise blood glucose levels and worsen diabetes.[34] When drugs are the only thing in your toolbox, people suffer needlessly, and the loss of life is very real.

The finding shocked doctors, and most still remain confused about this, as they have been encouraged to lower blood sugar, cholesterol, and blood pressure levels aggressively with medications and have almost no skill in educating and motivating patients and supporting them regarding aggressive dietary modifications. The comments posted in the scientific literature are quite ridiculous, with almost nobody admitting that diet, not drugs, should be the main approach to address type 2 diabetes.

In other words, medical care is ineffective in reducing diabetes-related deaths. The scientific conclusion by a specially appointed panel of elementary school children concluded: *It's the food, stupid.*

My mantra to diabetic patients is this: Don't just *treat* your diabetes—get rid of it. Flood your body with protective nutrients, lose the weight, and restore your health as fast as you can.

I have treated several hundred diabetic patients with the Nutritarian approach, and the vast majority became nondiabetic within four months, even before they lost all their excess weight. In a published pilot study of patients with type 2 diabetes, 90 percent of those following my nutrient-dense, plant-rich, low-glycemic diet were able to eliminate all of their medications for diabetes, and their mean HbA1c levels went from 8.2 to 5.8 percent, which is within the normal range.[35] And the benefits of the Nutritarian diet are not limited to reversing diabetes, as blood pressure and cholesterol are predictably normalized as well.[36]

Some suggested diets for diabetics rely on meat and other animal products as the major source of calories to keep glycemic effects down. This strategy has serious drawbacks. Diets high in animal protein promote inflammation and oxidative stress and lead to further weight gain, diabetes, and heart disease,[37] and they have been linked to cancer and premature death.[38] As we have seen, a Nutritarian diet permits animal products only in small amounts and restricts the consumption of high-glycemic sweeteners and baked goods. But this doesn't mean giving up good food. You will find the menus and recipes in the coming chapters to include a wide portfolio of delicious foods.

When following the Nutritarian eating style, you will eliminate foods that are empty of nutrients, such as sugar, sweeteners, white flour, and processed foods. My menus for diabetes reversal are similar to those for my regular Nutritarian program except that dried fruit is more limited, and only one fresh fruit is served with each meal. If you desire more fruit, then you can add lower-sugar fruit, such as berries, pomegranates, grapefruit, kumquats, guava, passion fruit, dragon fruit, and loquats. In other words, even diabetics can have more than one fruit with a meal if they choose low-glycemic fruits such as strawberries. All grains that are ground into flour are also omitted with this diet, even if they are whole grains; only intact grains that are cooked in water are acceptable. Beans are the most favored high-carbohydrate food because they are so low glycemic and so slowly absorbed.

An important reminder: It is critical to follow blood sugar levels carefully and reduce medications appropriately so your levels do not get too low. Err on the side of caution and let your sugar levels run a bit high to prevent a hypoglycemic reaction when changing to a Nutritarian diet. Medications should be cut by about one-third on day one and then by about half of the original amount by the end of the first week. Remove the medications that cause weight gain first, meaning insulin, sulfonylureas, and thiazolidinediones, leaving metformin on board until more weight loss is achieved.

SAVING THE LIVES OF TYPE 1 DIABETICS

Conventional medical care contributes to increased morbidity and mortality also in type 1 diabetes. People with this kind of diabetes not only die more than ten years before other Americans, but they also suffer much more with serious medical issues.

Most type 1 diabetics are given instruction to increase their insulin dose depending on their carbohydrate intake to adequately control their blood sugar levels. This results in a needless "death sentence" because they are not aware of the dangers of excessive use of insulin and the importance of maintaining a diet that requires only healthful amounts of insulin. Most endocrinologists are concerned with the dangers only of the high glucose and not of the health risks of the excess insulin given to control it.

I strongly advocate that type 1 diabetes is still compatible with a long, healthy lifespan and need not increase cardiovascular risk. The dangerous standard American diet (SAD) creates the morbidity and increased early mortality in type 1 diabetes, not the lack of insulin production. A Nutritarian diet needs to be prescribed to every person who has type 1 diabetes to help save their life.

A NUTRITARIAN DIET

- Protects against the effects of metabolic toxins that create damage
- Effectively lowers blood glucose levels
- Prevents sugar highs and lows

Howard Spinowitz, DO
A Case Study in Type 1 Diabetes

I was diagnosed with type 1 diabetes right after medical school during my first year of residency in internal medicine. I gained 50 pounds during the first month on insulin and have been on it ever since—but I've had a highly problematic course on insulin, with both severe highs and lows. While the term has fallen into disuse, my diabetes has often been described as extremely "brittle" by a few docs over the years.

Before I read *The End of Diabetes* in January 2013, I weighed about 185 pounds, although I was pretty physically fit due to weight lifting and cardio exercises three to four times per week. I was on an insulin pump, averaging 80–100 units per day. After following Dr. Fuhrman's recommendations for a couple of weeks, my insulin usage dropped dramatically, and my wife agreed with me that the whole family should switch over to this healthy way of eating.

I currently weigh 146 pounds and am a lean, sinewy mass of muscle! I feel like I'm 25 years old (despite turning 47 later this year), and I've discontinued the insulin pump in favor of 13 units per day of a long-acting insulin injection; additionally, I use about 12–15 units per day of a fast-acting mealtime insulin. While I've been teased by a few fellow physicians at my hospital for eating so healthfully (go figure!), I tirelessly advocate the Nutritarian eating plan, and a few doctors have bought Dr. Fuhrman's books, changed their diets, and noticeably improved both their appearances and their outlooks. Whenever possible, I recommend his books to sick patients. Additionally, my wife and I have tried our best to make a healthy impact on our community and educators.

- Enables the type 1 diabetic to use less than half as much insulin

- Prevents conditions associated with diabetes such as kidney failure and blindness

- Permits a favorable healthspan and lifespan

AUTOIMMUNE DISEASES AND CANCER

A Nutritarian diet predictably reverses and resolves autoimmune diseases and also is effective in reversing many early-stage cancers and prolonging lifespan in later-stage cancers. I link cancer and auto-immune disease together for a few reasons:

1. Both are affected by enhancing immune system health.

2. The same nutritional protocol has been shown to be effective in both (juicing greens, reducing glycemic carbohydrates, and eliminating animal protein).

3. Autoimmune diseases are linked to an increased risk of cancer in later life.

4. Chemotherapeutic agents that are carcinogenic and increase the risk of later-life cancer are used as therapy for both.

REVERSE AUTOIMMUNE DISEASE

Some of the cases in this book feature people who have reversed their autoimmune disease, and I have published some of these medical histories.[39] Many of my patients over the years have made full recoveries from

- Autoimmune hepatitis

- Inflammatory bowel disease (Crohn's and ulcerative colitis)

- Lupus or SLE (systemic lupus erythematosus) and discoid lupus

- Mixed connective tissue disease

- Multiple sclerosis

- Psoriasis and psoriatic arthritis

- Rheumatoid arthritis

- Scleroderma

- Sjögren's syndrome

In fact, over the past thirty years of my treating such individuals with the Nutritarian approach, it was rare when people did not make a full recovery—gradually discontinuing their medications and eventually getting totally well. This is particularly important because not only are the medications usually not completely effective, but their toxicity, side effects, and increased risk of later-life cancer are significant.[40]

I always enjoyed caring for patients with autoimmune diseases because they had such a fierce desire to get well; they were typically very eager to do everything I asked of them. I remember a patient with psoriatic arthritis who told me, "I will eat nothing but sawdust if it will make me well."

One 16-year-old girl with severe lupus was waiting for a kidney transplant, as her creatinine level was an extremely high 4.2 mg/dL when she saw me. She had very little kidney function left. But after nine months on a Nutritarian diet, she made a complete recovery and was able to stop all medications. Incredibly, her kidney function returned to normal, with a creatinine of 0.8 mg/dL.

Another young woman with lupus saw me regularly for almost a year as we adjusted and perfected her diet and slowly weaned her off all medications. Even after that year, she continued to have recurrent high fevers, and I was concerned that her disease activity was too strong and resistant to nutritional care, so I referred her back to her rheumatologist to be put back on the immunosuppressive drugs methotrexate and hydroxychloroquine (Plaquenil), which are commonly used to treat lupus. When I saw her one year later, she looked great and her disease was gone. She told me that she never went back to the rheumatologist, as I had suggested, and just continued on the Nutritarian diet, along with some intermittent fasting. About nine months later, the remaining vestiges of lupus were gone. I had given up on her after one year, but unusually, it took almost two years for her to get better, and *she* did not give up; she did not want to go back on those dangerous drugs.

One young man I cared for with ulcerative colitis was having about fifteen bloody bowel movements a day when he came to me. His situation was so severe I had to admit him to the hospital, where he was started on intravenous steroids, and no food. Surgeons were consulted about removing his entire large bowel; but instead, since he was no longer bleeding after a week, he left the hospital and we gave nutritional care a shot. I continued to have him fast with water only for a week more after his hospital discharge, and then the following week, he had only heated vegetable juices while we lowered the oral prednisone he was still taking. During subsequent weeks, I added steamed zucchini and then boiled and creamed root vegetables to his diet. We slowly advanced his diet, and over time, he made a complete recovery and got his healthy life back.

It is always thrilling to see people make recoveries from what the general medical community considers permanent illnesses that require medical treatment for the rest of their lives.

HIGH-VEGETABLE, PLANT-BASED DIETS: EFFECTIVE FOR PREVENTING AND TREATING CANCER

As we saw in Chapter 5, you can eat to prevent cancer, to slow cancerous progression, and, possibly, to beat cancer. No treatment—not even this "gold-standard" program of nutritional excellence—is consistently effective for "curing cancer," as cancer is a bizarre end-stage disease that responds to interventions in unpredictable ways. But dietary protocols that include vegetable juicing and high intake of cruciferous vegetables offer potential for treatment and for increasing the survival of cancer patients. Other plant-based regimens have also been shown to reduce the occurrence of cancer and slow or halt its progression.

Dr. Chris Miller
Reversed Her Lupus and Is Back at Work Again

At age 36, I was a practicing emergency physician in a busy emergency department when inflammation took over my body. I had joint swelling and pain in multiple fingers, strange rashes, then chest pains, and then even kidney involvement. I was promptly diagnosed with systemic lupus erythematosus (SLE, also known as lupus) and placed on heavy-duty immunosuppressive medications, including high-dose prednisone. Despite the medications, my disease continued to progress, and soon I was on six medications. It seemed the more medications my doctors gave me, the more complications I had, and the more I needed to take. It was a terrifying cycle, and I felt depressed and hopeless. I even had to quit my job in the emergency room, as my immune system was so suppressed, it wasn't safe for me to be around other sick people.

As I continued to have increased pain and fatigue, it became apparent that the medications were not working. My doctor wanted to try experimental drugs and at one point told me to go to the emergency room because I was having so much inflammation around my lungs, I couldn't even sit or lie down to rest because of the severe pain. I had gained weight from the prednisone, had no energy, was losing my hair, and felt sad and scared.

I began to search for other treatment options for lupus, and I'm so grateful that the first name that came up in my search was Dr. Joel Fuhrman, who had been helping patients with lupus and other autoimmune

diseases for decades. I quickly signed up for a weekend retreat that was just a few weeks away, right before Thanksgiving. I will never forget the first time I heard Dr. Fuhrman speak that weekend. He explained the science behind inflammation and chronic disease, the role of nutrient-dense foods to heal the body, and the cases and studies to support it. I couldn't believe it. I was stunned—how could I not have known any of this, during all of my years of medical training?

After the weekend of following the Nutritarian diet, hearing about the science of disease reversal, and learning techniques to prepare nutrient-rich meals, I felt inspired. For the first time since my diagnosis, I had hope of a future and of healing my disease. For me, this was the greatest gift of all. I flew home with my husband, and we promptly emptied our cupboards and fridge of all processed foods, animal products, and oils—all gone. We restocked them with my new foods: legumes, nuts, seeds, whole grains, and lots and lots of vegetables and fruits. I learned to make (and love!) fresh vegetable juices, blended vegetable-bean soups, and a variety of salads. I juiced for a few days at a time, and I even did some water-fasting.

Although I began to lose weight right away, it took some time for me to actually have less joint pain and other symptoms. Because of food sensitivities, I was taught to eliminate certain foods to quiet the inflammation, and after a certain period of time, I would reintroduce them. Over the years, I have been able to slowly taper off my medications. The hallmark antibodies of lupus have been cut in half, and the other improvements in my lab results have documented the reversal. My kidney function has returned to normal. Also, important to me, my hair has grown back, I no longer have rashes, and my energy is often through the roof. I even ran my first half marathon in over eight years!

I've been able to start working again, and now I teach my patients about the importance of eating nutrient-dense foods and removing proinflammatory ones. I'm proud to be a Nutritarian, I love eating these healing foods, and I can't imagine eating any other way. While it hasn't always been easy, it has been a fun journey and one that I'm so thankful to be on. I'm especially grateful to Dr. Fuhrman for giving me hope, guidance, and encouragement along the way. And I am so very appreciative to not only have my life back, but to live a better, more compassionate version of it.

Though I insist the evidence is overwhelming that the Nutritarian program would prevent more than 90 percent of all common cancers, I am not claiming that it can cure all cancer or even most cancer. However, scores of my patients who had life-threatening cancers have had spectacular results with this protocol, and many more with early stages of prostate and breast cancers have seen their conditions reverse.

For example, in 1997 my patient Pam had metastatic ovarian cancer that had moved to her lungs; 4 liters of fluid were extracted from her lungs so she could breathe. She was given only a few months to live, even with treatment. But today, twenty-two years later, she is thriving.

Likewise, twenty years ago, Irene reversed and eliminated her Grade 2, Stage 4 non-Hodgkin's lymphoma with the Nutritarian protocol and no other medical intervention, and it never came back. "In addition to reversing my lymphoma," she said, "I dropped over 40 pounds; I am also feeling extremely energetic and healthier in many ways as I build a protective environment within. It is no wonder that all of my vital readings indicate that I am in terrific shape, from my cholesterol, to my blood pressure, to my chemical makeup . . . It is also great to be slim and agile. Obviously, I will continue to follow the special diet that you designed for me. Once again, thank you, thank you, thank you!"

My thirty-plus years of utilizing this approach with cancer patients has made it clear that a Nutritarian diet saves lives.

The fact that dietary excellence can not only reduce cancer risk dramatically, but also prolong life in people who have cancer is well-established in scientific studies. In 2007, the World Cancer Research Fund (WCRF) and the American Institute for Cancer Research (AICR) released eight recommendations related to body fatness, physical activity, and diet aimed at preventing the most common cancers worldwide. Today, they offer ten recommendations.[41] The most important dietary advice is to be as lean as possible within the normal range of body weight, to mostly eat foods of plant origin, and to limit alcohol intake. Breast cancer risk was reduced by 60 percent in women who met at least five recommendations compared with those who met none.[42]

What about other cancers? Greater adherence to the WCRF/AICR dietary guidelines was associated with significantly less breast, endometrial, colorectal, lung, kidney, stomach, oral, liver, and esophageal cancers.[43] Similarly, a larger follow-up study following 469,000 people for eleven years found that just a 3 percent increase in the consumption of animal protein calories was associated with a 15 percent higher risk of bladder cancer whereas just a 2 percent increase in plant protein was associated with a 23 percent lower risk.[44]

AICR recommendation no. 10 is that cancer survivors follow the recommendations for cancer prevention. The same diet that can help prevent cancer in the first place can be used to help save lives after diagnosis. Adherence to the guidelines for cancer prevention was found to be associated with lower mortality among older female cancer survivors, including specifically breast cancer and other cancers in general.[45]

THE PROSTATE CANCER LIFESTYLE TRIAL

Other like-minded physicians have seen outcomes similar to those I have experienced. Notably, studies conducted by Dr. Dean Ornish (the Prostate Cancer Lifestyle Trial) established that diet and lifestyle intervention is effective at halting the progression of prostate cancer.[46] Ninety-three men with prostate cancer who had decided not to undergo treatment were recruited for the study. The control group did not change their diets and exercise habits, and the intervention group consumed a high-fiber, vegetable-rich diet for one year. PSA levels, which can indicate the presence of prostate cancer when they are high, decreased by 4 percent in the intervention group and increased by 6 percent in the control group. Six patients in the control group were treated because of evidence of disease progression. None of the patients in the intervention group required treatment.

The researchers also documented enhanced inhibition of cultured prostate cancer cell growth by serum from intervention group patients compared with controls; that is, only the blood of the participants on the vegetable-rich diet had cancer-fighting ability when added to cells growing in a petri dish. Two years after the end of the trial, thirteen patients in the control group and only two in the intervention group had required conventional treatment.[47] Choosing to treat lower-risk prostate cancers with a diet and lifestyle intervention allows men to delay or avoid the side effects of radiation and prostatectomy.

US POLYP PREVENTION TRIAL AND WOMEN'S HEALTHY EATING AND LIVING STUDY

When interventions to radically increase vegetable intake are utilized, cancer markers and outcomes improve. Amazing and life-altering results are possible only with significant and committed dietary changes.

The US Polyp Prevention Trial used more moderated dietary improvements and saw more moderate results.[48] Participants (who had at least one previous colorectal adenoma) were assigned to either a control group or a dietary intervention group. Those in the intervention group were advised to limit fat to 20 percent of calories, consume at least 18 grams of fiber per day, and eat 3.5 servings of fruits and vegetables for every 1,000 calories. Of the 821 in the intervention group who completed the study, 210 were "super compliers," meaning they met at least nine of the twelve dietary goals (three goals, once every four years). The super compliers had a 30 percent reduction in adenoma (polyp) recurrence and a 50 percent reduction in multiple or advanced adenoma recurrence compared with controls.

Also, when the data were analyzed for increase in bean intake, those with the greatest increase during the intervention had the lowest risk of advanced adenoma recurrence—a 65 percent decrease in risk compared with those who remained at baseline intake.[49] Higher intake of several different classes of flavonoids was also associated with a lower risk of recurrence.[50]

In the Women's Healthy Eating and Living (WHEL) Study, women eating a healthier and earlier dinner had fewer hot flashes and other negative menopausal symptoms, as well as fewer cancer deaths. Women assigned to the higher fruit, vegetable, and fiber group ate more vegetable and fruit servings and more fiber, and this group of women with breast cancer had 31 percent fewer incidences of cancer recurrence over seven years of follow-up.[51] Further analysis of the hot flash–negative women and of women taking the estrogen modulator tamoxifen to treat their cancer showed that those with higher vegetable consumption also had a lower risk of breast cancer recurrence.

LIVE FOR 100 YEARS

The average age of death should be 100 years old, not 80. Most people should live between 90 and 110 years, and they would if they ate super healthfully and avoided exposure to dangerous carcinogens. The Blue Zones around the world may have the most centenarians, yet none of these areas has an average death rate higher than 90 years of age because the dietary patterns in these areas are not scientifically designed to extend human lifespan. Rather, eating styles in the Blue Zones are haphazard and based on the foods available locally in those areas.

A Nutritarian diet, on the other hand, is specifically designed to comprehensively include all the foods that are the most lifespan-promoting from various agricultural zones and climates. It is considerably healthier than a Blue Zone diet because it checks off the box of every dietary element and practice that slows aging, prevents common diseases that cause death, and extends lifespan. The bottom line is that this "gold standard" should be expected to enable a longer lifespan.

SLEEP AND OVERNIGHT FASTING

Adequate sleep is a prerequisite for good health. In addition to leaving you feeling rested and alert, good-quality sleep is essential for your immune system to work properly. There is even some evidence that poor sleep impairs the immune system's ability to eliminate small, newly established tumors before they become dangerous.[52]

Michael de Marillac
Recovered from Near Death with Heart Problems

I was admitted to the hospital on December 8, 2015, with severe heart failure. I had uncontrolled diabetes on admission, with a blood sugar level so high that an accurate measure was not possible. My troponin levels indicated I had had a heart attack, but my main problem was the severe hypertrophic cardiomyopathy that had worsened over the previous year, in conjunction with the cardiac arrhythmias, with a heart rate between 165 and 220 beats per minute that could not be controlled. I almost died in the hospital and was brought back during a code with a defibrillator before I even made it to the operating room. When I had a procedure to help control my heart rate, the anesthetist told my family that my chances were slim, and if during this procedure my heart stopped again, it was unlikely that I would wake up.

I was told it was a miracle that I had survived long enough even to get to the hospital alive. My left ventricle ejection fraction was only at 20 percent, and I was on oxygen, still short of breath and barely surviving. I had some very close calls in the ICU, with crash teams constantly working on me to keep me alive. After the failed attempts to bring my heart under control, one of the doctors touched me on the shoulder and said they had done all that they could. I was waiting to die. My nurses cried and I felt dreadful. That was the turning point for me. I obtained permission from my electrocardiologist to eat Dr. Fuhrman's Nutritarian diet in the hospital.

I had known about the Nutritarian diet for a few years, and actually had bought a few of Dr. Fuhrman's books and had a set of videos, but I did not follow it until here I was, almost dead. It was nothing short of a miracle, because hospital rules do not permit special food to be brought in, yet they did it in my case, and that was what saved my life.

My diabetes was the first thing to be brought under control. I was weaned off huge amounts of insulin within a few months, as my blood test confirmed I was no longer diabetic, and I no longer had high blood pressure. Eventually, my poor vision and diabetic eye damage cleared. Slowly and steadily, I healed.

Now more than three years later, my left ventricle has remodeled in a good way—it has shrunk down in size, and my ejection fraction is well in the normal range. I am entirely off all medications, including beta-blockers, warfarin, and twenty other drugs. My cardiologist

announced last year that my heart problems are history, and he has even cleared me for scuba diving. He said he is glad I am getting out there and enjoying my life, because I didn't have a life before. I could not walk, as my severe diabetes and heart problems had crippled me—I had to learn to walk again.

Now, I am back at work full time. My blood pressure is perfect, and I walk an hour a day. I have completely reversed my type 2 diabetes and all retinal damage. My eye specialist was about to schedule me for eye surgery, as my intraocular pressure was too high, but for reasons he did not understand (but I do!), the high eye pressure returned to the normal range.

My medical bill was more than $100,000 as I was in the intensive coronary care unit for so long. Once my condition was more stable, I was also given an angiogram and asked to give permission for insertion of a stent if they found any blockage. I refused to sign; I said if there was any blockage, it could be reversed with exceptional nutrition. The medical resident looked at me as if I were insane. The cardiologist for that procedure showed me the video, and it was clear.

One of the most startling things that happened was that one of my treating cardiologists came to my bedside and, with great emotion, confided in me about the death of his parents from diabetes. He said that he was heading the same way and felt ashamed that, as a doctor, he was supposed to be an example to his patients but was failing himself. I told him—and the rest of the medical staff who treated me—about Dr. Fuhrman and his healthy eating style. It got them thinking hard, but I knew some did not know what to think, because they said I must have a very strong will to do this.

It was painful to be so sick, but getting well by following Dr. Fuhrman's Nutritarian diet was the easiest thing I have ever done. Now I am a new person with a new excitement about life.

Melatonin, which is a hormone produced in response to darkness and during sleep, is not only a promoter of sleep but also an anti-oxidant and an inhibitor of cancer cell growth.[53] To get good-quality sleep, you need to make your bedroom as dark as possible. Do not put clocks with lights or night lights in the bedroom. Use either light-blocking curtains or a sleep mask to reduce exposure to outside light. A dark room at bedtime promotes melatonin production and good sleep. Also, screens of smartphones, computers, and other electronic devices emit blue light, which is a signal for daytime and therefore diminishes melatonin production. Light exposure at night has been associated with an increased risk of cancer, breast cancer in particular.[54]

It is also important to stop eating several hours before bed and to have at least a thirteen-hour overnight fast between dinner and breakfast. As I noted earlier, prolonged nightly fasting has been linked to a reduction in breast cancer recurrence. In the Women's Healthy Eating and Living Study, women with breast cancer whose nightly fast was less than thirteen hours compared with thirteen hours or more had a 36 percent increase in breast cancer recurrence over a follow-up period of seven years. Plus, the women who had a longer nightly fasting period got more sleep.[55] When you modify your lifestyle to accommodate the triad of healthy eating, adequate sleep, and finishing dinner early so you can go to bed on an empty stomach, you maximize your protection against cancer.

CHAPTER SEVEN: QUICK SUMMARY

REVERSE HEART DISEASE AND DIABETES

Heart disease is a food-created issue, and a superior diet can restore your health relatively quickly. If you strictly follow the meal plans in this book, you can significantly lower your blood pressure and cholesterol in fewer than three weeks. In the vast majority of cases, you will not need medication anymore, ever. The Nutritarian dietary protocol works.

A Nutritarian diet is the most effective program to lower blood pressure and reverse heart disease because it is designed to reduce body fat, slow aging, improve nutrient bioavailability, reduce inflammation, and achieve nutritional completeness. It is lower-glycemic rather than a low-fat vegan diet and has more micronutrients and antioxidants to protect cells and to protect against cardiac arrhythmias. It also has all the benefits of restricting animal products, without the deficiencies that doing so imposes.

THE QUESTION OF NUTS

Radical fat-avoiding vegans often advocate against consuming longevity-promoting foods like walnuts or taking supplemental DHA and EPA, stating it is harmful or not needed with a vegan diet. But the inclusion of a reasonable amount of seeds and nuts and a conservative amount of DHA-EPA get equally impressive or superior results when it comes to plaque reduction, disease reversal, and weight reduction without the added risks of depression, memory loss, and later-life dementia.

Scores of studies show an increase in cardiovascular deaths when nuts and seeds are avoided. When we look at large studies that involve thousands of participants and go on for decades, and that have hard endpoints, we see the same thing: The lack of sufficient nuts or seeds in the diet increases premature death, particularly cardiovascular death. This is seen clearly in the Adventist Health Studies, which have studied a population that includes many people who eat a plant-based diet.

ADDRESSING TYPE 2 AND TYPE 1 DIABETES
WITH A NUTRITARIAN DIET

The Nutritarian diet is high in micronutrients and fiber and is caloric and glycemic favorable. It is designed to address the specific and urgent health challenges of type 2 diabetics and to resolve them quickly. This targeted dietary program allows most people with type 2 diabetes to eliminate all their diabetes medication within ninety days.

Conventional medical care contributes to the increased morbidity and mortality in type 1 diabetes. Most endocrinologists are concerned only about the dangers of high glucose levels, not about the health risks of the excess insulin given to control them. A Nutritarian diet protects against the effects of metabolic toxins that create damage, effectively lowers blood glucose levels, prevents glucose highs and lows, and enables people with type 1 diabetes to use less than half as much insulin.

AUTOIMMUNE DISEASES RESPOND

One of the most rewarding aspects of my nutrition-based medical practice over the past thirty-plus years has been the consistent positive outcomes from aiding those with autoimmune disease. I have seen numerous cases of complete remission from a broad assortment of autoimmune diseases and the lack of response to this portfolio of nutritional recommendations was rare.

PREVENT AND TREAT CANCER, LIVE LONGER

Eat a high-vegetable, plant-based diet to prevent cancer, to slow cancerous progression, and, possibly, to beat cancer. No treatment is consistently effective for curing cancer, but dietary protocols that include vegetable juicing and high intake of cruciferous vegetables offer the most potential for treatment and for increasing the survival of cancer patients. Other plant-based regimens have been shown to reduce the occurrence of cancer and slow or halt its progression.

If everyone ate super healthfully and were not exposed to dangerous carcinogens, most people would live 90 to 110 years. The Nutritarian diet is designed to include every dietary element and practice that slows aging, prevents diseases that cause death, and extends lifespan.

CHAPTER EIGHT

How to Cook, Eat, Live

This chapter provides four sets of menus to address a variety of health concerns:

1. General menus for optimal health and ideal body weight (three weeks of menus)

2. Menus designed specifically for reversing type 2 diabetes (one week of menus)

3. Emergency menus for radical weight reduction, advanced heart disease, or out-of-control diabetes (one week of menus)

4. Menus designed specifically to attack autoimmune disease or cancer (one week of menus)

Before you pick a menu, it is worthwhile to review all the options. The goal is for you to understand the reasons for the slight modifications among menus so you can best devise the right one to enable you (and/or a loved one) to enjoy eating healthfully for the rest of your life. You will be able to modify the general menu to meet your individual needs. Please note that you can also add more food and more nuts and seeds if you desire weight gain or if you are losing too much weight.

It is essential to keep in mind that it takes a full three months for a person's food preferences and addictive attraction to unhealthy foods to change. I urge you to remain patient and positive. The benefits are worth the wait. If you want to enjoy eating this way to the fullest and to live healthfully in the long term, you need to abstain from your addictive food triggers and do so 100 *percent*. And although you might miss your illicit pleasures at first, I promise you that the cravings and the tastes for these addictive, life-shortening foods *will* go away.

Create a "food business plan" that will help you to not be tempted to eat dangerous conventional foods and to keep the right type of healthful food choices in your home. Many of us find this works best when confining most of the food shopping and preparation to two days a week, such as Wednesday and Sunday, and then eating leftovers on the other days.

At my Eat to Live Retreat in San Diego, we initially see individuals who are addicted to foods and struggling with their weight wanting more salt and more sugar in their food; but as the weeks progress, they notice their taste buds changing, and the flavors of the delicious and nutritious foods offered start bursting forth. They simply do not desire added salt and sugar anymore.

The transformation is exciting and is not just about taste. We see an improvement in overall clarity of thought and enthusiasm about life. And by the time our guests go home, they not only love eating a Nutritarian diet; they love how it makes them feel. Most of the people who come to the retreat continue to progress and lose more weight after they leave, because they have learned the skills for their continued success.

Having seen such success with patients for decades, I know that when you do this at home, *it is necessary to follow and hold to the plan as it is prescribed.* If you do, the results and lifelong benefits will follow.

When people come to see me, I encourage them to let me decide what they will eat, reminding them, "What you thought you should eat, what you felt like eating, and what you liked to eat is what got you into this problem to begin with." But I promise them that they will love eating this way if they can just keep it going strictly for the first three months.

There is no question that your taste and food preferences will change. You will become more comfortable eating high-nutrient foods, and it will become second nature because you will like the food and you will like the way you feel. Putting nutrient-dense foods in your body improves every aspect of your physical life: digestion, energy, sleep, focus, and so much more.

Remember: This Nutritarian program is a high-nutrient eating style, not a calorie-counting formula. If you eat nutrient-dense foods, you don't have to count calories in the same way. Keep your focus on increasing your intake of disease-protective nutrients with natural, whole foods. When you eat so high on the nutrient-density line, you are satisfied with fewer calories and you are no longer driven to overeat.

HOW MUCH TO EAT

How much you should eat varies widely from person to person. Most of us have been trained to overeat throughout our lives, and the concept of how much we need to eat has been grossly distorted by the addictive nature of modern foods. Most people chronically overeat, even those who are not overweight. When you eat healthfully, you have fewer physical symptoms driving you to overeat, but it is still possible to eat recreationally and for emotional reasons or just to overeat out of habit.

Ideally, you want to eat when you are hungry and not eat when you are not hungry. Eating when you are not really hungry causes obesity. Also, eat until satisfied, but not until you are full. You should never feel uncomfortably full after a meal. And you don't want to eat so much that you are not hungry again for the next meal. In general, you need to eat enough to maintain a favorable musculature without getting overly thin. But since almost everyone is overweight, a person at the right weight might be viewed as being too thin. Remember, excess fat on the body is never a favorable thing, so even people who look thin may have too much fat on them. We gain healthy weight (muscle and bone) in the gym, and we lose weight (fat) in the kitchen. Exercise builds and maintains favorable musculature and strength, and eating healthy foods when you are hungry keeps you strong and lean.

Nutritarians eat to achieve superior health. They are at their ideal weight, or if they are not, they are moving closer to achieving it every day. If you are significantly overweight or obese, you should be losing at least 2 pounds a week on your way to recovery. By adjusting your food intake to maintain this rate of weight loss, you will be lowering your insulin resistance, lowering inflammatory proteins, and lowering unfavorable fat storage hormones.

HERE'S WHAT YOU SHOULD EAT

1. A large salad every day

2. At least a half-cup serving of beans/legumes in soup, salad, or a dish once daily

3. At least three fresh fruits every day

4. At least 1 ounce of raw nuts and seeds every day

5. At least one large (double-size) serving of green vegetables daily, steamed or as an ingredient in a soup or entrée

All the menus that follow require your pledge to avoid the six most deadly food habits listed below, especially during the first three months when you are still fragile and your new food preferences are being formed.

DON'T EAT

1. Barbecued meats, processed meats, or red meat

2. Fried foods

3. Dairy (cheese, ice cream, butter, milk)

4. Soft drinks, sugar, and other sweeting agents or artificial sweeteners

5. White flour products

6. Oil

Even a little bit of oil or a small amount of flour can derail your weight loss. Sugar, white flour, and oil halt fat breakdown. You need to give your body a real chance to change its biochemistry and build up its nutrient stores; then you will see how much better your life can be when you are well-nourished, and you can let the Nutritarian diet do its magic.

I always say the magic is in the last 5 percent—that is, you are already doing this 95 percent, so going all the way to 100 percent is only a touch more. But when you make that commitment and achieve that degree of excellence, you lose your food addictions more quickly and you see a miraculous transformation occur in your body. Dabbling in dangerous foods, even once in a while, can lead to your continuing to desire them.

THE SALAD IS THE MAIN DISH

The general dietary approach I use to plan my meals is this: You can skip a meal, substitute a vegetable juice, or just have some fruit in place of a meal, but don't skip lunch. Lunch is the most important meal. Have a big salad with lunch every day.

Here are my general guidelines for each meal:

Breakfast—A small to moderate amount of an intact grain such as quinoa or steel cut oats, 1 or 2 tablespoons of seeds (flax, chia, or hemp), and some fresh or frozen fruit

Lunch—A big salad and a bowl of vegetable bean soup or a bean stew or chili and a piece of fruit for dessert

Dinner—Some raw vegetables with a dip or salsa, a steamed or wokked green vegetable, a vegetable medley or vegetable stew, and some frozen fruit or a healthful fruit-based dessert

I prefer that people stay on a vegan diet (avoiding all animal products) and then supplement to get the tiny bit of extra nutrients needed when not eating animal products. Some people, however, want to remain on the Nutritarian program and still consume animal products; in this case, I want them to measure that amount so it stays restricted to less than 6 ounces a week and not more than 2 ounces per serving.

You can use wild fish and certified organic poultry as a minor component or flavoring agent in recipes. For example, in the recipes provided in Chapter 9, you could add a small amount of wild salmon to the Roasted Vegetable Salad with Balsamic Vinaigrette, or chicken to Chickpea, Greens, and Farro Stew. Instead of the baked tofu in the Broccoli and Snow Pea Stir-Fry, you could use 8 ounces of cooked shrimp or chicken (2 ounces per serving).

I used to think that a bit of sardines or shellfish was a favorable addition to a Nutritarian diet, but I have changed my mind, given the findings of extensive microplastic contamination in seafood. I feel safer eating vegan the vast majority of the time and go off it only on rare special occasions. I find this way of eating extremely satisfying, delicious—and intellectually, emotionally, and environmentally peaceful.

When you review these recipes and menus, keep in mind that most people eat a smaller variety of dishes and more leftovers. In other words, learn to cook in larger amounts so you can eat leftovers and have less food prep to do. These menus provide a variety of recipes to demonstrate a variety of options, but you can stretch one week of menus out for two to three weeks of options.

GENERAL MENUS FOR GREAT HEALTH AND IDEAL WEIGHT

WEEK 1

DAY 1

BREAKFAST
Steel Cut Oats and Cherries*

LUNCH
Salad with lettuce, arugula, tomatoes, green onions, edamame, and Seedy Ranch Dressing*

Portabella Pizza*

Orange, clementines, or other citrus fruit

DINNER
G-BOMB Thai Vegetable Curry*

Water-sautéed green leafy vegetable such as kale, collard greens, or cabbage sprinkled with lightly toasted unhulled sesame seeds

Fresh or thawed frozen berries

DAY 2

BREAKFAST
Cranberry Smoothie*

Easy trail mix (walnuts mixed with raisins/ currants or sunflower seeds mixed with unsweetened dried cherries or blueberries)

LUNCH
Raw vegetables with Aquafaba Hummus*

Leftover G-BOMB Thai Vegetable Curry*

Thawed frozen peaches or other fruit

DINNER
Eggplant Meatballs*

Zucchini noodles tossed with Garlic Nutter*

Melon or other fruit

Asterisks indicate recipes that are provided in Chapter 9.

DAY 3

BREAKFAST

Mixed fruit plate topped with hemp seeds

One slice bread (100 percent whole grain or sprouted-grain) with Nutritarian Cream Cheese*

LUNCH

Mexican Burrito Bowl*

Apple or other fruit

DINNER

Mac and Peas*

Leafy green vegetable such as bok choy, spinach, or Swiss chard steamed or water-sautéed with onions and mushrooms

No-Bake Brownies* or choice of fruit

DAY 4

BREAKFAST

Teff Porridge with Walnuts and Berries*

LUNCH

Salad with lettuce, shredded cabbage, black beans, tomato, red onion, and Bing Cherry Walnut Vinaigrette*

Steamed or roasted green beans or other green vegetable

DINNER

Broccoli and Snow Pea Stir-Fry with Pineapple and Baked Tofu*

Brown quinoa

Melon or other fruit

DAY 5

BREAKFAST

Oatmeal Raisin Cookie Breakfast Bowl*

Fresh or thawed frozen berries

LUNCH

Salad vegetables with leftover Aquafaba Hummus* or Bing Cherry Walnut Vinaigrette*

Avocado Toast with Shredded Brussels Sprouts*

Pear or other fruit

DINNER

Dr. Fuhrman's Famous Anticancer Soup*

Steamed broccoli or other green vegetable

Orange, clementines, or other citrus fruit

DAY 6

BREAKFAST
Clean Green Juice*

Mini Corn Muffins*

Banana or other fruit

LUNCH
Leftover Dr. Fuhrman's Famous Anticancer Soup*

Salad with mixed greens, watercress, cherry tomatoes, green onions, other vegetables, and Lemon Chia Dressing*

Thawed frozen cherries or other fresh or frozen fruit

DINNER
Veggie-Bean Burgers* with lettuce, tomato, and red onion with half a 100% whole grain or sprouted grain bun or pita (or can be served with a collard or cabbage wrap)

Crispy Onion Rings*

Pistachio Gelato* with blackberries

DAY 7

BREAKFAST
Nutty Collard Fruit Wraps*

Leftover Mini Corn Muffins*

LUNCH
Cannellini Beans and Greens*

Italian Stewed Tomatoes*

Grapes or other fruit

DINNER
Kale, Cabbage, and Mushroom Salad*

Garlic-Infused Quinoa with Tomatoes and Poblano Peppers*

Berries sprinkled with chopped almonds and unsweetened shredded coconut

WEEK 2

DAY 1

BREAKFAST
Apple Chai Steel Cut Oats*

LUNCH
Salad with mixed greens, shredded red cabbage, tomatoes, radishes, other vegetables, and Hemp Seed and Herb Dressing*

Leftover Cannellini Beans and Greens*

Fresh or thawed frozen berries

DINNER
Bean Pasta with Roasted Red Pepper Alfredo*

Steamed asparagus or other green vegetable

Thawed frozen peaches or other fruit

DAY 2

BREAKFAST

One slice 100% whole grain or sprouted grain bread with Whipped Pistachio Butter* or Chocolate Hemp Seed Butter*

Fresh or thawed frozen berries

LUNCH

Roasted Vegetable Salad with Balsamic Vinaigrette*

Pear or other fruit

DINNER

Corn and Red Lentil Chowder*

Salad with leftover Hemp Seed and Herb Dressing*

Cherry Apricot Oatmeal Cookies* or choice of fruit

DAY 3

BREAKFAST

Mango Ginger Smoothie*

Dried unsulfured figs or apricots

LUNCH

Sugar snap peas or salad with Curried Peanut Butter Dressing*

Leftover Corn and Red Lentil Chowder*

Fresh or thawed frozen mango or other fruit

DINNER

Mexican Cauliflower Rice and Beans*

Water-sautéed mushrooms, peppers, and onions

Fresh or thawed frozen berries

DAY 4

BREAKFAST

Butternut Breakfast Soup*

Fresh or thawed frozen berries

LUNCH

Herbed "Cheese" and Greens Wraps*

Leftover Cherry Apricot Oatmeal Cookies*

Apple or other fruit

DINNER

Kale, Chickpea, and Grain Bowl*

Salad with Walnut Vinaigrette Dressing*

Melon or other fruit

DAY 5

BREAKFAST

2 pieces of fruit

¼ cup walnuts or almonds

LUNCH

Salad with lettuce, shredded cabbage, tomatoes, and sautéed mushrooms and leftover Walnut Vinaigrette Dressing*

Five-Seed Crackers* with Cheezy Bean Dip*

Fresh fruit

DINNER

Mushroom and Wheat Berry Soup*

Steamed vegetable such as cauliflower, broccoli, or asparagus with Garlic Nutter*

Orange, clementines, or other citrus fruit

DAY 6

BREAKFAST

Mushroom and Kale Frittata*

Fresh or thawed frozen berries

LUNCH

Leftover Mushroom and Wheat Berry Soup*

Leftover Five-Seed Crackers* and Cheezy Bean Dip*

Fresh or thawed frozen pineapple or other fruit

DINNER

Green Pizza*

Salad with baby greens, arugula, tomatoes, red onion, other vegetables, and White Bean Dressing*

Melon or other fruit

DAY 7

BREAKFAST

Mango Blueberry Crisp*

Bok Choy Ginger Juice*

LUNCH

Broccoli Quiche with Aquafaba*

Salad with mixed greens, shredded red cabbage, tomatoes, beans, and leftover White Bean Dressing*

DINNER

Nutritarian Borscht*

Steamed string beans, chopped with thawed frozen green peas topped with toasted chopped almonds

Dark Chocolate Mousse* with raspberries

WEEK 3

DAY 1

BREAKFAST

Nutritarian Granola*
with berries and
nondairy milk

LUNCH

Salad with romaine,
spinach, red onion,
white beans, orange
segments, sesame
seeds, and Orange
Sesame Dressing*

Sweet and Sour
Cabbage and Split Pea
Soup*

Fresh fruit

DINNER

Chickpea, Greens, and
Farro Stew*

Steamed Brussels
sprouts or other
vegetable tossed with
lemon and walnuts

Kiwi or other fruit

DAY 2

BREAKFAST

Blueberry Chia Soaked
Oats*

LUNCH

Salad with leftover
Orange Sesame
Dressing*

Leftover Chickpea,
Greens, and Farro
Stew*

Fresh fruit

DINNER

Roasted Cauliflower
Tacos* served with
avocado slices and
lime wedges

Vanilla Nice Cream*
with thawed frozen
cherries

DAY 3

BREAKFAST

Mixed fruit salad
topped with hemp
seeds and almonds

LUNCH

Curried Egg-less Salad
with Cashews and
Dried Apricots* served
on 100% whole grain or
sprouted grain bread
or a bed of greens

Sliced tomatoes and
red onion

Thawed frozen cherries

DINNER

Broccoli Fra Diavolo*
with bean pasta

Endive leaves or salad
with Nutritarian Caesar
Dressing*

Pear or other fruit

DAY 4

BREAKFAST

Cherry Chocolate Smoothie*

LUNCH

Salad with romaine, arugula, tomatoes, green onions, and sautéed mushrooms with leftover Nutritarian Caesar Dressing*

Tofu Crackers*

Fresh or thawed frozen berries

DINNER

Black Bean and Butternut Squash Chili*

Steamed broccoli or other green vegetable tossed with flavored vinegar and pine nuts

Apple slices sprinkled with Ceylon cinnamon

DAY 5

BREAKFAST

Banana Pancakes* with blueberries

Vanilla Almond Milk*

Papaya or other fruit

LUNCH

Leftover Black Bean and Butternut Squash Chili*

Kohlrabi Apple Slaw*

Fresh fruit

DINNER

Asian Ginger Lime Zoodles*

Mushrooms and onions water-sautéed with choice of green vegetable

Fresh or thawed frozen pineapple

DAY 6

BREAKFAST

Huevos Rancheros (Mexican Scrambled Tofu)*

Melon or other fruit

LUNCH

Greens and Berries Salad with Raspberry Dressing*

Tofu Crackers*

Fresh fruit

DINNER

Zucchini Ravioli with Cauliflower Chickpea "Ricotta"* with Intense Marinara Sauce*

Carrot Cake Bites*

Grapes or other fruit

DAY 7

BREAKFAST

Cinnamon-Spiced Sweet Potato Breakfast Topped with Pecan Granola and Blueberry Sauce*

Fresh or thawed frozen berries

LUNCH

Sweet and Sour Cabbage and Split Pea Soup*

Steamed asparagus or other vegetable

Fresh fruit

DINNER

Korean Vegetable and Mushroom Lettuce Wraps*

Fresh or thawed frozen mango

DIABETES-REVERSAL MENUS

These menus are a bit lower in grains and fruit; and the desserts have fewer dates and tropical fruits, so more berries, grapefruit, and kiwis are utilized. Other very-low-sugar fruits, besides all berries, include loquats, passion fruit, guava, kumquats, grapefruit, kiwi, and dragon fruit. These recipes also include no white potato or bread.

DAY 1

BREAKFAST

Berries and grapefruit with lemon, topped with ground flaxseeds and crushed walnuts

LUNCH

Herbed "Cheese" and Greens Wraps*

Edamame, Corn, and Tomato Salad with Balsamic Dressing*

DINNER

Mexican Cauliflower Rice and Beans*

Steamed green leafy vegetable such as kale or collard greens

Melon or other fruit

DAY 2

BREAKFAST

Steel cut oatmeal with nondairy milk, berries, and chopped raw almonds

LUNCH

Salad with baby greens, cabbage, tomatoes, red onion, beans, and Creamy Ginger Dressing*

Strawberry Banana Ice Cream*

DINNER

Quinoa or other intact whole grain with steamed zucchini, red peppers, and onion seasoned with Dukkah Spice Blend* or other no-salt seasoning blend

Thawed frozen wild blueberries

Asterisks indicate recipes that are provided in Chapter 9.

DAY 3

BREAKFAST
Clean Green Juice*

Romaine leaves with Whipped Pistachio Butter* or raw almond butter

Fresh or thawed frozen berries

LUNCH
Salad with romaine, broccoli sprouts, grape tomatoes, scallions, and Russian Fig Dressing*

Steamed green beans or other green vegetable

Fresh fruit

DINNER
Mushroom and Wheat Berry Soup*

Steamed broccoli or steamed artichokes

Melon

DAY 4

BREAKFAST
Blueberry Chia Soaked Oats*

LUNCH
Korean Vegetable and Mushroom Lettuce Wraps*

Raw vegetables with Aquafaba Hummus*

Apple or other fruit

DINNER
Nutritarian Borscht*

Five-Seed Crackers*

Orange or clementines

DAY 5

BREAKFAST
Cranberry Smoothie*

Leftover Five-Seed Crackers*

LUNCH
Salad with baby greens, arugula, tomato, shallots, beans, pomegranate arils, and White Bean Dressing*

Steamed cauliflower or other vegetable seasoned with flavored vinegar or a no-salt seasoning blend

DINNER
Broccoli and Snow Pea Stir-Fry with Pineapple and Baked Tofu*

Spiced lentils and quinoa

Kiwi and berry fruit cup

DAY 6

BREAKFAST
Teff Porridge with
Walnuts and Berries*

LUNCH
Edamame and Green
Pea Soup*

Napa Cabbage Slaw*

Fresh or thawed frozen
mango

DINNER
Black Bean and
Butternut Squash Chili*

Bok choy or other
leafy green vegetable
seasoned with flavored
vinegar or no-salt
seasoning

Thawed frozen cherries
or other fruit

DAY 7

BREAKFAST
Mushroom and Kale
Frittata*

Thawed frozen
peaches or partially
thawed frozen jackfruit

LUNCH
Salad with romaine,
spinach, shredded
cabbage, tomatoes,
red onion, and Orange
Sesame Dressing*

Leftover Black Bean
and Butternut Squash
Chili*

DINNER
Broccoli Quiche with
Aquafaba*

Brussels sprouts or
other green vegetable

Vanilla Nice Cream*
with raspberries

EMERGENCY MENUS: TIME-RESTRICTED EATING AND INTERMITTENT FASTING

This is a sample of the more calorie-restricted diet that I use in my medical practice for people with life-threatening heart disease, diabetics who need to lower their blood sugar levels into the normal range quickly, and individuals who need to lower blood pressure rapidly or get rid of severe migraines. It delivers the highest level of nutrient density with fewer calories.

The emergency menus incorporate the concept of intermittent fasting. They include two meals and a vegetable juice each day. A key component of these menus is the incorporation of a restricted eating window to accelerate both the cleansing and repair mechanisms in your cells and weight loss. If possible, consume the two daily meals and the juice within an eight-hour window to ensure a long overnight fast. Aim to have brunch between 8:00 and 11:00 a.m. and dinner between 3:00 and 5:00 p.m. During the fasting periods after dinner and before brunch, drink only water.

Vegetable juices are included because they are an easy way to consume a large amount of vegetables, thereby receiving a hefty dose of phytochemicals without getting too full. This speeds up the time it takes for plant-derived nutrients to get into the cells. In other words, it could take six months to change a low skin carotenoid score to an adequate one, but with juicing the time is halved.

A juice is different from a smoothie. A smoothie retains the fiber (pulp), and juicing removes it. More vegetables can go into a serving of juice, delivering more nutrients that can be comfortably digested. Using lots of raw greens, with raw scallion and onion and cooked mushrooms, greatly enhances the immune benefits too.

I don't use any rice or wild rice anymore because of the potential for arsenic contamination. I also don't use white potato here because of its higher glycemic response. People with conditions that require more calories or more protein can use more dried soybeans, soaked and made into soups, and more hemp seeds.

DAY 1

BRUNCH

Super Immunity Juice*

Salad with romaine, arugula, tomato, sliced red onion, hemp seeds, and Nutritarian Caesar Dressing*

Fresh or thawed frozen berries

DINNER

Cannellini Beans and Greens*

Quinoa, thawed frozen peas, and chopped scallion

Melon or grapefruit

DAY 2

BRUNCH

Super Immunity Juice*

Asian Ginger Lime Zoodles*

Fresh or thawed frozen berries

DINNER

Salad with mixed greens, shredded cabbage, tomato, red onion, and Leftover Nutritarian Caesar Dressing*

Chickpea, Greens, and Farro Stew*

Fresh or thawed frozen mango or other fruit

DAY 3

BRUNCH

Bok Choy Ginger Juice*

Leftover Chickpea, Greens, and Farro Stew*

Melon or other fruit

DINNER

Salad with romaine, arugula, tomato, scallions, orange segments, and Orange Sesame Dressing*

Green beans or other vegetable

Fresh or thawed frozen blackberries and strawberries

DAY 4

BRUNCH

Bok Choy Ginger Juice*

Steel Cut Oats and Cherries*

Walnuts or lightly toasted raw almonds

DINNER

Mexican Burrito Bowl*

Kale or other green leafy vegetable

Fresh or thawed frozen pineapple

Asterisks indicate recipes that are provided in Chapter 9.

DAY 5

BRUNCH

Clean Green Juice*

Herbed "Cheese" and Greens Wraps*

Raw veggies with Aquafaba Hummus*

Apple or other fruit

DINNER

Salad with mixed greens, shredded cabbage, tomato, scallions, and Seedy Ranch Dressing*

Corn and Red Lentil Chowder*

Fresh or thawed frozen berries

DAY 6

BRUNCH

Clean Green Juice*

Salad with romaine, watercress, tomato, red onion, and leftover Seedy Ranch Dressing*

Pear or other fruit

DINNER

Bean Pasta with Roasted Red Pepper Alfredo*

Brussels sprouts or other green vegetable

Kiwis and fresh or thawed frozen red raspberries

DAY 7

BRUNCH

Super Immunity Juice*

Korean Vegetable and Mushroom Lettuce Wraps*

Fresh or thawed frozen berries

DINNER

Raw veggies with Aquafaba Hummus*

Dr. Fuhrman's Famous Anticancer Soup*

Orange or other citrus fruit

The Seven Most Important Dietary Steps to Prevent and Fight Cancer

1. Vegetables (not fruits) should be the largest part of your diet. They should include raw salad vegetables; raw solid vegetables such as broccoli and snow pea pods; and cooked vegetables, which should be steamed lightly, lightly sautéed in water, wokked, or cooked in soups.

2. Broccoli sprouts are the richest natural source of sulforaphane (an isothiocyanate). Baby greens, microgreens, and sprouts have more anticancer phytonutrients compared with the mature version. Use these in salads. Chew all your salads and raw vegetables to a liquid in your mouth before swallowing. Chew open every plant cell in your mouth for maximal absorption.

3. Include in your salads large amounts of green, leafy vegetables with lots of green lettuces and choices from the cruciferous family (cabbage, baby bok choy, Brussels sprouts, red cabbage, collards, turnip greens, Chinese cabbage, arugula, and watercress). You can lightly steam or wok them, too.

4. Drink fresh-squeezed vegetable juice two times a day made with carrots, beets, tomatoes, and greens: kale, collards, bok choy, wheatgrass, and cabbage and lettuces. This provides the broadest spectrum of cancer-fighting nutrients. Only use organic vegetables for juicing.

5. Beans contain powerful cancer-fighting compounds, especially the darker colored and reddish beans. Plus, they are low-glycemic and high in fiber and resistant starch. Use them in a carrot juice-based or tomato juice-based soup with added mushrooms and cruciferous vegetables.

6. Use organic raw fruits, especially those with high antioxidant content and free radical–absorptive capacity, such as all berries, pomegranates, kiwis, kumquats, black grapes, cherries, papayas, and passion fruit. Frozen fruits and vegetables are acceptable.

7. Use only raw, unsalted seeds and nuts and avocado in your diet as your source of fat. Do not use any animal fats or oils. You can lightly toast nuts and seeds in a toaster oven on the light setting.

AUTOIMMUNE DISEASE OR CANCER MENUS

Attack autoimmune disease or cancer with this menu. For people who have serious autoimmune disease, such as rheumatoid arthritis, lupus, or psoriasis, or for those with cancer, we are looking to strengthen the immune system and remove toxic wastes as quickly as possible. I often increase the juicing to an 8- to 10-ounce glass twice a day because the concentration of nutrients from colorful plants in the body's tissues and cells is needed to enhance or normalize immune function.

Even though in autoimmune disease the immune system is overly active, we can calm it down with better phytochemical exposure and removal of toxic excitatory compounds. Plus, enhancing function of T-suppressor cells is essential to normalizing autoimmune activity and reducing the uncontrolled release of antibodies.

Depending on the severity of the case and the amount and type of medications used, I usually start decreasing medications within a month; depending on the case, many people can be off all medications within three to six months.

DAY 1

BREAKFAST

Super Immunity Juice*

Mixed berries topped with a mix of ground flax, chia, and hemp seeds

LUNCH

Salad with mixed greens, arugula, tomatoes, red onion, and Walnut Vinaigrette Dressing*

Steamed bok choy topped with lightly toasted sesame seeds

Orange, clementine, or other citrus fruit

DINNER

Super Immunity Juice*

Raw vegetables with Cheezy Bean Dip*

Dr. Fuhrman's Famous Anticancer Soup*

Fresh or thawed frozen mango or other fruit

Asterisks indicate recipes that are provided in Chapter 9.

DAY 2

BREAKFAST
Clean Green Juice*

Cherry Berry Smoothie Bowl*

LUNCH
Salad with romaine, broccoli sprouts, tomato, scallions, and leftover Walnut Vinaigrette Dressing*

Leftover Dr. Fuhrman's Famous Anticancer Soup*

Melon or other fruit

DINNER
Clean Green Juice*

Artichoke-Stuffed Portabella Mushrooms*

Steamed broccoli or other cruciferous vegetable

Berries with shredded coconut and chopped walnuts

DAY 3

BREAKFAST
Super Green Smoothie*

Nutty Collard Fruit Wraps*

LUNCH
Greens and Berries Salad with Raspberry Dressing*

Edamame with no-salt seasoning

DINNER
Clean Green Juice*

Kale or collard greens water-sautéed with mushrooms, onions, and garlic

Quinoa seasoned with Dukkah Spice Blend*

Fresh fruit

DAY 4

BREAKFAST
Bok Choy Ginger Juice*

Steel cut oatmeal with nondairy milk, berries, and walnuts

LUNCH
Tofu and Sun-Dried Tomato Burgers* on a bed of greens or collard or cabbage wrap topped with sautéed mushrooms, lettuce, tomato, and sliced red onion

Apple slices with Chocolate Hemp Seed Butter*

DINNER
Bok Choy Ginger Juice*

Salad with baby greens, spinach, arugula, tomatoes, and Lemon Chia Dressing*

Soybean and Red Lentil Soup*

Melon

DAY 5

BREAKFAST

Super Immunity Juice*

Berries and other fruit topped with raw nuts and/or seeds

LUNCH

Salad with romaine, watercress, tomatoes, scallions, and leftover Lemon Chia Dressing* or flavored vinegar

Leftover Soybean and Red Lentil Soup*

DINNER

Super Immunity Juice*

Bean pasta with sautéed mushrooms and Intense Marinara Sauce* topped with toasted pine nuts

Steamed Brussels sprouts or other cruciferous vegetable

Thawed frozen cherries or other fruit

DAY 6

BREAKFAST

Super Green Smoothie*

Blueberry Chia Soaked Oats*

LUNCH

Kale, Cabbage, and Mushroom Salad*

Pear or apple

DINNER

Clean Green Juice*

G-BOMB Thai Vegetable Curry*

Pomegranate arils or blackberries

DAY 7

BREAKFAST

Clean Green Juice*

Mushroom and Kale Frittata*

Fresh or thawed frozen berries

LUNCH

Salad with mixed greens, broccoli sprouts, shredded cabbage, tomato, red onion, and Seedy Ranch Dressing*

Edamame and Green Pea Soup*

Grapes or cherries

DINNER

Bok Choy Ginger Juice*

Broccoli Fra Diavolo*

Riced cauliflower with chopped almonds

Orange

Remember, you don't have to make any fancy recipes to make this work; you can use lots of frozen foods, such as frozen artichoke hearts, broccoli florets, asparagus, and peas. Just sprinkle some herbal seasonings or Dukkah Spice Blend on top, and you have a dish. You can use frozen fruits and even boxed Nutritarian soups, jarred sauces, and all-natural oil-free salad dressings (check out DrFuhrman.com) if you have no time to cook.

CHAPTER NINE

The Nutritarian Diet Recipes

Losing weight, feeling strong and healthy, reversing disease, and adding years to your life is well within your grasp. My goal is to provide the why and the how. Chapters 1 through 7 presented the "why"; this chapter of delicious Nutritarian recipes gives you the "how." Your job is to make it happen. I know you can do it!

In the recipes that follow:

A favorite dish—done Nutritarian style.

†*If you like it spicy, you can adjust these recipes to dial up the heat!*

DRINKS AND SMOOTHIES

BREAKFAST

Apple Chai Steel Cut Oats, 282

Banana Pancakes, 282

Blueberry Chia Soaked Oats, 283

Cherry Berry Smoothie Bowl, 283

Cinnamon-Spiced Sweet Potato Breakfast Topped with
Pecan Granola and Blueberry Sauce, 284

Huevos Rancheros (Mexican Scrambled Tofu),*† 285

Mango Blueberry Crisp, 286

Mini Corn Muffins, 286

Mushroom and Kale Frittata, 287

Nutritarian Granola, 288

Nutty Collard Fruit Wraps, 288

Oatmeal Raisin Cookie Breakfast Bowl, 289

Steel Cut Oats and Cherries, 289

Teff Porridge with Walnuts and Berries, 290

SALAD DRESSINGS, DIPS, AND SPREADS

Bing Cherry Walnut Vinaigrette, 291

Creamy Ginger Dressing,* 291

Curried Peanut Butter Dressing,* 292

Hemp Seed and Herb Dressing, 292

Lemon Chia Dressing, 292

Nutritarian Caesar Dressing, 293

Orange Sesame Dressing, 293

Russian Fig Dressing, 294

Seedy Ranch Dressing, 294

Walnut Vinaigrette Dressing, 295

White Bean Dressing, 295

Aquafaba Hummus, 296

Cheezy Bean Dip,† 296

Chocolate Hemp Seed Butter, 297

Garlic Nutter, 297

Nutritarian Cream Cheese, 298

California Creamed Kale, 321

Cannellini Beans and Greens,† 322

Eggplant Meatballs,* 322

Farro and Mushroom Risotto, 323

Garlic-Infused Quinoa with Tomatoes and Poblano Peppers, 324

G-BOMB Thai Vegetable Curry,*† 324

Intense Marinara Sauce, 325

Italian Stewed Tomatoes,* 326

Kale, Chickpea, and Grain Bowl, 326

Mac and Peas, 327

Mexican Burrito Bowls,*† 328

Mexican Cauliflower Rice and Beans,*† 329

Roasted Radishes and Turnips, 330

Zucchini Ravioli with Cauliflower Chickpea "Ricotta",* 331

BURGERS, PIZZA, AND QUICK FOOD

Avocado Toast with Shredded Brussels Sprouts, 332

Crispy Onion Rings, 333

Five-Seed Crackers, 334

Green Pizza,* 335

Herbed "Cheese" and Greens Wraps, 336

Korean Vegetable and Mushroom Lettuce Wraps, 337

Lentil Walnut Burritos with Peppers, Onions, and Salsa,* 338

Portabella Pizza,* 339

Roasted Cauliflower Tacos,*† 340

Tofu and Sun-Dried Tomato Burgers, 340

Tofu Crackers and Tofu Jerky Pizza, 341

Sweet Potato Toast with Roasted Broccoli, 342

Veggie-Bean Burgers, 343

DESSERTS

Almond Blondies, 344

Blueberry Cherry Crumble, 345

Carrot Cake Bites, 346

DRINKS AND SMOOTHIES

Cherry Chocolate Smoothie

SERVES 2

1 cup unsweetened soy, hemp, or almond milk

2 teaspoons unsweetened natural cocoa powder

2 Medjool or 4 regular dates, pitted (see Note)

5 ounces baby kale or a mixture of baby kale and spinach

¼ cup walnuts

1 tablespoon pure vanilla bean powder or alcohol-free vanilla extract

2 cups frozen cherries

1 cup frozen wild blueberries

Blend ingredients in a high-powered blender until smooth. Add water if needed to achieve desired consistency.

Note: For diabetic or weight-loss diets, omit the dates.

PER SERVING: CALORIES 353; PROTEIN 8G; CARBOHYDRATES 63G; SUGARS 43G; TOTAL FAT 12G; SATURATED FAT 1.2G; SODIUM 125MG; FIBER 9.3G; BETA-CAROTENE 6,639MCG; VITAMIN C 102MG; CALCIUM 408MG; IRON 3.3MG; FOLATE 50MCG; MAGNESIUM 98MG; POTASSIUM 945MG; ZINC 1.3MG; SELENIUM 1.7MCG

Cranberry Smoothie

SERVES 2

½ cup fresh or frozen cranberries (see Note)

1 cup blueberries or other berries

1 banana

2 cups chopped kale

1 cup unsweetened soy, hemp, or almond milk

4 walnut halves

1 tablespoon chia seeds

Blend ingredients in a high-powered blender until smooth.

Note: Do not use dried cranberries, as they contain added sugar.

PER SERVING: CALORIES 210; PROTEIN 6G; CARBOHYDRATES 37G; SUGARS 16G; TOTAL FAT 6.6G; SATURATED FAT 0.6G; SODIUM 125MG; FIBER 7.8G; BETA-CAROTENE 6,230MCG; VITAMIN C 96MG; CALCIUM 394MG; IRON 2.5MG; FOLATE 41MCG; MAGNESIUM 77MG; POTASSIUM 627MG; ZINC 1MG; SELENIUM 4.3MCG

Mango Ginger Smoothie

SERVES 2

1 cup frozen mango
1 cup frozen butternut squash
2 cups shredded green cabbage or chopped kale
1 banana
⅛ cup unsweetened shredded coconut

1 lemon, juiced
⅛ cup raw almonds
1 tablespoon ground flaxseed
1 tablespoon chia seed
1 tablespoon fresh ginger, minced
1 teaspoon minced fresh turmeric, (or ¼ teaspoon dried turmeric)

Blend ingredients in high-powered blender until smooth.

PER SERVING: CALORIES 293; PROTEIN 7G; CARBOHYDRATES 47G; SUGARS 24G; TOTAL FAT 11.9G; SATURATED FAT 4.1G; SODIUM 22MG; FIBER 10.9G; BETA-CAROTENE 3,531MCG; VITAMIN C 88MG; CALCIUM 142MG; IRON 2.4MG; FOLATE 111MCG; MAGNESIUM 120MG; POTASSIUM 909MG; ZINC 1.2MG; SELENIUM 6.6MCG

Peachy Green Smoothie

SERVES 1

1 cup frozen peaches
½ frozen banana
1 cup unsweetened soy, hemp, or almond milk
2 cups kale or a mixture of kale and spinach

1 tablespoon chia, hemp, or ground flax seeds
¼ teaspoon vanilla bean powder or alcohol-free vanilla extract
Dash cinnamon (optional)

Blend ingredients in a high-powered blender until smooth.

PER SERVING: CALORIES 299; PROTEIN 16G; CARBOHYDRATES 39G; SUGARS 20G; TOTAL FAT 11G; SATURATED FAT 1.6G; SODIUM 61MG; FIBER 6.4G; BETA-CAROTENE 5,497MCG; VITAMIN C 83MG; CALCIUM 96MG; IRON 3.8MG; FOLATE 89MCG; MAGNESIUM 180MG; POTASSIUM 879MG; ZINC 1.9MG; SELENIUM 1.3MCG

Super Green Smoothie

SERVES 2

2 cups chopped green cruciferous vegetables (kale, collards, mustard greens, or turnip greens)

⅓ cup broccoli sprouts or radish sprouts

6 walnut halves

2 tablespoons hemp seeds

8 ounces frozen strawberries or raspberries

1 cup carrot or beet juice

Squeeze of lemon

Blend ingredients in a high-powered blender until smooth.

PER SERVING: CALORIES 127; PROTEIN 4G; CARBOHYDRATES 29G; SUGARS 10G; TOTAL FAT 0.8G; SATURATED FAT 0.1G; SODIUM 113MG; FIBER 5.2G; BETA-CAROTENE 17,256MCG; VITAMIN C 150MG; CALCIUM 143MG; IRON 2.7MG; FOLATE 52MCG; MAGNESIUM 55MG; POTASSIUM 869MG; ZINC 0.7MG; SELENIUM 2.3MCG

Vanilla Almond Milk

SERVES 6

½ cup raw almonds

½ cup hemp seeds

6 cups water

4 Medjool or 8 regular dates, pitted

2 tablespoons vanilla bean powder or alcohol-free vanilla extract

Blend ingredients in a high-powered blender until smooth.

PER SERVING: CALORIES 191; PROTEIN 7G; CARBOHYDRATES 16G; SUGARS 12G; TOTAL FAT 12.9G; SATURATED FAT 1.1G; SODIUM 13MG; FIBER 2.8G; BETA-CAROTENE 16MCG; CALCIUM 55MG; IRON 1.6MG; FOLATE 23MCG; MAGNESIUM 137MG; POTASSIUM 351MG; ZINC 1.8MG; SELENIUM 0.4MCG

Bok Choy Ginger Juice

SERVES 2

1 small head bok choy (about 8 stalks)

1 cucumber

4 stalks celery

1 green apple, cored

½ lime, peeled

2-inch piece ginger, or to taste

Wash all ingredients. Run ingredients through juicer.

PER SERVING (ESTIMATED; WILL VARY DEPENDING ON JUICER): CALORIES 134; PROTEIN 8G; CARBOHYDRATES 21G; SUGARS 17G; TOTAL FAT 1.3G; SATURATED FAT 0.2G; SODIUM 341MG; BETA-CAROTENE 11,569MCG; VITAMIN C 204MG; CALCIUM 507MG; IRON 4.1MG; FOLATE 320MCG; MAGNESIUM 113MG; POTASSIUM 1,584MG; ZINC 1.3MG; SELENIUM 2.9MCG

Clean Green Juice

SERVES 2

5 kale leaves
6 celery stalks
1 cucumber

1 green apple, cored and quartered
½ lemon, peeled
1-inch piece ginger, or to taste

Wash all ingredients. Run ingredients through juicer.

PER SERVING (ESTIMATED; WILL VARY DEPENDING ON JUICER): CALORIES 131; PROTEIN 7G;
CARBOHYDRATES 27G; SUGARS 12G; TOTAL FAT 1.7G; SATURATED FAT 0.2G; SODIUM 171MG;
BETA-CAROTENE 15,823MCG; VITAMIN C 217MG; CALCIUM 294MG; IRON 3.4MG; FOLATE 109MCG;
MAGNESIUM 88MG; POTASSIUM 1,298MG; ZINC 1.1MG; SELENIUM 2.1MCG

Super Immunity Juice

SERVES 2

6–7 ounces cruciferous leafy greens,
such as 4 stalks bok choy, ⅓ head
cabbage, 10 collard leaves
20 kale leaves

5 carrots and/or 3 beets
2 heads lettuce, 6 stalks celery, or
1 large cucumber

Wash all ingredients. Juice and store in airtight container until ready to use.
Use within 48 hours.

PER SERVING (ESTIMATED; WILL VARY DEPENDING ON JUICER; NUTRITION FACTS BASED
ON A COMBINATION OF BOK CHOY, CARROTS, AND CUCUMBER): CALORIES 65; PROTEIN 5G;
CARBOHYDRATES 13G; SUGARS 9G; TOTAL FAT 0.8G; SATURATED FAT 0.1G; SODIUM 202MG;
BETA-CAROTENE 13,154MCG; VITAMIN C 104MG; CALCIUM 274MG; IRON 2.4MG; FOLATE 166MCG;
MAGNESIUM 70MG; POTASSIUM 1,038MG; ZINC 0.9MG; SELENIUM 1.6MCG

Apple Chai Steel Cut Oats

SERVES 1

If you are short on time in the morning, make these tasty oats the night before and then eat cold or reheat in the morning before serving.

1 cup water
¼ cup steel cut oats
½ cup diced apple
2 tablespoons raisins or dried currants
½ tablespoon ground flaxseeds

½ teaspoon pure vanilla bean powder or alcohol-free vanilla extract
½ teaspoon cinnamon
¼ teaspoon cardamom
⅛ teaspoon cloves
⅛ teaspoon nutmeg

In a saucepan, bring water to a boil and stir in all ingredients. Reduce heat, cover, and simmer 12–15 minutes or until oats are tender and water is absorbed, stirring occasionally.

PER SERVING: CALORIES 283; PROTEIN 8G; CARBOHYDRATES 55G; SUGARS 7G; TOTAL FAT 4.9G; SATURATED FAT 0.3G; SODIUM 17MG; FIBER 9.7G; BETA-CAROTENE 19MCG; VITAMIN C 4MG; CALCIUM 62MG; IRON 2.8MG; FOLATE 6MCG; MAGNESIUM 28MG; POTASSIUM 260MG; ZINC 0.3MG; SELENIUM 1.1MCG

Banana Pancakes

SERVES 3

1 cup old-fashioned rolled oats
⅓ cup unsweetened soy, hemp, or almond milk
2 medium bananas

½ teaspoon pure vanilla bean powder or alcohol-free vanilla extract
Optional add-ins: Blueberries or raisins/currants

Add oats, nondairy milk, bananas, and vanilla to a blender and blend until smooth. Mix in blueberries, raisins, or currants. Wipe or spray a nonstick skillet with a small amount of oil and heat over medium heat. Pour or scoop batter onto skillet using about ¼ cup per pancake. Cook on both sides until golden.

Makes about 6 medium-size pancakes.

PER SERVING: CALORIES 178; PROTEIN 4G; CARBOHYDRATES 36G; SUGARS 10G; TOTAL FAT 2.7G; SATURATED FAT 0.4G; SODIUM 28MG; FIBER 4.7G; BETA-CAROTENE 20MCG; VITAMIN C 7MG; CALCIUM 79MG; IRON 7MG; FOLATE 16MCG; MAGNESIUM 24MG; POTASSIUM 283MG; ZINC 0.1MG; SELENIUM 0.8MCG

Blueberry Chia Soaked Oats

SERVES 1

½ cup old-fashioned oats

2 tablespoons chia seeds

1 cup unsweetened soy, hemp, or almond milk

2 tablespoons raisins or dried currants

½ cup fresh or thawed frozen blueberries (or other fruit)

½ teaspoon cinnamon

Combine the oats, chia seeds, nondairy milk, and raisins. Soak for at least 60 minutes or overnight. Stir in blueberries and cinnamon.

PER SERVING: CALORIES 334; PROTEIN 9G; CARBOHYDRATES 58G; SUGARS 7G; TOTAL FAT 9.4G; SATURATED FAT 0.9G; SODIUM 193MG; FIBER 10.5G; BETA-CAROTENE 24MCG; VITAMIN C 8MG; CALCIUM 590MG; IRON 12.4MG; FOLATE 8MCG; MAGNESIUM 62MG; POTASSIUM 247MG; ZINC 0.8MG; SELENIUM 5.8MCG

Cherry Berry Smoothie Bowl

SERVES 2

1 cup frozen cherries

1 cup frozen blueberries

1 ½ cups chopped kale or baby greens

1 cup unsweetened soy, hemp, or almond milk

2 tablespoons raw cashews

1 tablespoon chia seeds

1 tablespoon unsweetened, natural cocoa powder

1 Medjool or 2 regular dates, pitted

Optional toppings: Cocoa nibs, chopped raw nuts, seeds, unsweetened dried coconut

Blend all ingredients except toppings in a high-powered blender. Transfer to a bowl and add your choice of toppings.

PER SERVING: CALORIES 231; PROTEIN 7G; CARBOHYDRATES 39G; SUGARS 22G; TOTAL FAT 8.3G; SATURATED FAT 1.2G; SODIUM 119MG; FIBER 8.1G; BETA-CAROTENE 5,073MCG; VITAMIN C 64MG; CALCIUM 389MG; IRON 3.3MG; FOLATE 30MCG; MAGNESIUM 99MG; POTASSIUM 564MG; ZINC 1.4MG; SELENIUM 5.4MCG

Cinnamon-Spiced Sweet Potato Breakfast Topped with Pecan Granola and Blueberry Sauce

SERVES 4

FOR THE SWEET POTATO MIXTURE
2 cups baked sweet potatoes in chunks with skins
1 teaspoon cinnamon
1 teaspoon pure vanilla bean powder or alcohol-free vanilla extract
¼ teaspoon allspice
⅛ teaspoon cardamom
1 cup unsweetened soy, hemp, or almond milk
¼ cup unsweetened coconut flakes
2 tablespoons ground flaxseed

FOR THE TOPPING
¾ cup old-fashioned oats
½ cup coarsely chopped pecans
4 regular or 2 Medjool dates, pitted and chopped
1 teaspoon cinnamon
1 teaspoon vanilla extract

FOR THE BLUEBERRY SAUCE
2 Medjool dates (soaked for 1 hour in water, then drained and chopped)
2 cups frozen blueberries
½ cup water

Preheat oven to 350°F. To prepare the sweet potato mixture, blend the baked sweet potatoes, cinnamon, vanilla, allspice, and cardamom until smooth. In a separate bowl, mix nondairy milk, coconut, and ground flaxseed and chill for 10 minutes. Add to the sweet potato mixture and blend.

To prepare the topping, combine the oats, pecans, dates, cinnamon, and vanilla in a medium bowl. Evenly distribute on a parchment-lined baking sheet and bake 10–12 minutes, stirring occasionally.

Prepare the blueberry sauce by blending the dates, blueberries, and water in a medium saucepan. Cook on medium-low heat, stirring frequently until reduced by half. In four small bowls, layer the sweet potato mixture, the topping, and the blueberry sauce. Refrigerate until ready to eat.

Note: For a nice presentation as well as a portable breakfast, layer mixture in 12-ounce Mason jars. Seal and refrigerate.

PER SERVING: CALORIES 389; PROTEIN 7G; CARBOHYDRATES 59G; SUGARS 18G; TOTAL FAT 16.2G; SATURATED FAT 4.4G; SODIUM 92MG; FIBER 12.5G; BETA-CAROTENE 11,555MCG; VITAMIN C 22MG; CALCIUM 222MG; IRON 6MG; FOLATE 15MCG; MAGNESIUM 79MG; POTASSIUM 731MG; ZINC 1.8MG; SELENIUM 2.8MCG

Huevos Rancheros (Mexican Scrambled Tofu)

SERVES 6

FOR THE HUEVOS RANCHEROS
1 large onion, diced
1 red bell pepper, diced
8 ounces mushrooms (cremini, portabella, or button), sliced
½ cup roughly chopped garlic
3 tablespoons white wine or water
1 block (14–16 ounces) extra-firm tofu, drained
⅓ cup nutritional yeast
3 Roma tomatoes
12 ounces chopped spinach
1 ½ cups cooked or 1 (15-ounce) can no-salt-added or low-sodium black beans, drained
1 cup chopped cilantro leaves, plus more for garnish
Squeeze of lemon
Tabasco sauce or cayenne pepper to taste

FOR THE MEXICAN CHILI SAUCE (OR USE A BOTTLE OF LOW-SODIUM SALSA, SUCH AS DR. FUHRMAN'S TEX MEX SALSA)
1 ½ cups chopped Roma tomatoes
½ teaspoon pure chipotle chili powder
1 ½ teaspoons pure ancho chili powder
1 ½ teaspoons dried oregano
2 cloves garlic, peeled
2 teaspoons ground cumin
2 teaspoons lemon juice
1 teaspoon cider vinegar
2 Medjool or 4 regular dates, chopped
Freshly ground black pepper

In a large deep skillet, sauté the onions, peppers, mushrooms, and garlic in the white wine, covered, over medium-high heat for 10 minutes or until the onions are translucent and lightly browned. Crumble in the tofu, add the Mexican Chili Sauce (instructions below) and the nutritional yeast, stir and cook an additional 5 minutes. Add the tomatoes, spinach, black beans, and cilantro and continue cooking until heated through. Finish with a squeeze of lemon juice and adjust heat level by adding Tabasco or cayenne. Serve sprinkled with a little chopped cilantro.

To make the Mexican Chili Sauce, combine all ingredients in a high-powered blender and process until smooth (makes about 1½ cups).

PER SERVING: CALORIES 267; PROTEIN 21G; CARBOHYDRATES 35G; SUGARS 10G; TOTAL FAT 6.1G; SATURATED FAT 0.7G; SODIUM 83MG; FIBER 11.3G; BETA-CAROTENE 4,233MCG; VITAMIN C 63MG; CALCIUM 268MG; IRON 5.9MG; FOLATE 233MCG; MAGNESIUM 157MG; POTASSIUM 1,173MG; ZINC 4.4MG; SELENIUM 20.2MCG

Mango Blueberry Crisp

SERVES 6

This fruity recipe makes a great breakfast or dessert.

6 cups frozen mango chunks
2 cups frozen blueberries
¾ cup walnuts or pecans
¼ cup old-fashioned rolled oats

¼ cup raisins or dried currants
¼ cup water
½ teaspoon cinnamon
¼ teaspoon nutmeg

Preheat oven to 350°F. Line a 9-by-13-inch pan with parchment paper. Add the mangoes and spread them evenly. Sprinkle the blueberries over the mangoes. In a food processor, chop and mix the remaining ingredients. Sprinkle evenly over the top. Bake uncovered for 20 minutes.

PER SERVING: CALORIES 238; PROTEIN 4G; CARBOHYDRATES 40G; SUGARS 27G; TOTAL FAT 9.4G; SATURATED FAT 1G; SODIUM 5MG; FIBER 5.8G; BETA-CAROTENE 1,072MCG; VITAMIN C 62MG; CALCIUM 39MG; IRON 1.7MG; FOLATE 87MCG; MAGNESIUM 41MG; POTASSIUM 411MG; ZINC 0.6MG; SELENIUM 1.7MCG

Mini Corn Muffins

SERVES 12

1 cup unsweetened soy, hemp, or almond milk
1 tablespoon chia seeds
1 teaspoon apple cider vinegar
1 cup cooked chickpeas
¾ cup cornmeal

1 ½ tablespoons nutritional yeast
1 tablespoon low-sodium baking powder
3 Medjool dates or 6 regular dates, pitted
¼ cup currants

Preheat oven to 350°F. Combine nondairy milk, chia seeds, and vinegar and let stand for 10 minutes. Meanwhile, in a high-powered blender, blend chickpeas, cornmeal, nutritional yeast, and baking powder. Place in a bowl. Add the milk / chia seed mixture to the blender along with the dates and blend until smooth. Add to the bowl with the chickpeas and cornmeal and mix thoroughly. Stir in currants. Line a mini muffin pan with paper liners or lightly spray with cooking spray and dust with cornmeal. Divide batter among muffin cups. Bake 25 minutes. Cool on a cooling rack. Makes 24 muffins.

PER SERVING: CALORIES 84; PROTEIN 3G; CARBOHYDRATES 16G; TOTAL FAT 1.4G; SATURATED FAT 0.2G; SODIUM 8MG; FIBER 2.6G; BETA-CAROTENE 15MCG; CALCIUM 72MG; IRON 1MG; FOLATE 30MCG; MAGNESIUM 29MG; ZINC 0.6MG; SELENIUM 2.2MCG

Mushroom and Kale Frittata

SERVES 6

3 tablespoons ground flaxseeds
¾ cup water
16 ounces mushrooms, chopped
1 cup chopped kale
3 green onions, chopped
1 leek, thinly sliced crosswise
¼ cup unsulfured, unsalted sun-dried tomatoes

¼ cup nutritional yeast
2 tablespoons garbanzo bean flour
1 tablespoon no-salt Italian seasoning
1 tablespoon garlic powder
1 tomato, chopped
1 tablespoon fresh oregano

Preheat oven to 350°F. Whisk together ground flaxseeds and water in a medium bowl. Let sit for 10 minutes. Add mushrooms, kale, green onions, leek, sun-dried tomatoes, nutritional yeast, garbanzo bean flour, Italian seasoning, and garlic powder and mix well. Scoop frittata mixture into lightly oiled muffin tins and top with chopped tomatoes and oregano. Bake 30–35 minutes until starting to set. Remove from oven, loosen sides with a knife, and cool for 10 minutes. Remove from tins with a spoon and place on a cooling rack. They will continue to firm up as they cool.

Note: You can also make this frittata in a medium baking dish. Add an additional 10 minutes to baking time.

PER SERVING: CALORIES 127; PROTEIN 26G; CARBOHYDRATES 68G; SUGARS 5G; TOTAL FAT 3.1G; SATURATED FAT 0.3G; SODIUM 28MG; FIBER 5.9G; BETA-CAROTENE 1,440MCG; VITAMIN C 22MG; CALCIUM 55MG; IRON 34.9MG; FOLATE 42MCG; MAGNESIUM 49MG; POTASSIUM 505MG; ZINC 1.9MG; SELENIUM 8.9MCG

Nutritarian Granola

SERVES 10

½ cup raw almond or cashew butter

1 medium apple, peeled and quartered

1 ripe banana

1 ½ teaspoons ground cinnamon

¼ teaspoon ground nutmeg

1 ½ teaspoons pure vanilla bean powder

4 cups old-fashioned rolled oats

1 cup chopped raw walnuts or pecans

½ cup raw pumpkin seeds

¼ cup unhulled sesame seeds

⅓ cup unsweetened shredded coconut

1 cup dried currants

Preheat oven to 225°F. Place the nut butter, apple, banana, cinnamon, nutmeg, and vanilla in a high-powered blender and blend until smooth and creamy. In a large bowl, mix the oats, nuts, seeds, and coconut. Add the blended mixture and toss to combine. Transfer mixture to two parchment-lined baking pans and spread out thinly over the pans so the granola can bake evenly. Bake 20 minutes, stirring occasionally. After baking, stir in currants.

Allow to cool, then store in an airtight container. Enjoy with nondairy milk and fruit, or for an easy on-the-run breakfast, add a scoop to a bowl of berries.

PER SERVING: CALORIES 403; PROTEIN 11G; CARBOHYDRATES 45G; SUGARS 14G; TOTAL FAT 22.9G; SATURATED FAT 4.8G; SODIUM 5MG; FIBER 7.7G; BETA-CAROTENE 18MCG; VITAMIN C 3MG; CALCIUM 69MG; IRON 10.6MG; FOLATE 23MCG; MAGNESIUM 109MG; POTASSIUM 388MG; ZINC 2.1MG; SELENIUM 4.5MCG

Nutty Collard Fruit Wraps

SERVES 1

1 collard leaf, washed and dried

1 tablespoon raw nut or seed butter (almond, cashew, sunflower)

Sprinkle of rolled oats

Sprinkle of raisins / dried currants or blueberries

Sprinkle of cinnamon and/or nutmeg

½ banana or several apple slices (depending on the size of the collard leaf)

Lay the collard leaf flat and remove the bottom half of the stem. Thinly spread the leaf with nut or seed butter and sprinkle with oats, raisins, and cinnamon. Cut the banana lengthwise to make it thinner and add to the leaf, or use apple slices. Fold the short end of the collard leaf over the filling and tightly roll up the leaf until you reach the end. Lay the wrap seam side down on a plate.

PER SERVING: CALORIES 198; PROTEIN 5G; CARBOHYDRATES 27G; SUGARS 8G; TOTAL FAT 9.5G; SATURATED FAT 0.8G; SODIUM 11MG; FIBER 5.4G; BETA-CAROTENE 1,399MCG; VITAMIN C 18MG; CALCIUM 115MG; IRON 1.7MG; FOLATE 80MCG; MAGNESIUM 67MG; POTASSIUM 467MG; ZINC 0.7MG; SELENIUM 1.5MCG

Oatmeal Raisin Cookie Breakfast Bowl

SERVES 1

½ cup old-fashioned oats

1 cup unsweetened soy, hemp, or almond milk

½ frozen banana (see Note)

1 tablespoon no-oil-added, unsalted peanut butter

2 tablespoons raisins

1 tablespoon chia seeds

½ teaspoon pure vanilla bean powder or alcohol-free vanilla extract

½ teaspoon cinnamon

Optional toppings: Chopped walnuts or pecans, raspberries, blueberries, unsweetened dried coconut

Blend all ingredients except toppings in a high-powered blender. Transfer to a bowl and add your choice of toppings.

Note: Peel bananas and seal in a plastic bag before freezing.

PER SERVING: CALORIES 393; PROTEIN 12G; CARBOHYDRATES 57G; SUGARS 19G; TOTAL FAT 15.8G; SATURATED FAT 1.8G; SODIUM 193MG; FIBER 10.4G; BETA-CAROTENE 17MCG; VITAMIN C 6MG; CALCIUM 628MG; IRON 3.8MG; FOLATE 47MCG; MAGNESIUM 140MG; POTASSIUM 594MG; ZINC 2.3MG; SELENIUM 15.3MCG

Steel Cut Oats and Cherries

SERVES 2

2 cups water

½ cup steel cut oats

2 Medjool or 4 regular dates, pitted, soaked in ¼ cup hot water for ½ hour

1 cup frozen cherries, thawed

1 tablespoon ground flaxseeds

1 teaspoon ground cinnamon

1 teaspoon pure vanilla bean powder or alcohol-free vanilla extract

In a saucepan, bring water to a boil and stir in oats. Reduce heat, cover and simmer 12–15 minutes or until oats are tender and water is absorbed, stirring occasionally. Meanwhile, in a high-powered blender, purée the dates, soaking water, cherries, flaxseeds, cinnamon, and vanilla until very smooth. Pour into the oats about 5 minutes before the end of the cooking time.

Note: If desired, top with additional chopped thawed frozen cherries.

PER SERVING: CALORIES 295; PROTEIN 9G; CARBOHYDRATES 58G; SUGARS 23G; TOTAL FAT 4.9G; SATURATED FAT 0.2G; SODIUM 12MG; FIBER 9.5G; BETA-CAROTENE 427MCG; VITAMIN C 1MG; CALCIUM 74MG; IRON 2.7MG; FOLATE 11MCG; MAGNESIUM 37MG; POTASSIUM 297MG; ZINC 0.4MG; SELENIUM 0.9MCG

Teff Porridge with Walnuts and Berries

SERVES 4

Teff is a tiny whole grain with a mild, nutty flavor that is popular in Ethiopian cooking.

1 cup teff

1 ½ cups unsweetened soy, hemp, or almond milk

1 ½ cups water

½ cup chopped, pitted dates

½ teaspoon cinnamon

¼ cup chopped walnuts or pecans

½ cup berries

Place the teff in a medium saucepan and lightly toast until kernels start to pop, stirring constantly. Add nondairy milk and water to the saucepan and bring to a boil. Reduce heat, cover and simmer for 10 minutes. Stir in the dates and cinnamon and continue cooking, stirring frequently to prevent sticking, for another 5 to 10 minutes until water is absorbed and teff is soft. Remove from heat and stir in nuts and fruit. To reheat, add additional nondairy milk if needed to achieve desired consistency.

PER SERVING: CALORIES 303; PROTEIN 9G; CARBOHYDRATES 54G; SUGARS 15G; TOTAL FAT 7.1G; SATURATED FAT 0.7G; SODIUM 80MG; FIBER 6.4G; BETA-CAROTENE 11MCG; VITAMIN C 2MG; CALCIUM 302MG; IRON 4.5MG; FOLATE 13MCG; MAGNESIUM 117MG; POTASSIUM 374MG; ZINC 2.1MG; SELENIUM 3.1MCG

SALAD DRESSINGS, DIPS, AND SPREADS

Bing Cherry Walnut Vinaigrette

SERVES 8

3 cups fresh pitted or thawed frozen cherries

1 cup tart cherry juice or pomegranate juice

¾ cup walnuts

2 tablespoons ground chia seeds

½ cup Dr. Fuhrman's Pomegranate Balsamic Vinegar or other balsamic vinegar

¼ cup unsweetened dried cherries, or 6 regular or 3 Medjool dates, pitted

Blend all ingredients in a high-powered blender until smooth and creamy, adding more cherry juice if needed to facilitate blending.

PER SERVING: CALORIES 156; PROTEIN 3G; CARBOHYDRATES 19G; SUGARS 12G; TOTAL FAT 8.3G; SATURATED FAT 0.8G; SODIUM 8MG; FIBER 2.8G; BETA-CAROTENE 305MCG; VITAMIN C 2MG; CALCIUM 42MG; IRON 1MG; FOLATE 21MCG; MAGNESIUM 35MG; POTASSIUM 215MG; ZINC 0.6MG; SELENIUM 2MCG

Creamy Ginger Dressing

SERVES 2

¼ cup unsweetened almond milk

1 teaspoon finely grated ginger

1 tablespoon reduced-sodium miso

2 tablespoons unhulled sesame seeds

1 tablespoon rice vinegar

Blend ingredients until smooth.

PER SERVING: CALORIES 67; PROTEIN 2G; CARBOHYDRATES 4G; TOTAL FAT 5.1G; SATURATED FAT 0.7G; SODIUM 185MG; FIBER 1.3G; BETA-CAROTENE 3MCG; CALCIUM 151MG; IRON 1.5MG; FOLATE 10MCG; MAGNESIUM 37MG; POTASSIUM 83MG; ZINC 0.8MG; SELENIUM 3MCG

Curried Peanut Butter Dressing

SERVES 2

2 tablespoons no-oil-added, unsalted peanut butter

1 teaspoon curry powder

½ teaspoon coconut aminos

1 teaspoon lime juice

2 teaspoons rice vinegar

3 tablespoons warm water

Combine all ingredients with a whisk. Add more water if needed to achieve a creamy consistency.

PER SERVING: CALORIES 96; PROTEIN 4G; CARBOHYDRATES 4G; SUGARS 1G; TOTAL FAT 7.8G; SATURATED FAT 1.1G; SODIUM 58MG; FIBER 1.6G; BETA-CAROTENE 7MCG; VITAMIN C 1MG; CALCIUM 15MG; IRON 0.7MG; FOLATE 24MCG; MAGNESIUM 31MG; POTASSIUM 124MG; ZINC 0.6MG; SELENIUM 1.3MCG

Hemp Seed and Herb Dressing

SERVES 4

¼ cup hemp seeds

¼ cup raw almonds

½ cup water

2 tablespoons Dr. Fuhrman's Riesling Reserve Vinegar or apple cider vinegar

1 Medjool or 2 regular dates, pitted

1 clove garlic

¼ teaspoon dried oregano

¼ teaspoon dried basil

¼ teaspoon black pepper

Blend ingredients in a high-powered blender until smooth and creamy. Add additional water as needed to adjust consistency.

PER SERVING: CALORIES 121; PROTEIN 5G; CARBOHYDRATES 6G; SUGARS 3G; TOTAL FAT 8.8G; SATURATED FAT 0.8G; SODIUM 2MG; FIBER 2.5G; BETA-CAROTENE 2MCG; CALCIUM 37MG; IRON 1.7MG; FOLATE 5MCG; MAGNESIUM 87MG; POTASSIUM 93MG; ZINC 1.3MG; SELENIUM 0.5MCG

Lemon Chia Dressing

SERVES 2

½ cup water

2 Medjool or 4 regular dates, pitted

3 tablespoons raw cashews

2 tablespoons chia seeds

3 tablespoons Dr. Fuhrman's Lemon Basil Vinegar or fresh lemon juice

Blend all ingredients until smooth.

PER SERVING: CALORIES 192; PROTEIN 5G; CARBOHYDRATES 28G; SUGARS 17G; TOTAL FAT 8.8G; SATURATED FAT 1.3G; SODIUM 6MG; FIBER 5.6G; BETA-CAROTENE 22MCG; VITAMIN C 9MG; CALCIUM 88MG; IRON 1.9MG; FOLATE 11MCG; MAGNESIUM 86MG; POTASSIUM 316MG; ZINC 1.3MG; SELENIUM 8.2MCG

Nutritarian Caesar Dressing

SERVES 6

⅓ cup raw cashews

2 tablespoons hemp seeds

6 ounces firm silken tofu

3 large cloves garlic

2 medium celery stalks, chopped

½ cup water

¼ cup freshly squeezed lemon juice

½ teaspoon low-sodium white miso

2 teaspoons Dijon mustard

4 regular or 2 Medjool dates, pitted

1 teaspoon kelp granules

2 tablespoons nutritional yeast

Freshly ground pepper to taste

Blend all ingredients in a high-powered blender until smooth, adding some nondairy milk if needed to adjust consistency. Taste and adjust seasonings and refrigerate until ready to use. Makes about 2½ cups.

PER SERVING: CALORIES 91; PROTEIN 5G; CARBOHYDRATES 9G; SUGARS 4G; TOTAL FAT 4.3G; SATURATED FAT 0.8G; SODIUM 62MG; FIBER 1.6G; BETA-CAROTENE 37MCG; VITAMIN C 5MG; CALCIUM 27MG; IRON 1.1MG; FOLATE 10MCG; MAGNESIUM 39MG; POTASSIUM 190MG; ZINC 1.2MG; SELENIUM 2.5MCG

Orange Sesame Dressing

SERVES 4

6 tablespoons unhulled sesame seeds, divided

2 navel oranges, peeled

¼ cup Dr. Fuhrman's Blood Orange Vinegar or white wine vinegar

¼ cup raw cashews

Lightly toast the sesame seeds in a dry skillet over medium-high heat for about 3 minutes, shaking the pan frequently. In a high-powered blender, combine oranges, vinegar, cashews, and 4 tablespoons of the sesame seeds. Toss with the salad, sprinkling remaining sesame seeds on top of the salad.

PER SERVING: CALORIES 162; PROTEIN 5G; CARBOHYDRATES 15G; SUGARS 7G; TOTAL FAT 10.5G; SATURATED FAT 1.6G; SODIUM 4MG; FIBER 3.4G; BETA-CAROTENE 61MCG; VITAMIN C 41MG; CALCIUM 165MG; IRON 3MG; FOLATE 39MCG; MAGNESIUM 81MG; POTASSIUM 246MG; ZINC 1.6MG; SELENIUM 6.4MCG

Russian Fig Dressing

SERVES 4

½ cup no-salt-added or low-sodium pasta sauce
⅓ cup raw almonds
2 tablespoons raw sunflower seeds

3 tablespoons Dr. Fuhrman's Black Fig Vinegar or balsamic vinegar
1 tablespoon raisins

Blend all ingredients in a high-powered blender until smooth.

PER SERVING: CALORIES 127; PROTEIN 4G; CARBOHYDRATES 10G; SUGARS 5G; TOTAL FAT 8.6G; SATURATED FAT 0.7G; SODIUM 13MG; FIBER 2.5G; BETA-CAROTENE 127MCG; VITAMIN C 0.7MG; CALCIUM 48MG; IRON 1.1MG; FOLATE 20MCG; MAGNESIUM 54MG; POTASSIUM 243MG; ZINC 0.67MG; SELENIUM 3MCG

Seedy Ranch Dressing

SERVES 4

½ cup hemp seeds
¼ cup raw cashews
½ cup unsweetened soy, hemp, or almond milk
3 tablespoons fresh lemon juice
1 ½ tablespoons nutritional yeast

1 teaspoon coconut aminos
1 small clove garlic
¼ teaspoon black pepper
½ teaspoon dried parsley
½ teaspoon dried dill

Blend all ingredients except parsley and dill in a high-powered blender until creamy and smooth. Add additional nondairy milk if needed to adjust consistency. Add parsley and dill and pulse for just a few seconds to combine.

PER SERVING: CALORIES 182; PROTEIN 10G; CARBOHYDRATES 7G; SUGARS 1G; TOTAL FAT 14G; SATURATED FAT 1.6G; SODIUM 83MG; FIBER 2G; BETA-CAROTENE 50MCG; VITAMIN C 6MG; CALCIUM 89MG; IRON 2.5MG; FOLATE 29MCG; MAGNESIUM 174MG; POTASSIUM 319MG; ZINC 3.3MG; SELENIUM 1.8MCG

Walnut Vinaigrette Dressing

SERVES 4

¼ cup balsamic vinegar
½ cup water
¼ cup walnuts
¼ cup raisins

1 teaspoon Dijon mustard
1 clove garlic
¼ teaspoon dried thyme

Blend all ingredients in a high-powered blender until smooth.

PER SERVING: CALORIES 84; PROTEIN 1G; CARBOHYDRATES 11G; SUGARS 8G; TOTAL FAT 4.2G; SATURATED FAT 0.4G; SODIUM 21MG; FIBER 0.8G; BETA-CAROTENE 3MCG; VITAMIN C 1MG; CALCIUM 20MG; IRON 0.6MG; FOLATE 7MCG; MAGNESIUM 16MG; POTASSIUM 119MG; ZINC 0.3MG; SELENIUM 0.9MCG

White Bean Dressing

SERVES 6

This works great as a salad dressing or a sauce for cooked vegetables.

1 (15-ounce) can no-salt-added or low-sodium great northern or other white bean, undrained
2 tablespoons hemp seeds
2 tablespoons lemon juice
1 tablespoon rice vinegar

1 ½ teaspoons onion powder
½ teaspoon garlic powder
1 teaspoon dried basil
1 teaspoon dried oregano
1 teaspoon reduced-sodium white miso paste

Blend ingredients in a high-powered blender until smooth. Add water if needed to adjust consistency.

PER SERVING: CALORIES 89; PROTEIN 5G; CARBOHYDRATES 13G; TOTAL FAT 2G; SATURATED FAT 0.2G; SODIUM 27MG; FIBER 5.1G; BETA-CAROTENE 3MCG; VITAMIN C 2MG; CALCIUM 47MG; IRON 1.9MG; FOLATE 68MCG; MAGNESIUM 58MG; POTASSIUM 273MG; ZINC 0.9MG; SELENIUM 0.8MCG

Aquafaba Hummus

SERVES 6

Aquafaba ("water-bean") is the starchy liquid found in canned beans or the liquid left over from cooking your own. It acts as a great binder in recipes.

1 clove garlic

1 (15-ounce) can no-salt-added or low-sodium chickpeas, drained, with liquid reserved

3 tablespoons fresh lemon juice

¼ cup unhulled sesame seeds

1 bulb roasted garlic

½ teaspoon chili powder

½ teaspoon cumin

Roast the bulb of garlic at 300°F for 20 minutes. Squeeze out the soft, cooked garlic cloves from the bulb and blend all ingredients along with ¼ cup of the reserved chickpea liquid (called aquafaba) in a high-powered blender until very smooth and creamy. Add additional liquid if needed to adjust consistency.

PER SERVING: CALORIES 114; PROTEIN 5G; CARBOHYDRATES 16G; SUGARS 2G; TOTAL FAT 4.2G; SATURATED FAT 0.5G; SODIUM 9MG; FIBER 4.1G; BETA-CAROTENE 42MCG; VITAMIN C 6MG; CALCIUM 93MG; IRON 2.3MG; FOLATE 78MCG; MAGNESIUM 44MG; POTASSIUM 189MG; ZINC 1.2MG; SELENIUM 4.6MCG

Cheezy Bean Dip

SERVES 4

1 ½ cups cooked or 1 (15-ounce) can no-salt-added or low-sodium white beans, drained

2 tablespoons unhulled sesame seeds

¼ cup nutritional yeast

1 small clove garlic

1 tablespoon Dijon mustard

1 tablespoon apple cider vinegar

1 tablespoon water

½ teaspoon smoked paprika

Pinch cayenne pepper

Blend ingredients in a high-powered blender until smooth. Add additional water if needed to adjust consistency.

PER SERVING: CALORIES 154; PROTEIN 11G; CARBOHYDRATES 21G; TOTAL FAT 3G; SATURATED FAT 0.5G; SODIUM 52MG; FIBER 7G; BETA-CAROTENE 80MCG; CALCIUM 114MG; IRON 3.6MG; FOLATE 59MCG; MAGNESIUM 71MG; POTASSIUM 416MG; ZINC 2.9MG; SELENIUM 3.8MCG

Chocolate Hemp Seed Butter

SERVES 8

1 cup hemp seeds
¼ cup unsweetened soy, hemp, or
almond milk

1 ½ tablespoons unsweetened natural
cocoa powder
3 Medjool or 6 regular dates, pitted

Blend ingredients in a high-powered blender until smooth and creamy. Add additional nondairy milk if needed to adjust consistency.

PER SERVING: CALORIES 139; PROTEIN 7G; CARBOHYDRATES 9G; SUGARS 6G; TOTAL FAT 10G; SATURATED FAT 1G; SODIUM 7MG; FIBER 1.7G; BETA-CAROTENE 9MCG; CALCIUM 37MG; IRON 1.8MG; FOLATE 24MCG; MAGNESIUM 150MG; POTASSIUM 318MG; ZINC 2.1MG; SELENIUM 0.1MCG

Garlic Nutter

SERVES 8

Use to season cooked vegetables or add extra flavor to soups and sauces. Spread it on a wrap or pita sandwich. Make a salad dressing by adding tomato sauce, vinegar, and some basil.

2 bulbs garlic
¾ cup raw cashews
¼ cup hemp seeds

⅓ cup water or unsweetened
nondairy milk
1 tablespoon nutritional yeast

Preheat oven to 350°F. Roast garlic in a small baking dish for about 25 minutes or until soft. Cool, and then squeeze roasted garlic from skins. Blend garlic and remaining ingredients in a high-powered blender until smooth.

To make Garlic Dijon Dressing: Add balsamic vinegar, Dijon mustard, and additional water or nondairy milk to achieve desired consistency.

To make Garlic Marinara Dressing: Add low-sodium pasta sauce and balsamic or fig vinegar.

To make Pesto: Add 3 cups of basil, 1 clove raw garlic, and lemon juice.

PER SERVING: CALORIES 122; PROTEIN 5G; CARBOHYDRATES 9G; SUGARS 1G; TOTAL FAT 8.5G; SATURATED FAT 1.4G; SODIUM 5MG; FIBER 1G; BETA-CAROTENE 1MCG; VITAMIN C 4MG; CALCIUM 31MG; IRON 1.4MG; FOLATE 15MCG; MAGNESIUM 73MG; POTASSIUM 178MG; ZINC 1.5MG; SELENIUM 3.1MCG

Nutritarian Cream Cheese

SERVES 6

2 tablespoons agar flakes (see Note)
1 cup water, divided
½ cup raw cashews
1 tablespoon hemp seeds
2 tablespoons nutritional yeast

1 tablespoon lemon juice
1 teaspoon arrowroot powder
½ teaspoon reduced-sodium miso
¼ teaspoon garlic powder
¼ teaspoon onion powder

Mix agar flakes and ½ cup of the water in a small pan. Slowly bring to a boil and simmer until flakes dissolve, about 3–5 minutes. Remove from heat. Blend remaining ½ cup water and all other ingredients in a high-powered blender until smooth. Add agar mixture and blend again until well-dispersed. Refrigerate 1–2 hours until firm.

Note: If using agar powder instead of flakes, reduce amount to 2 teaspoons. To make without agar, use only enough water to achieve desired consistency, about 2 tablespoons. It will not firm up or "gel."

PER SERVING: CALORIES 76; PROTEIN 3G; CARBOHYDRATES 5G; SUGARS 1G; TOTAL FAT 5.1G; SATURATED FAT 0.9G; SODIUM 13MG; FIBER 1G; VITAMIN C 1MG; CALCIUM 11MG; IRON 1MG; FOLATE 8MCG; MAGNESIUM 41MG; POTASSIUM 88MG; ZINC 1.2MG; SELENIUM 2.4MCG

Whipped Pistachio Butter

SERVES 8

1 cup raw unsalted, shelled pistachios

¼–½ cup coconut water or more as needed to facilitate blending

In a high-powered blender, blend the pistachios and just enough coconut water so that the mixture moves. Blend on high speed until smooth and fluffy. Refrigerate for several hours so it can thicken.

PER SERVING: CALORIES 88; PROTEIN 3G; CARBOHYDRATES 5G; SUGARS 1G; TOTAL FAT 7G; SATURATED FAT 0.9G; SODIUM 8MG; FIBER 1.7G; BETA-CAROTENE 38MCG; VITAMIN C 1MG; CALCIUM 18MG; IRON 0.6MG; FOLATE 8MCG; MAGNESIUM 20MG; POTASSIUM 176MG; ZINC 0.3MG; SELENIUM 1.2MCG

Dukkah Spice Blend

SERVES 13

Dukkah is a Middle Eastern seasoning made with nuts, seeds, and spices. Sprinkle it on salads, soups, or vegetables for great flavor and extra crunch.

½ cup hazelnuts
¼ cup blanched almonds
3 tablespoons sunflower seeds
1 tablespoon fennel seeds
3 tablespoons coriander seeds
3 tablespoons unhulled white sesame seeds

1 tablespoon cumin
1 tablespoon paprika
1 tablespoon turmeric
2 tablespoons unhulled black sesame seeds

In a food processor or blender pulse chop the hazelnuts and almonds to small pieces. In a pan over low heat, lightly toast the hazelnuts, almonds, and sunflower seeds (about 3 minutes), stirring occasionally. Place in a blender. Add the fennel, coriander, and white sesame seeds to the same pan and lightly toast (about 2 minutes), stirring occasionally. Add to the blender and pulse a few times until mixture is chopped a bit more, but not a powder. Add cumin, paprika, turmeric, and black sesame seeds and pulse a few more times to combine. Store refrigerated in an airtight container.

PER SERVING: CALORIES 91; PROTEIN 3G; CARBOHYDRATES 4G; TOTAL FAT 7.9G; SATURATED FAT 0.7G; SODIUM 3MG; FIBER 2.4G; BETA-CAROTENE 142MCG; VITAMIN C 1MG; CALCIUM 68MG; IRON 1.8MG; FOLATE 16MCG; MAGNESIUM 44MG; POTASSIUM 138MG; ZINC 0.7MG; SELENIUM 2.9MCG

Curried Egg-less Salad with Cashews and Dried Apricots

SERVES 6

1 pound extra-firm tofu, drained and pressed (see Note)

1 (12.3-ounce) package firm silken tofu

1 tablespoon curry powder

2 tablespoons lemon juice

1 orange, peeled

Pinch cayenne pepper

½ cup chopped unsulfured dried apricots, divided

¼ cup minced celery

¼ cup chopped scallions

¼ cup minced red pepper

½ cup currants

¼ cup chopped cilantro

¼ cup minced English cucumber

1 cup chopped romaine

½ cup chopped cashews (toast on lowest setting in toaster oven for 3 minutes)

6 slices (100% whole grain) bread or pitas (optional)

While the extra-firm tofu is being pressed, purée the silken tofu, curry powder, lemon juice, orange, cayenne, and ¼ cup of the dried apricots in a high-powered blender until smooth. Crumble the pressed tofu into a large mixing bowl and stir in the purée. Add the celery, scallions, red pepper, currants, cilantro, cucumber, and remaining ¼ cup apricots and mix well. Cover and refrigerate for at least an hour to let the flavors mingle and the apricots soften more. Serve on a bed of chopped romaine, with the cashews sprinkled on top.

Note: Wrap tofu in paper towels, place something heavy on top, and let sit for 30 minutes to remove excess moisture.

PER SERVING: CALORIES 361; PROTEIN 19G; CARBOHYDRATES 50G; SUGARS 19G; TOTAL FAT 12G; SATURATED FAT 1.8G; SODIUM 148MG; FIBER 9.1G; BETA-CAROTENE 4,819MCG; VITAMIN C 23MG; CALCIUM 140MG; IRON 5.5MG; FOLATE 89MCG; MAGNESIUM 77MG; POTASSIUM 704MG; ZINC 1.4MG; SELENIUM 3.3MCG

Edamame, Corn, and Tomato Salad with Balsamic Dressing

SERVES 4

2 cups frozen edamame, thawed

1 cup frozen corn kernels, thawed

1 cup halved cherry or grape tomatoes

½ cup chopped red bell pepper

½ cup finely chopped red onion

2 tablespoons chopped parsley

2 tablespoons balsamic vinegar

2 tablespoons rice vinegar

1 tablespoon low-sodium ketchup

1 teaspoon stone-ground mustard

In a large bowl, combine edamame, corn, tomatoes, pepper, onion, and parsley. Whisk together balsamic and rice vinegar, ketchup, and mustard. Toss vinegar mixture with salad. Refrigerate for 1 hour to blend flavors. Toss again before serving.

PER SERVING: CALORIES 169; PROTEIN 11G; CARBOHYDRATES 24G; SUGARS 8G; TOTAL FAT 4.6G; SATURATED FAT 0.6G; SODIUM 29MG; FIBER 6.5G; BETA-CAROTENE 612MCG; VITAMIN C 41MG; CALCIUM 67MG; IRON 2.5MG; FOLATE 279MCG; MAGNESIUM 75MG; POTASSIUM 657MG; ZINC 1.5MG; SELENIUM 0.9MCG

Greens and Berries Salad with Raspberry Dressing

SERVES 2

5 cups mixed baby greens

5 cups chopped romaine lettuce

1 cup halved cherry tomatoes

¼ small red onion, sliced

¼ cup shredded carrots

1 cup fresh raspberries

¼ cup chopped almonds, lightly toasted

FOR THE DRESSING

1 ¼ cups frozen raspberries

½ apple, peeled, cored, and quartered

4 regular or 2 Medjool dates, pitted

½ clove garlic

½ teaspoon Dijon mustard

¼ cup water

1 tablespoon Dr. Fuhrman's Riesling Reserve Vinegar or apple cider vinegar

1 teaspoon fresh lime juice

Combine all salad ingredients, except raspberries and almonds. Add raspberries and toss lightly. Sprinkle almonds on top. Combine dressing ingredients in a high-powered blender. Pour desired amount over salad.

PER SERVING: CALORIES 347; PROTEIN 11G; CARBOHYDRATES 59G; SUGARS 34G; TOTAL FAT 11.6G; SATURATED FAT 0.9G; SODIUM 97MG; FIBER 19.5G; BETA-CAROTENE 16,175MCG; VITAMIN C 77MG; CALCIUM 259MG; IRON 5.2MG; FOLATE 371MCG; MAGNESIUM 172MG; POTASSIUM 1,577MG; ZINC 2.5MG; SELENIUM 2.8MCG

Kale, Cabbage, and Mushroom Salad

SERVES 2

FOR THE DRESSING (SEE NOTE)
½ cup water
¼ cup rice vinegar
¼ cup unhulled sesame seeds
1 teaspoon lemon zest
2 teaspoons coconut aminos

FOR THE SALAD
8 ounces mushrooms, sliced
3 cups chopped kale, tough stems removed
3 cups thinly sliced cabbage, or baby bok choy
½ cup chopped scallions
¼ cup grated carrot
½ cup pomegranate seeds (optional)

Blend dressing ingredients in a high-powered blender. Heat 2 tablespoons water in a sauté pan and water-sauté mushrooms until softened and tender. Allow to cool. Place chopped kale in a large salad bowl, add 2 tablespoons of the dressing, and using your fingers, massage the kale for a minute or two until it starts to wilt. Add sautéed mushrooms, cabbage, scallions, and carrot. Toss with desired amount of dressing. Garnish with pomegranate seeds, if desired.

Note: If you don't have time to make a dressing, you can use any of Dr. Fuhrman's no-oil bottled salad dressings in this recipe.

PER SERVING: CALORIES 256; PROTEIN 13G; CARBOHYDRATES 35G; SUGARS 12G; TOTAL FAT 10.6G; SATURATED FAT 1.5G; SODIUM 309MG; FIBER 10.3G; BETA-CAROTENE 10,606MCG; VITAMIN C 171MG; CALCIUM 389MG; IRON 6.2MG; FOLATE 145MCG; MAGNESIUM 135MG; POTASSIUM 1,308MG; ZINC 2.9MG; SELENIUM 18.3MCG

Kohlrabi Apple Slaw

SERVES 4

3 small kohlrabi bulbs (about 1 pound), peeled, cored, and cut into 1/4-inch matchsticks (you can also use jicama instead of kohlrabi)

1 apple, cored and cut into 1/4-inch matchsticks

1 shallot, thinly sliced

1 carrot, peeled and grated

1/4 cup chopped walnuts

Black pepper to taste

FOR THE DRESSING

1 orange, juiced (about 1/2 cup juice)

1/4 cup unsweetened soy, hemp, or almond milk

1/4 cup raw almonds

1 tablespoon balsamic vinegar

Combine kohlrabi, apple, shallot, carrot, and walnuts in a medium bowl. Blend dressing ingredients in a high-powered blender. Toss salad with desired amount of dressing. Season with black pepper.

PER SERVING: CALORIES 159; PROTEIN 5G; CARBOHYDRATES 20G; SUGARS 11G; TOTAL FAT 8.1G; SATURATED FAT 0.7G; SODIUM 45MG; FIBER 6.5G; BETA-CAROTENE 1,307MCG; VITAMIN C 82MG; CALCIUM 94MG; IRON 1.2MG; FOLATE 42MCG; MAGNESIUM 57MG; POTASSIUM 608MG; ZINC 0.5MG; SELENIUM 1.4MCG

Napa Cabbage Slaw

SERVES 4

FOR THE DRESSING

3 tablespoons no-oil-added, unsalted peanut butter

6 tablespoons warm water

1 Medjool or 2 regular dates, pitted

1 tablespoon apple cider vinegar

1 teaspoon reduced-sodium miso

1/4 teaspoon grated ginger

FOR THE SALAD

2 heads napa cabbage, finely chopped

2 cups finely chopped baby kale

3 green onions, chopped

1 cup cooked beans, any variety

1 cup fresh or defrosted frozen corn kernels

1 ripe avocado, cubed

1/4 cup fresh parsley

To make the dressing, blend the peanut butter, water, dates, vinegar, miso, and ginger. Place the salad ingredients in a large bowl and combine. Toss with desired amount of dressing. Refrigerate at least 1 hour and toss again before serving.

PER SERVING: CALORIES 289; PROTEIN 13G; CARBOHYDRATES 39G; SUGARS 8G; TOTAL FAT 12.4G; SATURATED FAT 2.2G; SODIUM 107MG; FIBER 10.4G; BETA-CAROTENE 3,729MCG; VITAMIN C 99MG; CALCIUM 167MG; IRON 3.1MG; FOLATE 296MCG; MAGNESIUM 112MG; POTASSIUM 1,210MG; ZINC 1.9MG; SELENIUM 2.6MCG

Roasted Vegetable Salad
with Balsamic Vinaigrette

SERVES 4

2 medium zucchini, sliced
1 red bell pepper, sliced
1 yellow or orange bell pepper, sliced
2 cups quartered mushrooms
1 red onion, sliced
1 tablespoon balsamic vinegar
1 teaspoon Dr. Fuhrman's VegiZest Seasoning (or other no-salt seasoning blend, adjusted to taste)
5 ounces mixed salad greens

FOR THE DRESSING (SEE NOTE)

¼ cup water
2 tablespoons balsamic vinegar
2 tablespoons raw almond butter
¼ teaspoon onion powder
¼ teaspoon garlic powder
⅛ teaspoon dried oregano
⅛ teaspoon dried basil

Preheat oven to 350°F. Toss zucchini, peppers, mushrooms, and onions with balsamic vinegar and no-salt seasoning. Place on a nonstick silicon baking sheet and bake for 20 minutes or until tender, stirring occasionally.

To make the dressing, whisk water, vinegar, and almond butter together until mixture is smooth and almond butter is evenly dispersed. Mix in remaining ingredients. Top salad greens with roasted vegetables and desired amount of dressing.

Note: This dressing works well on any salad. In a rush? Substitute Dr. Fuhrman's bottled Almond Balsamic Salad Dressing.

PER SERVING: CALORIES 127; PROTEIN 6G; CARBOHYDRATES 17G; SUGARS 8G; TOTAL FAT 5.3G; SATURATED FAT 0.5G; SODIUM 31MG; FIBER 4.4G; BETA-CAROTENE 2,735MCG; VITAMIN C 149MG; CALCIUM 90MG; IRON 1.9MG; FOLATE 106MCG; MAGNESIUM 71MG; POTASSIUM 797MG; ZINC 1.2MG; SELENIUM 5.4MCG

Sunrise Salad

SERVES 4

8 Brussels sprouts, halved

2 cups mushrooms

4 cups greens, such as romaine, spinach, kale, or chard

1 ½ cups or 1 (15-ounce) can no-salt-added or low-sodium black beans, drained

½ onion, sliced

1 green bell pepper, sliced

1 cup low-sodium salsa

⅓ cup unhulled sesame seeds

Dash turmeric

Dash ground black pepper

1 avocado, pitted and sliced (optional)

2 cups raspberries

Steam Brussels sprouts until tender, about 10 minutes. Lightly sauté mushrooms using a small amount of water if needed to prevent sticking.

Divide greens among four plates. Top with beans, onion, bell pepper, Brussels sprouts, mushrooms, salsa, and sesame seeds. Season with turmeric and black pepper to taste. If desired, add avocado slices. Serve with ½ cup of raspberries on the side of each plate.

PER SERVING: CALORIES 241; PROTEIN 13G; CARBOHYDRATES 37G; SUGARS 8G; TOTAL FAT 7.2G; SATURATED FAT 1G; SODIUM 26MG; FIBER 15G; BETA-CAROTENE 2,873MCG; VITAMIN C 90MG; CALCIUM 195MG; IRON 5.3MG; FOLATE 227MCG; MAGNESIUM 129MG; POTASSIUM 959MG; ZINC 2.5MG; SELENIUM 9.5MCG

Black Bean and Butternut Squash Chili

SERVES 5

2 cups chopped onions

3 cloves garlic, chopped

2 ½ cups chopped (½-inch pieces) butternut squash

4 ½ cups or 3 (15-ounce) cans no-salt-added or low-sodium black beans, undrained

2 tablespoons chili powder (see Note)

2 teaspoons ground cumin

2 cups low-sodium or no-salt-added vegetable broth

1 ½ cups no-salt-added diced tomatoes, packaged in non-BPA packaging

1 bunch Swiss chard, tough stems removed, chopped

Add all ingredients except Swiss chard to a large pot. Bring to a boil, reduce heat, and simmer, uncovered, until squash is tender, about 20 minutes. Stir in Swiss chard and simmer until chard is tender, about 4 minutes.

Note: If you like it spicy, use a hot chili powder blend.

PER SERVING: CALORIES 300; PROTEIN 17G; CARBOHYDRATES 58G; SUGARS 6G; TOTAL FAT 1.8G; SATURATED FAT 0.4G; SODIUM 204MG; FIBER 18.4G; BETA-CAROTENE 4,928MCG; VITAMIN C 38MG; CALCIUM 146MG; IRON 6MG; FOLATE 276MCG; MAGNESIUM 180MG; POTASSIUM 1,231MG; ZINC 2.4MG; SELENIUM 3.8MCG

Butternut Breakfast Soup

SERVES 6

4 cups frozen butternut squash

2 medium apples, peeled, seeded, and chopped

4 cups (packed) kale, tough stems and center ribs removed and leaves chopped, or frozen, chopped

1 cup chopped onion

2 tablespoons Dr. Fuhrman's Pomegranate Balsamic Vinegar or other fruity vinegar

5 cups carrot juice

½ cup unsweetened soy, almond, or hemp milk

½ cup raw cashews

¼ cup hemp seeds

1 teaspoon cinnamon

½ teaspoon nutmeg

Place squash, apples, kale, onion, vinegar, and carrot juice in a soup pot. Bring to a boil, lower heat, cover, and simmer for 30 minutes or until kale is tender. Purée half of the soup with the nondairy milk, cashews, and hemp seeds in a high-powered blender. Return blended mixture to soup pot. Add cinnamon and nutmeg.

PER SERVING: CALORIES 310; PROTEIN 9G; CARBOHYDRATES 57G; SUGARS 18G; TOTAL FAT 8.3G; SATURATED FAT 1.3G; SODIUM 167MG; FIBER 9.7G; BETA-CAROTENE 28,816MCG; VITAMIN C 106MG; CALCIUM 266MG; IRON 4.2MG; FOLATE 70MCG; MAGNESIUM 158MG; POTASSIUM 1,524MG; ZINC 1.9MG; SELENIUM 8.4MCG

Chickpea, Greens, and Farro Stew

SERVES 4

3 cups or 2 (15-ounce) cans no-salt-added or low-sodium chickpeas with liquid, divided

4 cups low-sodium or no-salt-added vegetable broth, divided

1 medium onion, diced

1 medium carrot, diced

2 stalks celery, diced

8 cloves garlic, chopped

1 cup sliced mushrooms

1 ½ cups diced tomatoes

1 teaspoon Herbes de Provence (see Note)

¼ teaspoon black pepper

½ cup farro

3 cups finely chopped kale or collard greens

Blend one can of the chickpeas (including the liquid) with ½ cup of the vegetable broth until smooth. Place that and the remaining ingredients in a pot, bring to a boil, and then reduce heat, cover, and simmer for 30 minutes or until farro and vegetables are tender.

Note: Herbes de Provence, a combination of herbs typical of the Provence region of France, can be found in the spice section of most grocery stores. The stew can also be seasoned with a combination of dried thyme, savory, oregano, and fennel.

PER SERVING: CALORIES 377; PROTEIN 17G; CARBOHYDRATES 68G; SUGARS 10G; TOTAL FAT 4.6G; SATURATED FAT 0.4G; SODIUM 198MG; FIBER 16.1G; BETA-CAROTENE 6,282MCG; VITAMIN C 75MG; CALCIUM 198MG; IRON 6.3MG; FOLATE 253MCG; MAGNESIUM 92MG; POTASSIUM 897MG; ZINC 2.4MG; SELENIUM 5.5MCG

Corn and Red Lentil Chowder

SERVES 4

1 large onion, chopped

1 tablespoon curry powder

½ teaspoon turmeric

4 cups no-salt-added or low-sodium vegetable broth

2 cups water

1½ cups dry red lentils, rinsed and drained

1 tablespoon peeled and grated ginger

¼ teaspoon black pepper

3 cups fresh or frozen corn kernels, divided

2 cups finely chopped spinach or kale

Add onion, curry powder, turmeric, broth, water, lentils, ginger, and black pepper to a soup pot, bring to a boil, reduce heat, cover, and simmer for 20 minutes or until lentils are tender. Add 2 cups of the corn and cook for an additional 10 minutes, stirring occasionally. Place soup in a high-powered blender and blend until smooth and creamy. Return to soup pot, add remaining 1 cup of corn and the chopped spinach or kale and cook another 10 minutes, until greens have softened.

PER SERVING: CALORIES 392; PROTEIN 23G; CARBOHYDRATES 74G; SUGARS 4G; TOTAL FAT 2.9G; SATURATED FAT 0.5G; SODIUM 166MG; FIBER 11.9G; BETA-CAROTENE 3,176MCG; VITAMIN C 51MG; CALCIUM 123MG; IRON 7.5MG; FOLATE 203MCG; MAGNESIUM 92MG; POTASSIUM 877MG; ZINC 3.5MG; SELENIUM 7.4MCG

Dr. Fuhrman's Famous Anticancer Soup

SERVES 9

½ cup dried adzuki or other beans

5 cups water

5 pounds organic carrots, juiced (about 6 cups carrot juice; see Note)

2 bunches celery, juiced (about 2 cups celery juice; see Note)

6 medium zucchini

2 tablespoons Dr. Fuhrman's VegiZest (or other no-salt seasoning blend, adjusted to taste)

1 teaspoon Mrs. Dash salt-free seasoning or ⅓ teaspoon black pepper

½ cup dried split peas

4 medium onions

3 leeks, roots and 1 inch from top cut off

2 bunches kale, collard greens, or other greens

¾ cup raw cashews

¼ cup hemp seeds

10 ounces fresh mushrooms (shiitake, cremini, and/or white), chopped

Place the beans, 5 cups water, carrot juice, celery juice, zucchini, VegiZest, and Mrs. Dash seasoning in a large pot and bring to a boil, then reduce heat to a simmer. Add the dried split peas to 1½ cups of water in a separate small covered pot and cook on a low flame for 30 minutes or until tender. Meanwhile, blend the onions and leeks with a small amount of the soup liquid and add to the pot and then do the same for the kale (or other dark greens) and add that purée back to the pot. Remove the softened zucchini with tongs, place in the blender along with cashews and hemp seeds and blend until creamy and add back to soup pot. Add the chopped mushrooms and continue to simmer for an additional 2 hours. Take the small pot of cooked split peas and blend until smooth and add to the main pot.

Note: Freshly juiced organic carrots and celery will maximize the flavor of this soup.

PER SERVING: CALORIES 335; PROTEIN 16G; CARBOHYDRATES 55G; SUGARS 15G; TOTAL FAT 9G; SATURATED FAT 1.4G; SODIUM 174MG; FIBER 10.9G; BETA-CAROTENE 21,994MCG; VITAMIN C 136MG; CALCIUM 225MG; IRON 5.7MG; FOLATE 209MCG; MAGNESIUM 180MG; POTASSIUM 1,749MG; ZINC 3.3MG; SELENIUM 8.2MCG

Edamame and Green Pea Soup

SERVES 3

1 small onion, chopped
2 stalks celery, chopped
4 cloves garlic, chopped
2 cups frozen edamame
2 cups frozen green peas

2 cups finely chopped kale
⅛ teaspoon black pepper
4 cups no-salt-added or low-sodium vegetable broth

Place all ingredients in a soup pot. Bring to a boil, reduce heat, and simmer for 25 minutes. Blend in a high-powered blender until smooth and creamy. Return to pot and reheat.

PER SERVING: CALORIES 256; PROTEIN 18G; CARBOHYDRATES 34G; SUGARS 8G; TOTAL FAT 6.1G; SATURATED FAT 0.8G; SODIUM 327MG; FIBER 11.2G; BETA-CAROTENE 5,288MCG; VITAMIN C 80MG; CALCIUM 200MG; IRON 5.2MG; FOLATE 396MCG; MAGNESIUM 111MG; POTASSIUM 908MG; ZINC 2.5MG; SELENIUM 2.9MCG

Mushroom and Wheat Berry Soup

SERVES 6

¾ cup raw cashews, soaked in advance

8 cups no-salt-added or low-sodium vegetable broth, divided

1 onion, chopped

2 carrots, chopped

4 ribs celery, chopped

8 cloves garlic, chopped

1 cup organic wheat berries, rinsed and drained

1 teaspoon coconut aminos

1 teaspoon dried thyme

⅛ teaspoon black pepper, or to taste

¼ cup hemp seeds

1½ cups cooked or 1 (15-ounce) can no-salt-added or low-sodium cannellini beans

10 ounces mushrooms, sliced

In a large soup pot, bring 7 cups of the vegetable broth to a simmer. Add onion, carrots, celery, garlic, wheat berries, coconut aminos, thyme, and black pepper. Return to a simmer, cover pot, and cook for 30 minutes. Place remaining cup of vegetable broth, soaked and drained cashews, hemp seeds, and beans in a high-powered blender and blend until very smooth. Add blended mixture to soup pot along with the mushrooms, partially cover, and cook for an additional 15 minutes, stirring occasionally.

PER SERVING: CALORIES 342; PROTEIN 16G; CARBOHYDRATES 47G; SUGARS 4G; TOTAL FAT 11.9G; SATURATED FAT 1.8G; SODIUM 246MG; FIBER 9.7G; BETA-CAROTENE 1,764MCG; VITAMIN C 5MG; CALCIUM 91MG; IRON 5.4MG; FOLATE 106MCG; MAGNESIUM 168MG; POTASSIUM 785MG; ZINC 3.4MG; SELENIUM 31.4MCG

Nutritarian Borscht

SERVES 4

1 onion, chopped

8 cloves garlic, chopped

2 medium beets, washed, peeled, and cut into bite-size pieces

4–6 cups no-salt-added or low-sodium vegetable broth

2 medium carrots, sliced into rounds

2 cups chopped cabbage

1 medium zucchini, chopped

1 cup chopped fresh green beans

½ cup frozen corn kernels or 1 ear corn, kernels removed

2 tomatoes, chopped

2 tablespoons lemon juice

1 bunch fresh dill, chopped (or 1–2 teaspoons dried dill)

2 tablespoons chopped fresh parsley

FOR THE "SOUR CREAM"

1 cup cashews, soaked overnight

1 Medjool or 2 regular dates, soaked overnight with the cashews

2 tablespoons fresh lemon juice

1 teaspoon apple cider vinegar

1 teaspoon white miso

Place onion, garlic, beets, and vegetable broth in a soup pot. Bring to a boil, then reduce heat, cover, and simmer for 10 minutes. Add carrots, cabbage, zucchini, green beans, corn, and tomatoes and cook for 30 minutes. Remove from heat and add lemon juice, dill, and parsley. Let sit for 10 minutes. Serve with a dollop of cashew "sour cream."

To make the cashew "sour cream," drain the cashews and dates, reserving the soaking liquid. In a high-powered blender, purée the cashews, dates, lemon juice, vinegar, and miso plus ½ cup of the soaking liquid. Add additional liquid if needed to facilitate blending. Chill until ready to use. Leftover "sour cream" will keep for several days and can be used with other dishes.

PER SERVING: CALORIES 319; PROTEIN 10G; CARBOHYDRATES 39G; SUGARS 16G; TOTAL FAT 15.8G; SATURATED FAT 2.8G; SODIUM 299MG; FIBER 7G; BETA-CAROTENE 2,820MCG; VITAMIN C 46MG; CALCIUM 116MG; IRON 4.4MG; FOLATE 121MCG; MAGNESIUM 149MG; POTASSIUM 896MG; ZINC 2.8MG; SELENIUM 8.1MCG

Soybean and Red Lentil Soup

SERVES 8

1 cup dried soybeans, soaked overnight and drained

1 onion, chopped

2 carrots, chopped

2 stalks celery, chopped

4 cloves garlic, chopped

2 cups red lentils, rinsed

14 ounces crushed tomatoes

6 cups vegetable broth

2 teaspoons ground cumin

2 teaspoons ground coriander

1/4 teaspoon black pepper

4 cups thinly sliced kale

1/2 lemon, juiced

Place soybeans in a large soup pot along with 4 cups of water. Bring to a boil, reduce heat, partially cover, and simmer for 2½ hours. Add onion, carrots, celery, garlic, red lentils, tomatoes, vegetable broth, cumin, coriander, and black pepper to the soybeans, bring to a simmer and cook for 30 minutes, or until lentils are soft and vegetables are tender. Add additional vegetable broth if needed to adjust consistency. Add kale and heat until wilted. Remove from heat and stir in lemon juice.

PER SERVING: CALORIES 322; PROTEIN 22G; CARBOHYDRATES 46G; SUGARS 2G; TOTAL FAT 6.3G; SATURATED FAT 0.9G; SODIUM 143MG; FIBER 10.9G; BETA-CAROTENE 4,557MCG; VITAMIN C 51MG; CALCIUM 174MG; IRON 8.8MG; FOLATE 210MCG; MAGNESIUM 123MG; POTASSIUM 1,045MG; ZINC 3.4MG; SELENIUM 8.9MCG

Sweet and Sour Cabbage and Split Pea Soup

SERVES 4

6 unsulfured prunes, pitted

2 Granny Smith apples, cored and quartered

5 cups water, divided

⅓ cup dried split peas

1 large onion, chopped

1 cup chopped carrots

2 cups unsweetened soy, hemp, or almond milk

¼ cup hulled barley

½ head cabbage, coarsely chopped

1 teaspoon dried basil

1 teaspoon dried oregano

½ teaspoon dried thyme

Black pepper to taste

3 tablespoons lemon juice

½ cup raw walnuts, toasted, then finely chopped

1 teaspoon caraway seeds

Blend the prunes, apples, and 2 cups of the water in a high-powered blender until smooth and creamy. In a small pot, cook the split peas in 1½ cups of water, simmering for 20 minutes or until tender. Remove the split peas from the small pot and blend until smooth. Add blended prune mixture and blended split peas to a soup pot along with the remaining water, onion, carrots, nondairy milk, barley, cabbage, basil, oregano, thyme, and black pepper. Bring to a boil, reduce heat, cover, and simmer for 30 minutes or until barley and vegetables are tender. Stir in lemon juice, chopped walnuts, and caraway seeds.

PER SERVING: CALORIES 347; PROTEIN 13G; CARBOHYDRATES 52G; SUGARS 22G; TOTAL FAT 12.4G; SATURATED FAT 1.3G; SODIUM 106MG; FIBER 14.3G; BETA-CAROTENE 2,773MCG; VITAMIN C 56MG; CALCIUM 282MG; IRON 3.5MG; FOLATE 121MCG; MAGNESIUM 104MG; POTASSIUM 894MG; ZINC 2.1MG; SELENIUM 6.3MCG

Artichoke-Stuffed Portabella Mushrooms

SERVES 4

2 tablespoons no-salt-added or low-sodium vegetable broth or more as needed

1 small onion, sliced

½ red bell pepper, sliced

6 cloves garlic, minced

10 ounces frozen artichoke hearts, thawed and sliced

4 large portabella mushrooms, stems removed

¼ teaspoon garlic powder

¼ teaspoon dried basil

¼ teaspoon dried oregano

½ cup low-sodium pasta sauce

2 tablespoons pine nuts

FOR THE CASHEW "CHEESE"

1 cup raw cashews, soaked for at least 2 hours, then drained

2 tablespoons nutritional yeast

2 tablespoons freshly squeezed lemon juice

½ teaspoon coconut aminos

¼ teaspoon black pepper

¼ cup water

Preheat oven to 350°F. In a large skillet, heat vegetable broth and sauté onion and peppers until tender, about 4 minutes. Add garlic, cook for 30 seconds, then add sliced artichoke hearts and cook until heated through, about 2 minutes. Place mushrooms on a rimmed baking sheet, gill sides up, and sprinkle with garlic powder, basil, and oregano. Bake for 10 minutes. Top with pasta sauce and artichoke mixture, and sprinkle with pine nuts. Bake for an additional 10 minutes or until mushrooms are tender.

To make the Cashew "Cheese," place the ingredients in a high-powered blender and blend until smooth, adding small amounts of additional water if needed to adjust consistency. Top mushrooms with a dollop of Cashew "Cheese" before serving. Store leftover Cashew "Cheese" refrigerated in a sealed container for up to 4 days.

PER SERVING: CALORIES 217; PROTEIN 10G; CARBOHYDRATES 22G; SUGARS 7G; TOTAL FAT 11.7G; SATURATED FAT 1.8G; SODIUM 88MG; FIBER 6.7G; BETA-CAROTENE 369MCG; VITAMIN C 28MG; CALCIUM 48MG; IRON 2.7MG; FOLATE 135MCG; MAGNESIUM 96MG; POTASSIUM 800MG; ZINC 2.9MG; SELENIUM 19.9MCG

Asian Ginger Lime Zoodles

SERVES 4

FOR THE SAUCE
1 ½ cups water
4 Medjool or 8 regular dates, pitted
¼ cup no-oil-added, unsalted peanut butter
3 tablespoons hemp seeds
1 teaspoon minced ginger
1 small clove garlic
1 tablespoon lime juice
1 teaspoon red curry powder

½ teaspoon chili powder
½ teaspoon ground cumin
4 large zucchini, cut into noodles using a spiralizer or vegetable peeler
1 large carrot, cut into noodles using a spiralizer or vegetable peeler
1 cup shelled edamame
2 tablespoons chopped fresh cilantro
2 green onions, sliced

Blend water and dates in a high-powered blender, then add the peanut butter, hemp seeds, ginger, garlic, lime juice, red curry powder, chili powder, and ground cumin and blend until smooth and creamy. Heat 1 cup of the sauce in a large skillet, add the zucchini and carrot noodles, and cook for 2 minutes or until the zucchini starts to soften. Add additional sauce as needed to reach desired consistency. Stir in edamame, cilantro, and green onions.

PER SERVING: CALORIES 294; PROTEIN 15G; CARBOHYDRATES 32G; SUGARS 21G; TOTAL FAT 15.2G; SATURATED FAT 2.6G; SODIUM 65MG; FIBER 8.7G; BETA-CAROTENE 1,972MCG; VITAMIN C 64MG; CALCIUM 115MG; IRON 3.7MG; FOLATE 227MCG; MAGNESIUM 175MG; POTASSIUM 1,402MG; ZINC 3MG; SELENIUM 2.4MCG

Bean Pasta with Roasted Red Pepper Alfredo

SERVES 4

8 ounces bean pasta, cooked according to package instructions

5 ounces spinach

8 ounces mushrooms, any variety, sliced

FOR THE SAUCE (SEE NOTE)

1 red bell pepper, raw or roasted

½ cup water

½ cup raw cashews

¼ cup hemp seeds

¼ cup nutritional yeast

¼ teaspoon onion powder

¼ teaspoon garlic powder

¼ teaspoon ground turmeric

⅛ teaspoon nutmeg

Black pepper to taste

Dash red pepper flakes

Combine sauce ingredients in a high-powered blender and blend until smooth. Heat 2–3 tablespoons water in a large sauté pan and water-sauté mushrooms until cooked through. Add spinach and continue cooking until wilted. Add cooked pasta and desired amount of sauce and cook for 1–2 minutes.

Note: For a super-quick and easy meal, substitute Dr. Fuhrman's bottled Mushroom Alfredo Sauce.

PER SERVING: CALORIES 416; PROTEIN 26G; CARBOHYDRATES 52G; SUGARS 4G; TOTAL FAT 12.9G; SATURATED FAT 2.1G; SODIUM 39MG; FIBER 14.5G; BETA-CAROTENE 2,477MCG; VITAMIN C 49MG; CALCIUM 223MG; IRON 8.9MG; FOLATE 126MCG; MAGNESIUM 123MG; POTASSIUM 618MG; ZINC 6.1MG; SELENIUM 10.8MCG

Bolognese Sauce

SERVES 6

1 (8-ounce) package tempeh, broken into chunks

1 medium onion, chopped

2 medium carrots, finely chopped

8 ounces mushrooms, chopped

6 cloves garlic, chopped

2 tablespoons tomato paste

2 cups no-salt-added or low-sodium vegetable broth

28 ounces crushed tomatoes

2 teaspoons coconut aminos

1 teaspoon dried basil

1 teaspoon dried oregano

¼ teaspoon crushed red pepper flakes, or to taste

Place the tempeh in a food processor and pulse until it is crumbled. Heat 2–3 tablespoons of water or vegetable broth in a large sauté pan. Add the onion, carrots, mushrooms, garlic, and crumbled tempeh and sauté for 5 minutes, or until vegetables are tender, adding more liquid if needed. Add the tomato paste and cook, stirring for 1 minute, then add the vegetable broth, crushed tomatoes, coconut aminos, basil, oregano, and red pepper flakes. Bring to a simmer, cover, and cook for 20 minutes. Serve over your choice of bean pasta or quinoa.

PER SERVING (ANALYSIS DOES NOT INCLUDE PASTA OR QUINOA): CALORIES 149; PROTEIN 11G; CARBOHYDRATES 20G; SUGARS 9G; TOTAL FAT 4.7G; SATURATED FAT 1G; SODIUM 308MG; FIBER 4.1G; BETA-CAROTENE 1,923MCG; VITAMIN C 17MG; CALCIUM 117MG; IRON 3.6MG; FOLATE 42MCG; MAGNESIUM 69MG; POTASSIUM 819MG; ZINC 1.1MG; SELENIUM 4.7MCG

Broccoli and Snow Pea Stir-Fry with Pineapple and Baked Tofu

SERVES 4

14 ounces extra-firm tofu

FOR THE SAUCE (SEE NOTE)

1 cup warm water
¼ cup no-oil-added, unsalted peanut butter
1 teaspoon arrowroot powder
1 teaspoon minced ginger
2 teaspoons coconut aminos
1 teaspoon lime juice
¾ teaspoon red curry powder
¼ teaspoon chili powder
¼ teaspoon ground cumin

FOR THE STIR-FRY

4 cups broccoli florets
2 cups halved snow peas
1 red bell pepper, sliced
1 cup sliced mushrooms
2 cups cubed pineapple
4 scallions, sliced
1 tablespoon unhulled sesame seeds, lightly toasted

Preheat oven to 350°F. Wrap tofu in a clean absorbent towel and set something heavy on it to press out the liquid. Let sit for 20 minutes. Cut tofu into ½-inch pieces and place on baking sheet lined with parchment paper or a nonstick silicone mat. Bake for 30 minutes until golden and firm. Meanwhile, mash and whisk together water, peanut butter, arrowroot powder, ginger, coconut aminos, lime juice, and spices until smooth and well-combined. Pour into a bowl, add baked tofu, and marinate for 5 minutes, stirring occasionally.

Add ¼ cup water to a large pan or wok and when hot, add the broccoli florets, cover, and cook for 4 minutes, stirring occasionally and adding additional water as needed to prevent sticking. Add the snow peas, red bell pepper, and mushrooms, cover, and cook for an additional 4 minutes or until vegetables are crisp tender. Stir in pineapple, scallions, tofu, and desired amount of marinade and cook until heated through. Sprinkle with sesame seeds before serving.

Note: Dr. Fuhrman's Thai Curry Sauce also works well with this recipe if you don't have time to make your own marinade.

PER SERVING: CALORIES 306; PROTEIN 19G; CARBOHYDRATES 30G; SUGARS 15G; TOTAL FAT 15.7G; SATURATED FAT 2G; SODIUM 193MG; FIBER 7.5G; BETA-CAROTENE 1,158MCG; VITAMIN C 181MG; CALCIUM 289MG; IRON 4.7MG; FOLATE 146MCG; MAGNESIUM 133MG; POTASSIUM 875MG; ZINC 2.5MG; SELENIUM 19.3MCG

Broccoli Fra Diavolo

SERVES 3

6 cups fresh broccoli florets

8 cloves garlic, chopped

1 ½ cups diced tomatoes

1 cup no-salt-added or low-sodium tomato or pasta sauce

⅛ teaspoon crushed red pepper flakes

1–2 teaspoons no-salt-added Italian seasoning

1 tablespoon Dr. Fuhrman's MatoZest (or other no-salt seasoning blend, adjusted to taste)

¼ cup nutritional yeast

Steam broccoli until just tender, about 10 minutes. In large saucepan over medium heat, sauté garlic in ¼ cup water for 3–4 minutes. Add tomatoes, tomato sauce, red pepper flakes, Italian seasoning, and MatoZest. Simmer for 5 minutes. Stir in steamed broccoli and nutritional yeast.

Note: Cauliflower also works well in this recipe.

PER SERVING: CALORIES 181; PROTEIN 14G; CARBOHYDRATES 29G; SUGARS 10G; TOTAL FAT 2.8G; SATURATED FAT 0.4G; SODIUM 100MG; FIBER 10.5G; BETA-CAROTENE 1,417MCG; VITAMIN C 180MG; CALCIUM 174MG; IRON 4.4MG; FOLATE 154MCG; MAGNESIUM 94MG; POTASSIUM 1,116MG; ZINC 3.4MG; SELENIUM 6.3MCG

Broccoli Quiche with Aquafaba

SERVES 4

1 large onion, sliced
1 cup chopped mushrooms
5 cups small broccoli florets
14 ounces firm tofu
½ cup chickpea aquafaba (see Note)
¼ cup unsweetened soy, hemp, or almond milk
¼ cup nutritional yeast
2 tablespoons raw cashew butter

2 tablespoons arrowroot powder
1 teaspoon coconut aminos or reduced-sodium soy sauce
1 teaspoon paprika
1 teaspoon Dijon mustard
½ teaspoon garlic powder
½ teaspoon turmeric
¼ teaspoon ground black pepper

Preheat oven to 375°F. Heat 2–3 tablespoons of water in a large pan and add sliced onion and mushrooms. Water-sauté until onions are tender, adding small amounts of additional water as needed to prevent sticking. Add broccoli and a few more tablespoons of water, cover and cook for 5 minutes or until broccoli is almost tender. Blend remaining ingredients in a high-powered blender for at least 1 minute to whip up the aquafaba. Mix with onions, mushrooms, and broccoli and place in an 8-inch cake pan that has been wiped with olive oil. Bake for 35–40 minutes or until top is golden. Allow to cool for 10 minutes before cutting.

Note: Aquafaba is the liquid found in canned beans and other legumes or left over from cooking your own dried beans. It can function as a good egg replacement, as in this recipe.

PER SERVING: CALORIES 240; PROTEIN 19G; CARBOHYDRATES 23G; SUGARS 4G; TOTAL FAT 9.5G; SATURATED FAT 1.5G; SODIUM 131MG; FIBER 6.8G; BETA-CAROTENE 562MCG; VITAMIN C 105MG; CALCIUM 234MG; IRON 3.3MG; FOLATE 88MCG; MAGNESIUM 65MG; POTASSIUM 542MG; ZINC 2.7MG; SELENIUM 6.1MCG

Buffalo Cauliflower

SERVES 4

1 cup almond flour
¼ cup nutritional yeast
1 teaspoon Dr. Fuhrman's MatoZest (or other no-salt seasoning blend, adjusted to taste)
1 teaspoon paprika

¼–½ teaspoon cayenne pepper, or to taste
⅔ cup water
1 tablespoon Dijon mustard
1 head cauliflower, cut into florets

Preheat oven to 350°F. Combine flour, nutritional yeast, and seasonings in a bowl. Using a whisk, add water gradually. Stir in mustard. Mixture should resemble a thick batter. Toss cauliflower florets with the coating. Place on baking sheet lined with parchment paper or a nonstick silicone mat and bake for 20–25 minutes until coating is dry and cauliflower is tender.

PER SERVING: CALORIES 235; PROTEIN 13G; CARBOHYDRATES 15G; SUGARS 4G; TOTAL FAT 15.4G; SATURATED FAT 1.3G; SODIUM 99MG; FIBER 7.7G; BETA-CAROTENE 177MCG; VITAMIN C 71MG; CALCIUM 109MG; IRON 2.1MG; FOLATE 101MCG; MAGNESIUM 111MG; POTASSIUM 640MG; ZINC 2.9MG; SELENIUM 3.1MCG

California Creamed Kale

SERVES 4

You can use the cashew cream sauce you make for this recipe with broccoli, spinach, or other steamed vegetables.

2 bunches kale, leaves removed from tough stems
¾ cup raw cashews
¼ cup hemp seeds
¾ cup unsweetened soy, hemp, or almond milk

¼ cup dehydrated onion flakes
1 tablespoon Dr. Fuhrman's VegiZest or nutritional yeast (or other no-salt seasoning blend, adjusted to taste)

Place kale in a large steamer pot. Steam 6–8 minutes or until soft. Meanwhile, place remaining ingredients in a high-powered blender and blend until smooth. Place kale in colander and press to remove excess water. Coarsely chop kale and mix in a bowl with the cream sauce. Optional: Top with a dollop of thick tomato sauce and chopped red onion or scallion.

PER SERVING: CALORIES 320; PROTEIN 16G; CARBOHYDRATES 32G; SUGARS 4G; TOTAL FAT 18.1G; SATURATED FAT 2.7G; SODIUM 97MG; FIBER 5.6G; BETA-CAROTENE 15,455MCG; VITAMIN C 206MG; CALCIUM 321MG; IRON 5.9MG; FOLATE 82MCG; MAGNESIUM 218MG; POTASSIUM 1,173MG; ZINC 3.5MG; SELENIUM 7MCG

Cannellini Beans and Greens

SERVES 4

1 ½ cups dried cannellini or other beans, soaked overnight, then drained

2 heads of garlic, cloves removed

3 sage leaves

6 cups water

1 tablespoon grated lemon zest

2 tablespoons fresh lemon juice

2 tablespoons pine nuts

Black pepper to taste

Pinch red pepper flakes, or more if you like it spicy

5 ounces spinach, kale, or other greens

Bring beans, half of the garlic, sage, and water to a boil in a soup pot. Reduce heat, cover, and simmer gently, stirring occasionally, until beans are tender, about 1½ hours. Using a slotted spoon, transfer beans to sauté pan and add lemon zest, lemon juice, and ¼ cup bean cooking liquid and toss, while cooking for 10 minutes, adding more cooking liquid if needed to reach desired consistency. Stir in pine nuts and season with black pepper.

Crush the remaining garlic. Heat 2–3 tablespoons water in a separate sauté pan, add crushed garlic and red pepper flakes, and cook until garlic is fragrant, about 30 seconds. Add spinach or other greens and cook until just wilted, about 3 minutes. Serve beans with greens on the side.

PER SERVING: CALORIES 305; PROTEIN 20G; CARBOHYDRATES 51G; SUGARS 2G; TOTAL FAT 3.8G; SATURATED FAT 0.4G; SODIUM 56MG; FIBER 12.8G; BETA-CAROTENE 2,037MCG; VITAMIN C 17MG; CALCIUM 253MG; IRON 9.4MG; FOLATE 368MCG; MAGNESIUM 191MG; POTASSIUM 1,632MG; ZINC 3.4MG; SELENIUM 11.2MCG

Eggplant Meatballs

SERVES 5

These delicious "meatballs" are great served on top of sautéed, spiralized zucchini "noodles."

1 tablespoon chia seeds

3 tablespoons water

¼ cup no-salt-added or low-sodium vegetable broth

1 small onion, chopped

8 cloves garlic, chopped

1 medium unpeeled eggplant, chopped

1 ½ cups cooked or 1 (15-ounce) can no-salt-added or low-sodium chickpeas, drained

¼ cup chopped fresh parsley

¼ cup unfortified nutritional yeast

1 cup whole wheat panko or bread crumbs

½ teaspoon dried oregano

½ teaspoon dried basil

Pinch red chili flakes, or to taste

2 cups no-salt-added or low-sodium pasta sauce

Preheat oven to 375°F. Whisk the chia seeds and water together in a small bowl and let sit for at least 10 minutes. Heat 2 tablespoons vegetable broth in a medium skillet and sauté the onion and garlic for 5 minutes or until the onion is translucent. Add the eggplant and continue cooking until eggplant is soft, about 12 minutes, stirring occasionally and adding additional vegetable broth as needed to prevent burning. Place the eggplant mixture in a mixing bowl and stir in the chia mixture, chickpeas, and parsley. Place in a food processor and pulse until chopped but not puréed. Return to the bowl and stir in nutritional yeast, panko, oregano, basil, and red chili flakes. Mix well, then roll into balls. Place on a baking sheet lined with parchment paper or a nonstick silicone mat and bake for 30 minutes, turning occasionally. Serve topped with pasta sauce. Makes about 30 balls.

PER SERVING: CALORIES 286; PROTEIN 13G; CARBOHYDRATES 47G; SUGARS 12G; TOTAL FAT 5.1G; SATURATED FAT 0.8G; SODIUM 209MG; FIBER 12.2G; BETA-CAROTENE 600MCG; VITAMIN C 10MG; CALCIUM 135MG; IRON 4.4MG; FOLATE 162MCG; MAGNESIUM 85MG; POTASSIUM 786MG; ZINC 2.9MG; SELENIUM 10.2MCG

Farro and Mushroom Risotto

SERVES 4

½ sweet onion, chopped
6 cloves garlic, minced
1 pound mushrooms, thinly sliced
1 ½ cups farro
3–4 cups no-salt-added or low-sodium vegetable broth
⅓ cup nutritional yeast
3 tablespoons chopped parsley

In a saucepan, heat 2–3 tablespoons water and sauté onions, garlic, and mushrooms until tender, adding additional water as needed. In another saucepan, toast farro for 3–4 minutes on low flame. Add ½ cup broth to the farro and stir until broth is absorbed. Continue adding broth ½ cup at a time, stirring intermittently, until all vegetable broth is absorbed and farro is al dente, about 15 minutes. Remove from heat, stir in mushroom mixture, nutritional yeast, and parsley.

PER SERVING: CALORIES 262; PROTEIN 13G; CARBOHYDRATES 45G; SUGARS 3G; TOTAL FAT 2.1G; SATURATED FAT 0.1G; SODIUM 78MG; FIBER 9.5G; BETA-CAROTENE 96MCG; VITAMIN C 6MG; CALCIUM 48MG; IRON 2.9MG; FOLATE 16MCG; MAGNESIUM 20MG; POTASSIUM 288MG; ZINC 1.9MG; SELENIUM 7.3MCG

Garlic-Infused Quinoa with Tomatoes and Poblano Peppers

SERVES 6

1 bulb garlic
1 poblano pepper
4 cups water

2 cups quinoa
2 tomatoes, chopped
½ red onion, chopped

Preheat oven to 325°F. Wrap garlic bulb and poblano pepper separately in foil. Roast garlic for 30 minutes and pepper for 15 minutes. Cut garlic cloves in half and squeeze out the soft garlic. Blend the cooked garlic with a cup of water and then stir in 3 more cups of water and heat to a boil. Add quinoa to garlic broth and simmer for 20 minutes. Dice the roasted poblano pepper and add to cooked quinoa along with chopped tomatoes and onion.

PER SERVING: CALORIES 229; PROTEIN 9G; CARBOHYDRATES 41G; SUGARS 2G; TOTAL FAT 3.6G; SATURATED FAT 0.4G; SODIUM 7MG; FIBER 5G; BETA-CAROTENE 203MCG; VITAMIN C 14MG; CALCIUM 43MG; IRON 2.8MG; FOLATE 115MCG; MAGNESIUM 120MG; POTASSIUM 469MG; ZINC 1.9MG; SELENIUM 5.6MCG

G-BOMB Thai Vegetable Curry

SERVES 4

FOR THE SAUCE
½ cup water
1 medium carrot
1 cup unsweetened coconut milk
¼ cup unsweetened shredded coconut
2 stalks lemongrass, tough outer segments removed
4 regular or 2 Medjool dates, pitted
6 cloves garlic
1-inch piece fresh ginger, peeled
½ bunch fresh basil leaves
1 tablespoon coconut aminos
1 tablespoon Thai green or red curry paste

FOR THE VEGETABLES
½ red bell pepper, seeded and thinly sliced
½ large eggplant, cut into 1-inch cubes
1 cup green beans, cut into 2-inch lengths
1 ½ cups sliced shiitake and button mushrooms
1 can bamboo shoots, sliced
1 pound tofu, cut into ½-inch cubes
8 ounces fresh baby greens
2 cups cooked quinoa or other intact whole grain
4 green onions, sliced on the diagonal into ½-inch pieces
¼ cup chopped raw cashews, lightly toasted

Reserving some extra basil leaves for garnish, blend sauce ingredients in a high-powered blender until smooth and creamy. Place bell pepper, eggplant, green beans, mushrooms, bamboo shoots, and tofu in a wok or large skillet. Steam-sauté, covered, for about 8 minutes or until vegetables are tender. Add sauce and bring to a simmer. Add greens just before serving. Serve on top of cooked quinoa, topped with sprigs of fresh herbs, green onions, and lightly toasted chopped cashews.

PER SERVING: CALORIES 410; PROTEIN 13G; CARBOHYDRATES 51G; SUGARS 16G; TOTAL FAT 20.9G; SATURATED FAT 13.5G; SODIUM 166MG; FIBER 11G; BETA-CAROTENE 5,434MCG; VITAMIN C 40MG; CALCIUM 127MG; IRON 5MG; FOLATE 155MCG; MAGNESIUM 169MG; POTASSIUM 1,192MG; ZINC 3.6MG; SELENIUM 13MCG

Intense Marinara Sauce

SERVES 6

1 yellow onion, quartered

36 ounces strained tomatoes, packaged in BPA-free containers, or homegrown garden tomatoes

6 cloves garlic

2 large carrots, roughly chopped

2 Medjool or 4 regular dates, pitted

1 tablespoon Italian spices

½ tablespoon MatoZest (or other no-salt-seasoning blend, adjusted to taste)

1 tablespoon onion powder

¼ teaspoon black pepper

Blend the onion in a high-powered blender until chopped, then add remaining ingredients. Blend partially, pour half into a pot, then blend the rest of the sauce until smooth. Add to the pot with the partially blended sauce and simmer uncovered on very low heat 4–6 hours or until desired thickness is achieved. The longer it cooks and the thicker it gets, the more concentrated the flavor.

PER SERVING: CALORIES 106; PROTEIN 4G; CARBOHYDRATES 25G; SUGARS 15G; TOTAL FAT 0.6G; SATURATED FAT 0.1G; SODIUM 245MG; FIBER 5.3G; BETA-CAROTENE 2,220MCG; VITAMIN C 20MG; CALCIUM 96MG; IRON 2.8MG; FOLATE 36UG; MAGNESIUM 48MG; POTASSIUM 688MG; ZINC 0.7MG; SELENIUM 1.8MCG

Italian Stewed Tomatoes

SERVES 4

8 medium tomatoes
¼ cup chopped celery
¼ cup chopped onion
¼ cup chopped green bell pepper

6 cloves garlic, chopped
2 tablespoons chopped fresh basil or
2 teaspoons dried basil

Place all ingredients in a large saucepan. Cover and cook over medium heat for 10 minutes or until tomatoes are softened, stirring occasionally to prevent sticking.

PER SERVING: CALORIES 53; PROTEIN 2G; CARBOHYDRATES 11G; SUGARS 7G; TOTAL FAT 0.5G; SATURATED FAT 0.1G; SODIUM 19MG; FIBER 3.4G; BETA-CAROTENE 1,186MCG; VITAMIN C 43MG; CALCIUM 35MG; IRON 0.8MG; FOLATE 43MCG; MAGNESIUM 31MG; POTASSIUM 640MG; ZINC 0.5MG; SELENIUM 0.2MCG

Kale, Chickpea, and Grain Bowl

SERVES 4

1 cup dry quinoa, rinsed (see Note)
1 ½ cups low-sodium or no-salt-added vegetable broth
1 ½ cups cooked or 1 (15-ounce) can no-salt-added or low-sodium chickpeas
1 cup finely chopped carrots
4 cups chopped kale or other greens
¼ cup very thinly sliced shallots or scallions
¼ cup chopped fresh parsley
¼ teaspoon black pepper

FOR THE DRESSING

1 ripe avocado, peeled and pitted
2 tablespoons fresh lemon juice
2 tablespoons water
2 tablespoons unhulled sesame seeds (lightly toasted in pan for 3 minutes)
1 clove garlic
¼ teaspoon ground turmeric

Place quinoa in a sauce pot with 1½ cups broth and the liquid drained from the can of chickpeas (called aquafaba) and bring to a boil. Reduce heat to a simmer, cover, and cook until liquid is absorbed, about 15 minutes. Transfer to a large bowl. Heat 2–3 tablespoons water in a large sauté pan, add carrots, and cook for 5 minutes, stirring occasionally and adding more water if needed to prevent sticking. Add kale and chickpeas, cover, and cook until kale is wilted and carrots are tender, about 3 minutes. Add kale and chickpea mixture, shallots, parsley, and black pepper to quinoa and toss. Blend dressing ingredients until smooth. Divide quinoa mixture among four bowls and drizzle with dressing.

Note: You can use other intact grains, such as freekeh, bulgur, or farro. Cook them according to package instructions.

PER SERVING: CALORIES 425; PROTEIN 16G; CARBOHYDRATES 69G; SUGARS 6G; TOTAL FAT 9.6G; SATURATED FAT 1.1G; SODIUM 62MG; FIBER 16.4G; BETA-CAROTENE 9,049MCG; VITAMIN C 94MG; CALCIUM 171MG; IRON 5.6MG; FOLATE 168MCG; MAGNESIUM 72MG; POTASSIUM 798MG; ZINC 1.8MG; SELENIUM 4.6MCG

Mac and Peas

SERVES 8

FOR THE SAUCE

½ cup steel cut oats

4 cups no-salt-added or low-sodium vegetable broth

6 cloves garlic

2 teaspoons reduced-sodium white miso

1 dime-size slice fresh turmeric (or about ¼ teaspoon ground)

1 cup nutritional yeast

½ teaspoon dried thyme

2 tablespoons fresh lemon juice, or to taste

2 teaspoons reduced-sodium yellow mustard

Freshly ground black pepper

FOR THE MACARONI

12 ounces elbow-shaped bean pasta, cooked according to package instructions

2 cups frozen peas

1 pound broccoli florets, steamed

FOR THE TOPPING

¼ cup ground raw almonds

¼ cup nutritional yeast

Pinch garlic powder

Preheat oven to 350°F. While the pasta is cooking, grind the steel cut oats in a blender to make a coarse flour. Combine the broth, oat flour, garlic, miso, turmeric, and nutritional yeast in a blender and purée until smooth. Transfer the purée to a saucepan, add the thyme, and cook over low heat, whisking often, until thick and bubbly. Stir in the lemon juice, mustard, and pepper, and remove from the heat. In a large mixing bowl, combine the drained pasta, sauce, peas, and broccoli. Pour into a large casserole dish or into individual ramekins. Whisk together topping ingredients and sprinkle on top. Bake 20 minutes (for the casserole) or 15 minutes (for the individual ramekins), until the tops are yellowed and firm. Let cool for 10 minutes before serving.

PER SERVING: CALORIES 372; PROTEIN 30G; CARBOHYDRATES 50G; SUGARS 3G; TOTAL FAT 5G; SATURATED FAT 0.6G; SODIUM 203MG; FIBER 17.5G; BETA-CAROTENE 649MCG; VITAMIN C 59MG; CALCIUM 209MG; IRON 8MG; FOLATE 79MCG; MAGNESIUM 72MG; POTASSIUM 287MG; ZINC 8.1MG; SELENIUM 3.1MCG

Mexican Burrito Bowls

SERVES 4

1 cup dry quinoa, rinsed (see Note)

2 ½ cups no-salt-added or low-sodium vegetable broth, divided

4 tablespoons chopped cilantro, divided

4 tablespoons fresh lime juice, divided

½ large onion, chopped

2 cloves garlic, minced

1 ½ cups cooked or 1 (15-ounce) can no-salt-added or low-sodium black beans, drained

¼ teaspoon chili powder

¼ teaspoon cumin

Pinch cayenne pepper, or to taste

1 cup shredded lettuce

TOPPING INGREDIENTS

Thawed frozen corn

Sliced avocado

Raw pumpkin seeds

No-salt-added salsa

Diced tomatoes

Sliced jalapeno peppers

Hot sauce or Sriracha sauce

Place quinoa and 2 cups of the vegetable broth in a saucepan, bring to a boil, reduce heat, and simmer for 20 minutes or until quinoa is tender and liquid is absorbed. Remove from heat and fluff with a fork. Stir in 2 tablespoons of the chopped cilantro and 2 tablespoons of the lime juice. While quinoa is cooking, heat 2–3 tablespoons water in a sauté pan and water-sauté onions until softened and tender. Add minced garlic and sauté another minute. Add black beans, remaining ½ cup vegetable broth, remaining 2 tablespoons cilantro, chili powder, cumin, and cayenne pepper. Bring to a boil, then reduce the heat and simmer for 15 minutes or until liquid is almost evaporated. Stir in remaining 2 tablespoons of lime juice. Divide quinoa between four bowls. Top each bowl with shredded lettuce, black bean mixture, and your choice of topping ingredients.

Note: You can use other intact grains, such as freekeh, bulgur, or farro. Cook them according to package instructions.

PER SERVING: CALORIES 346; PROTEIN 14G; CARBOHYDRATES 56G; SUGARS 3G; TOTAL FAT 8.5G; SATURATED FAT 1.2G; SODIUM 100MG; FIBER 12.4G; BETA-CAROTENE 633MCG; VITAMIN C 17MG; CALCIUM 76MG; IRON 4.3MG; FOLATE 223MCG; MAGNESIUM 152MG; POTASSIUM 860MG; ZINC 2.5MG; SELENIUM 5.1MCG

Mexican Cauliflower Rice and Beans

SERVES 4

1 medium head cauliflower, cut into florets (about 4 cups riced)

1 ½ cups cooked or 1 (15-ounce) can no-salt-added or low-sodium black beans

1 cup chopped onions

½ cup chopped red bell pepper

1 jalapeno pepper, seeded and chopped (see Note)

6 cloves garlic, minced

½ cup no-salt-added or low-sodium vegetable broth, or more as needed

1 ½ cups diced tomatoes

½ cup corn kernels, fresh or thawed frozen

2 teaspoons ground cumin

1 teaspoon chili powder

⅛ teaspoon ground black pepper

¼ cup chopped fresh cilantro

1 avocado, chopped

Grate cauliflower or pulse in a food processor until it resembles rice. Drain liquid from beans (called aquafaba) into a wok or large skillet. Add onions, red pepper, and jalapeno and water-sauté until they start to soften, about 2 minutes. Add riced cauliflower, garlic, and vegetable broth to the skillet. Cook for 6 minutes or until cauliflower is al dente, adding additional vegetable broth if needed to prevent sticking. Add remaining ingredients except for cilantro and avocado and cook for an additional 1–2 minutes or until heated through. Stir in cilantro. Serve topped with chopped avocado.

Note: Include the jalapeno seeds if you like it spicy.

PER SERVING: CALORIES 239; PROTEIN 11G; CARBOHYDRATES 38G; SUGARS 8G; TOTAL FAT 6.7G; SATURATED FAT 1G; SODIUM 84MG; FIBER 13.6G; BETA-CAROTENE 804MCG; VITAMIN C 116MG; CALCIUM 91MG; IRON 3.5MG; FOLATE 244MCG; MAGNESIUM 100MG; POTASSIUM 1,188MG; ZINC 1.8MG; SELENIUM 2.6MCG

Roasted Radishes and Turnips

SERVES 4

2 cups trimmed and halved radishes

2 cups trimmed and halved turnips

6 cloves garlic, minced

1 cup no-salt-added or low-sodium vegetable broth

1 tablespoon red wine vinegar

4 sprigs fresh dill

Black pepper to taste

Preheat oven to 350°F. Combine ingredients in an 8-by-9-inch baking dish. Cover with foil and bake for 25 minutes. Remove foil and bake for an additional 10 minutes.

PER SERVING: CALORIES 31; PROTEIN 1G; CARBOHYDRATES 7G; SUGARS 3G; TOTAL FAT 0.1G; SODIUM 87MG; FIBER 1.6G; BETA-CAROTENE 1MCG; VITAMIN C 18MG; CALCIUM 37MG; IRON 0.5MG; FOLATE 16MCG; MAGNESIUM 11MG; POTASSIUM 192MG; ZINC 0.3MG; SELENIUM 1MCG

Zucchini Ravioli with Cauliflower Chickpea "Ricotta"

SERVES 5

FOR THE CAULIFLOWER CHICKPEA "RICOTTA"

1 yellow onion, chopped

1 ½ cups cooked or 1 (15-ounce) can no-salt-added or low-sodium chickpeas

12 ounces frozen cauliflower rice (or ½ large head cauliflower pulsed into "rice")

¼ cup raw cashews

¼ cup hemp seeds

¼ cup nutritional yeast

6 cloves garlic

⅓ cup water

2 tablespoons lemon juice

2 tablespoons onion powder

½ tablespoon dulse flakes

½ cup chopped basil

TO ASSEMBLE THE RAVIOLI

3 zucchini

3 cups no-salt-added or low-sodium marinara sauce (see Note)

To make the cauliflower "ricotta," heat a dry pan over medium heat for 2 minutes, then add the chopped onion and stir for 3 minutes. Add the liquid from the can of chickpeas (called aquafaba) and cauliflower rice and sauté for an additional 6 minutes or until cauliflower is soft. Pulse the chickpeas in a food processor until crumbled (or mash with a fork) and set aside. In a high-powered blender, blend the cashews, hemp seeds, nutritional yeast, garlic, water, lemon juice, onion powder, and dulse flakes until super smooth and creamy, about 1–2 minutes. Place blended mixture in a large bowl along with chickpeas and cauliflower and mix until well combined. Stir in chopped basil.

To assemble the ravioli, preheat oven to 350°F. Cut ends off zucchini and slice lengthwise into very thin, slightly transparent slices, using a mandolin or vegetable peeler. Spread 1 cup of the marinara sauce on the bottom of a 9-by-11-inch baking pan. Lay two zucchini slices vertically so they overlap and then place another two overlapping slices across them to form a cross. Add a tablespoon of the ricotta in the center. Fold up all ends to form the ravioli. This should make 10–15 individual raviolis (depending on size) to place seam side down in the baking pan. Add remaining sauce on top and bake for 50 minutes. Serve warm.

Note: Use bottled marinara sauce or make your own.

PER SERVING: CALORIES 322; PROTEIN 17G; CARBOHYDRATES 43G; SUGARS 16G; TOTAL FAT 11.2G; SATURATED FAT 1.5G; SODIUM 77MG; FIBER 11.7G; BETA-CAROTENE 903MCG; VITAMIN C 58MG; CALCIUM 137MG; IRON 5MG; FOLATE 194MCG; MAGNESIUM 173MG; POTASSIUM 1,347MG; ZINC 4.2MG; SELENIUM 6.2MCG

Avocado Toast with Shredded Brussels Sprouts

SERVES 4

¼ cup chopped walnuts

4 cloves garlic, chopped

¾ pound Brussels sprouts, cut into ⅛-inch shredded slices

2 tablespoons raisins or currants

1 tablespoon nutritional yeast

Black pepper to taste

1 ripe avocado, mashed

4 slices (100% whole grain, sprouted) bread, lightly toasted

Lightly toast chopped walnuts in a small skillet over medium heat for 2–3 minutes. Heat 2 tablespoons water in a large skillet and sauté garlic for 1 minute; add shredded Brussels sprouts and cook for 2–3 minutes, until warm and just slightly wilted. Add a small amount of additional water if needed. Remove from heat and toss with toasted walnuts, currants (or raisins), nutritional yeast, and black pepper. Spread mashed avocado on toast, and then top with Brussels sprout mixture.

PER SERVING: CALORIES 233; PROTEIN 9G; CARBOHYDRATES 29G; SUGARS 6G; TOTAL FAT 10.7G; SATURATED FAT 1.4G; SODIUM 172MG; FIBER 7.7G; BETA-CAROTENE 405MCG; VITAMIN C 76MG; CALCIUM 92MG; IRON 2.8MG; FOLATE 113MCG; MAGNESIUM 58MG; POTASSIUM 625MG; ZINC 1.6MG; SELENIUM 10.1MCG

Crispy Onion Rings

SERVES 4

⅓ cup raw almond butter (at room temperature)

1 teaspoon coconut aminos

2 teaspoons balsamic vinegar

1–2 medium onions, sliced and separated into rings

½ cup almond flour (see Note)

¼ cup nutritional yeast

Preheat oven to 350°F. In a medium bowl, whisk together the almond butter (make sure it is at room temperature), coconut aminos, and vinegar. Use your fingers to rub and press the almond butter mixture onto the surface of the onions. Mix together the almond flour and nutritional yeast in a shallow dish. Dip each onion ring into the almond flour mixture. Place on a baking sheet lined with parchment paper or a nonstick silicone mat and bake 20–25 minutes or until crisp.

Note: You can purchase almond flour at most supermarkets or health food stores, or make it yourself by processing raw almonds in a food processor until finely ground.

PER SERVING: CALORIES 204; PROTEIN 12G; CARBOHYDRATES 11G; SUGARS 2G; TOTAL FAT 13.4G; SATURATED FAT 1.1G; SODIUM 61MG; FIBER 4.2G; BETA-CAROTENE 1MCG; VITAMIN C 2MG; CALCIUM 119MG; IRON 1.8MG; FOLATE 21MCG; MAGNESIUM 95MG; POTASSIUM 311MG; ZINC 2.5MG; SELENIUM 0.6MCG

Five-Seed Crackers

SERVES 12

½ cup ground flaxseeds

½ cup ground chia seeds

¼ cup unhulled sesame seeds

¼ cup hemp seeds

¼ cup chopped pumpkin seeds

½ cup chopped raw cashews

8 ounces frozen spinach, thawed, water squeezed out, chopped

¾ cup no-salt-added tomato sauce

¼ cup nutritional yeast

2 tablespoons apple cider vinegar

½ teaspoon garlic powder

½ teaspoon onion powder

1 teaspoon dried oregano

1 teaspoon chili powder

Mix all ingredients in a mixing bowl until a dough forms, adding a splash of water if needed. Place a piece of parchment paper on a large cutting board, place a third of the dough on the parchment paper, using moistened hands to handle the dough. Moisten another piece of parchment paper, place it on top of the dough and roll out to about ⅛-inch thickness. Remove the top piece of parchment paper and place the rolled-out dough (including the bottom layer of parchment) onto a rack in a dehydrator. Repeat process two more times with remaining dough.

Dehydrate at 125°F for 10 hours or until desired consistency. After about 5 hours, gradually loosen and remove bottom parchment paper. When done, break into pieces.

Note: To make in a conventional oven, spread dough evenly on a baking sheet lined with parchment paper or a nonstick silicone mat. Bake for 2 hours at lowest possible setting, then shut off the oven and leave in the oven for another 2–3 hours.

PER SERVING: CALORIES 167; PROTEIN 8G; CARBOHYDRATES 11G; SUGARS 1G; TOTAL FAT 11.5G; SATURATED FAT 1.5G; SODIUM 25MG; FIBER 6.6G; BETA-CAROTENE 1,404MCG; VITAMIN C 2MG; CALCIUM 127MG; IRON 2.8MG; FOLATE 43MCG; MAGNESIUM 127MG; POTASSIUM 305MG; ZINC 2.2MG; SELENIUM 9.4MCG

Green Pizza

SERVES 4

FOR THE CASHEW-HEMP CHEESE (SEE NOTE)
¾ cup raw cashews, soaked in water for at least 2 hours, drained
¼ cup hemp seeds
2 tablespoons nutritional yeast
2 tablespoons fresh lemon juice
¼ teaspoon garlic powder
¼ teaspoon black pepper
¼ cup water or more as needed

FOR THE PESTO (SEE NOTE)
4 cloves garlic
½ cup walnuts
¼ cup balsamic vinegar
½ cup water
½ tablespoon Dr. Fuhrman's VegiZest (or other no-salt seasoning blend, adjusted to taste)
½ tablespoon nutritional yeast
2 cups arugula
2 cups spinach

FOR THE PIZZA
4 (100% whole grain) tortillas or pitas
2 medium tomatoes, thinly sliced
Additional arugula and spinach for topping

Preheat oven to 350°F. To make the cashew-hemp cheese, place the cashews, hemp seeds, nutritional yeast, lemon juice, garlic powder, and black pepper in a food processor. Pulse until the cashews form a coarse meal. Scrape down the sides with a spatula. With the motor running, add the water and process for 10 seconds. Scrape down the sides again and continue processing until the mixture is smooth and thick, about 1–2 minutes, adding additional water if needed.

To make the pesto, add the garlic, walnuts, vinegar, water, VegiZest, and nutritional yeast to a food processor and blend at high speed. Turn to low, add the arugula and spinach and blend to a chunky consistency. Bake tortillas or pitas directly on the oven rack for 5 minutes or until just crisp. Spread a layer of pesto on each tortilla or pita, arrange sliced tomatoes on top, and add a few small dollops of cashew cheese. Bake for an additional 2–3 minutes or until toppings are warm. Remove from oven and top with arugula and spinach.

Note: You can use leftover cashew-hemp cheese and arugula pesto for other dishes. They will keep for up to 5 days in an airtight container in the refrigerator.

PER SERVING: CALORIES 353; PROTEIN 15G; CARBOHYDRATES 40G; SUGARS 6G; TOTAL FAT 16.3G; SATURATED FAT 2.4G; SODIUM 176MG; FIBER 8.8G; BETA-CAROTENE 1,757MCG; VITAMIN C 21MG; CALCIUM 122MG; IRON 4.9MG; FOLATE 84MCG; MAGNESIUM 105MG; POTASSIUM 507MG; ZINC 2.6MG; SELENIUM 4.4MCG

Herbed "Cheese" and Greens Wraps

SERVES 4

4 (100% whole grain) tortillas
4 cups spinach
1–2 cups baby arugula
2 medium tomatoes, sliced

FOR THE HERBED "CHEESE"

1 (12.3-ounce) package extra-firm silken tofu (see Note)
½ cup raw cashews
2 ½ tablespoons fresh lemon juice
½ teaspoon reduced-sodium white miso paste
2 tablespoons nutritional yeast
¼ teaspoon garlic powder
½ tablespoon finely chopped fresh parsley
½ tablespoon finely chopped fresh thyme
½ tablespoon finely chopped basil
1 shallot, minced

To make the Herbed "Cheese," place the tofu, cashews, lemon juice, miso, nutritional yeast, and garlic powder in a food processor or blender and process until smooth. Transfer to a bowl and stir in the herbs and shallot. To assemble the wraps, spread a layer of "cheese" on each tortilla. Add spinach, arugula, and sliced tomatoes, and roll up.

Note: Silken tofu is often sold in aseptic containers that do not need to be refrigerated. It is sometimes sold in a different section of grocery stores than regular tofu, which is packed in water and requires refrigeration.

PER SERVING: CALORIES 320; PROTEIN 19G; CARBOHYDRATES 37G; SUGARS 4G; TOTAL FAT 12.1G; SATURATED FAT 2G; SODIUM 271MG; FIBER 8.1G; BETA-CAROTENE 2,079MCG; VITAMIN C 23MG; CALCIUM 135MG; IRON 5.3MG; FOLATE 81MCG; MAGNESIUM 100MG; POTASSIUM 529MG; ZINC 2.3MG; SELENIUM 3.9MCG

Korean Vegetable and Mushroom Lettuce Wraps

SERVES 4

FOR THE VEGETABLES

1 medium onion, sliced

4 cups small broccoli florets

2 medium carrots, cut diagonally into ⅓-inch pieces

4 medium red bell peppers, seeded and cut into 1-inch squares

2 cups bok choy, cut into bite-size pieces

3 cups fresh mushrooms (shiitake, porcini, and/or cremini), stems removed

8 ounces fresh spinach

Romaine, Boston, or other lettuce leaves

FOR THE DRESSING

1 cup raw almonds

¼ cup unsweetened soy, almond, or hemp milk

¼ cup water

3 tablespoons unhulled sesame seeds

2 dates, pitted

2 cloves garlic, chopped

½-inch piece fresh ginger, peeled and chopped

Pinch hot pepper flakes or more to taste

Heat 2 tablespoons of water in a large sauté pan and water-sauté the onion, broccoli, carrots, and bell peppers for 5 minutes, adding more water as necessary to keep vegetables from scorching. Add the bok choy and mushrooms, cover, and simmer until vegetables are just tender. Remove the cover and cook off most of the water. Add the spinach and toss until wilted. Blend all dressing ingredients together in a high-powered blender until creamy. Add more water as necessary to adjust consistency. To serve, spoon vegetable filling onto a lettuce leaf, add a bit of dressing, and roll up.

PER SERVING: CALORIES 383; PROTEIN 15G; CARBOHYDRATES 37G; SUGARS 14G; TOTAL FAT 23.3G; SATURATED FAT 2.4G; SODIUM 143MG; FIBER 11.1G; BETA-CAROTENE 10,393MCG; VITAMIN C 271MG; CALCIUM 338MG; IRON 5.9MG; FOLATE 338MCG; MAGNESIUM 225MG; POTASSIUM 1,711MG; ZINC 3.3MG; SELENIUM 20.8MCG

Lentil Walnut Burritos with Peppers, Onions, and Salsa

SERVES 6

FOR THE LENTIL FILLING
1 cup walnuts
1 ¾ cups cooked brown lentils (see Note)
1 ½ teaspoons dried oregano
1 ½ teaspoons ground cumin
1 ½ teaspoons chili powder
2 tablespoons nutritional yeast
1 teaspoon coconut aminos
2 tablespoons water or as needed

FOR THE SALSA (OR USE A BOTTLED LOW-SODIUM SALSA SUCH AS DR. FUHRMAN'S TEX-MEX SALSA)
2 fresh tomatoes, chopped
1 small red onion, chopped
1 clove garlic, chopped
½ jalapeno chili pepper, seeded and minced
3 tablespoons fresh lime juice
1 tablespoon chopped cilantro

TO FINISH
1 large green bell pepper, thinly sliced
1 large onion, thinly sliced
6 (100% whole grain) tortillas

Place walnuts in a food processor and pulse several times to chop them. Add the cooked lentils, oregano, cumin, chili powder, nutritional yeast, and coconut aminos and pulse until mixture is thoroughly combined and crumbly. Add 1–2 tablespoons water as needed to facilitate mixing evenly. Stir together salsa ingredients. Heat 2–3 tablespoons water in a large skillet and water-sauté green bell pepper and onion until tender. To assemble burritos, spread lentil/walnut mixture on tortillas; top with sautéed peppers, onions, and salsa; and roll up.

Note: To cook dry lentils, bring 1 cup lentils and 2 cups water to a boil in a large saucepan. Reduce heat, cover, and cook for 25 minutes or until tender. Drain.

PER SERVING: CALORIES 371; PROTEIN 16G; CARBOHYDRATES 47G; SUGARS 5G; TOTAL FAT 15.1G; SATURATED FAT 1.7G; SODIUM 196MG; FIBER 13.4G; BETA-CAROTENE 371MCG; VITAMIN C 35MG; CALCIUM 101MG; IRON 5.5MG; FOLATE 139MCG; MAGNESIUM 67MG; POTASSIUM 528MG; ZINC 2MG; SELENIUM 2.9MCG

Portabella Pizza

SERVES 1

2 large portabella mushrooms, stems removed

¼ teaspoon garlic powder

¼ teaspoon dried basil

¼ teaspoon dried oregano

½ cup no-salt-added or low-sodium pasta sauce

⅓ cup thinly sliced onion

⅓ cup thinly sliced green or red bell pepper

2–3 tablespoons Nutritarian Parmesan (see Note)

Preheat oven to 350°F. Place mushrooms on a parchment-lined baking sheet, gill side up, and sprinkle with garlic powder, basil, and oregano. Bake for 6 minutes. Top with pasta sauce, onions, and peppers and a sprinkle of Nutritarian Parmesan. Bake for an additional 20 minutes or until vegetables are tender.

Note: To make Nutritarian Parmesan, place ¼ cup walnuts, almonds, or hemp seeds; ¼ cup nutritional yeast; ¼ teaspoon garlic powder; and ¼ teaspoon onion powder in a food processor and pulse until the texture of grated Parmesan is achieved. Store in an airtight container in the refrigerator.

PER SERVING: CALORIES 179; PROTEIN 11G; CARBOHYDRATES 26G; SUGARS 14G; TOTAL FAT 4.7G; SATURATED FAT 0.6G; SODIUM 59MG; FIBER 8G; BETA-CAROTENE 569MCG; VITAMIN C 30MG; CALCIUM 79MG; IRON 2.6MG; FOLATE 78MCG; MAGNESIUM 54MG; POTASSIUM 1,178MG; ZINC 3MG; SELENIUM 1.6MCG

Roasted Cauliflower Tacos

SERVES 6

1 head cauliflower, chopped into small florets

1 ½ teaspoons cumin

1 teaspoon chili powder

1 teaspoon paprika

12 corn or small whole grain tortillas

2 cups shredded cabbage

¼ cup cilantro

FOR THE SAUCE

¼ cup chopped raw almonds

1 bulb garlic, unpeeled

1 ½ cups chopped tomatoes

1 clove garlic, peeled

1 Medjool or 2 regular dates, pitted

3 tablespoons lime juice

¼ teaspoon paprika

½ teaspoon cumin

¼ teaspoon chipotle chili powder, or more to taste

Preheat oven to 375°F. Lightly oil a large baking sheet, add cauliflower, cumin, chili powder, and paprika and toss to combine. Roast for 25 minutes or until tender and lightly browned. On a separate small baking sheet, place almonds and bake for 5 minutes until lightly toasted. Remove almonds from pan. Place the bulb of unpeeled garlic on the pan and bake for 15 minutes or until garlic is soft. Let cool, cut open, and squeeze out the soft garlic paste. Place the roasted almonds and roasted garlic paste in a high-powered blender along with tomatoes, 1 clove raw peeled garlic, dates, lime juice, paprika, cumin, and chipotle chili powder. Blend until smooth and creamy. Warm tortillas in the microwave or oven. Fill each tortilla with 1–2 tablespoons of sauce, add roasted cauliflower, then top with shredded cabbage and cilantro. If desired, serve with sliced avocado, jalapeno slices, and lime wedges.

PER SERVING: CALORIES 200; PROTEIN 7G; CARBOHYDRATES 36G; SUGARS 7G; TOTAL FAT 5G; SATURATED FAT 0.6G; SODIUM 69MG; FIBER 7.6G; BETA-CAROTENE 460MCG; VITAMIN C 65MG; CALCIUM 108MG; IRON 2.2MG; FOLATE 80MCG; MAGNESIUM 81MG; POTASSIUM 655MG; ZINC 1.3MG; SELENIUM 4.3MCG

Tofu and Sun-Dried Tomato Burgers

SERVES 4

6 unsulfured, no-salt-added sun-dried tomatoes, soaked 60 minutes in enough warm water to cover

1 tablespoon chia seeds

9 ounces spinach

14-ounce block extra-firm tofu, frozen and thawed (see Note)

¼ cup finely chopped red onion

2 cloves garlic, finely chopped

½ cup whole wheat panko or bread crumbs

¼ cup unhulled sesame seeds, toasted

2 teaspoons coconut aminos

¼ teaspoon black pepper

Pinch red pepper flakes, or to taste

Preheat oven to 350°F. Drain the sun-dried tomatoes and chop, reserving the soaking water. Combine 3 tablespoons of the soaking water with chia seeds and allow to sit for at least 10 minutes. Heat a large wok or sauté pan, add the remaining tomato soaking water, and sauté spinach until wilted. Chop it finely and place in a large bowl. Crumble the thawed tofu into the bowl. Mix in the chia seeds, sun-dried tomatoes, red onion, garlic, panko, toasted sesame seeds, coconut aminos, black pepper, and red pepper flakes. Form into eight burgers and place on a baking pan lined with parchment paper. Bake for 15 minutes, carefully flip, and bake for another 10 minutes or until lightly browned. If desired, serve on a 100% whole grain pita with lettuce, tomato, red onion, and no-salt ketchup or Dr. Fuhrman's Nutritarian Ketchup.

Note: Freezing and thawing tofu gives it a "meaty" texture. Thaw tofu for about 6 hours.

PER SERVING: CALORIES 237; PROTEIN 16G; CARBOHYDRATES 21G; SUGARS 3G; TOTAL FAT 12.1G; SATURATED FAT 1.5G; SODIUM 278MG; FIBER 5G; BETA-CAROTENE 3,606MCG; VITAMIN C 21MG; CALCIUM 375MG; IRON 6.1MG; FOLATE 176MCG; MAGNESIUM 157MG; POTASSIUM 697MG; ZINC 2.5MG; SELENIUM 21.9MCG

Tofu Crackers and Tofu Jerky Pizza

SERVES 4

15 ounces extra-firm tofu, sliced very thin, no more than ⅛-inch thick

¼ cup aquafaba for brushing tofu (see Note)

¼ cup sesame, poppy, hemp, or flax seeds

Salt-free seasoning, if desired, such as Dr. Fuhrman's MatoZest or VegiZest

Place sliced tofu on dehydrator sheets. Brush lightly with aquafaba. Top with your choice of seeds and salt-free seasoning. Set dehydrator to 120°F. Dehydrate for 8 hours or until desired crispness. If using a conventional oven, place aquafaba-brushed tofu slices on a parchment-lined baking tray and bake at lowest setting for 3–4 hours or until desired crispness. As they cool, they will harden to cracker or jerky consistency.

For Tofu Jerky Pizza, spread with Intense Marinara Sauce or another low-salt tomato sauce.

Note: Aquafaba is the typically discarded cooking liquid found in retail cans and boxes of beans or the liquid left over from cooking dried beans.

PER SERVING: CALORIES 148; PROTEIN 12G; CARBOHYDRATES 4G; SUGARS 1G; TOTAL FAT 10.7G; SATURATED FAT 1.2G; SODIUM 9MG; FIBER 1.5G; VITAMIN C 1MG; CALCIUM 274MG; IRON 3.3MG; FOLATE 27MCG; MAGNESIUM 88MG; POTASSIUM 182MG; ZINC 1.9MG; SELENIUM 16.9MCG

Sweet Potato Toast with Roasted Broccoli

SERVES 4

1 large sweet potato, peeled and cut into 1-inch pieces
½ cup freshly squeezed orange juice
1 cup water
¼ teaspoon black pepper
1 head broccoli, cut into large florets

1 tablespoon fresh lemon juice
4 slices (100% whole grain) bread, lightly toasted
1 tablespoon unhulled sesame seeds, toasted

Combine sweet potato, orange juice, and water in a small saucepan. Bring to a boil, reduce heat, and simmer until sweet potato is very soft and liquid has evaporated, about 20 minutes. Mash and season with black pepper.

Preheat oven to 350°F. Place broccoli on a baking sheet lined with parchment paper or a nonstick silicone mat and roast until tender, about 20 minutes. Coarsely chop and toss with lemon juice. Spread toast with mashed sweet potato, top with roasted broccoli, and sprinkle with sesame seeds.

PER SERVING: CALORIES 184; PROTEIN 10G; CARBOHYDRATES 34G; SUGARS 8G; TOTAL FAT 2.8G; SATURATED FAT 0.5G; SODIUM 214MG; FIBER 7.6G; BETA-CAROTENE 3,325MCG; VITAMIN C 153MG; CALCIUM 140MG; IRON 2.5MG; FOLATE 125MCG; MAGNESIUM 77MG; POTASSIUM 750MG; ZINC 1.5MG; SELENIUM 18MCG

Veggie-Bean Burgers

SERVES 4

1 tablespoon ground flaxseeds
1 ½ cups cooked or 1 (15-ounce) can no-salt-added or low-sodium red kidney beans
½ medium onion, finely chopped
6 cloves garlic, minced
1 small carrot, grated
1 teaspoon cumin

1 teaspoon chili powder
½ teaspoon garlic powder
¼ teaspoon paprika
¼ teaspoon black pepper
¼ cup finely chopped kale
1–2 tablespoons rolled oats, if needed to adjust consistency

Preheat oven to 350°F. Combine flaxseeds and liquid from the canned beans (called aquafaba) in a small bowl and let stand for 10 minutes. Heat 2–3 tablespoons water in a sauté pan and water-sauté the onion, garlic, and carrot until the onions are translucent and the carrots begin to soften. Transfer to a large bowl and stir in the spices. Partially mash the beans with a fork or pulse in a food processor. Add to the vegetables and spices. Stir in the flax and aquafaba mixture and the kale. Form into four burgers and place on a baking pan lined with parchment paper or a nonstick silicone mat. (If mixture is too wet, add rolled oats to adjust consistency). Bake for 15 minutes, carefully flip, and bake for another 10 minutes or until lightly browned.

PER SERVING: CALORIES 116; PROTEIN 7G; CARBOHYDRATES 20G; SUGARS 2G; TOTAL FAT 1.4G; SATURATED FAT 0.2G; SODIUM 26MG; FIBER 6.5G; BETA-CAROTENE 1,566MCG; VITAMIN C 8MG; CALCIUM 49MG; IRON 2.8MG; FOLATE 95MCG; MAGNESIUM 45MG; POTASSIUM 402MG; ZINC 1MG; SELENIUM 2MCG

DESSERTS

Almond Blondies

SERVES 16

2 tablespoons ground flaxseeds

½ cup water

2 teaspoons baking powder

¾ cup unsweetened almond milk

⅓ cup raw almond butter

½ cup no-oil-added, unsalted peanut butter

¾ cup almond flour

1 ½ tablespoons maca powder

3 very ripe medium bananas

9 Medjool dates, pitted

1 tablespoon vanilla bean powder or 2 teaspoons alcohol-free vanilla extract

2 cups old-fashioned or steel cut oats

1 ½ cups whole wheat flour (or oat flour for gluten-free)

¾ cup raisins

Preheat oven to 350°F. Whisk together flaxseeds and water and allow to sit for 5 minutes. Add flaxseed mixture and all ingredients except oats, whole wheat flour, and raisins to a food processor and process until well-combined. Pour into a large bowl and mix in oats, flour, and raisins. Pour the batter into an 8-by-8-inch baking pan and spread evenly using a large spoon. Bake for 35 minutes or until top is golden and a toothpick inserted into the center comes out clean. Baking time may vary, so after 35 minutes, check every 5 minutes. Let cool and cut into squares.

PER SERVING: CALORIES 269; PROTEIN 8G; CARBOHYDRATES 41G; SUGARS 14G; TOTAL FAT 10.5G; SATURATED FAT 1.7G; SODIUM 15MG; FIBER 5.9G; BETA-CAROTENE 18MCG; VITAMIN C 6MG; CALCIUM 92MG; IRON 4MG; FOLATE 24MCG; MAGNESIUM 69MG; POTASSIUM 495MG; ZINC 1MG; SELENIUM 8.2MCG

Blueberry Cherry Crumble

SERVES 6

FOR THE CRUMB TOPPING
½ cup old-fashioned oats
½ cup almonds, lightly toasted
¼ cup sprouted lentil flour or almond flour
3 tablespoons raw almond butter, at room temperature
1 Medjool or 2 regular dates, pitted, soaked in warm water for 1 minute, then drained and mashed
1 teaspoon almond extract
1 teaspoon cinnamon

FOR THE FILLING
12 ounces fresh or frozen blueberries
8 ounces frozen cherries
½ teaspoon agar powder (see Note)
¼ cup water
1 tablespoon fresh lemon juice
1 Medjool date or 2 regular dates, pitted and chopped

Preheat oven to 350°F. To prepare topping, pulse oats and almonds in a high-powered blender until crumbly. Place in a medium bowl and stir in flour, almond butter, mashed date, almond extract, and cinnamon. Mix until completely combined. To make the filling, combine blueberries, cherries, agar, water, lemon juice, and date in a medium saucepan and cook over low heat until bubbly, about 10 minutes. Pour into an 8-by-8-inch or 1½-liter baking dish and top with crumb mixture. Bake for 15 minutes. Let cool for 10 minutes before serving.

Note: Agar is a vegetarian gelatin made from seaweed. It can be used as a thickening agent in a variety of recipes. If using agar flakes instead of powder, double the amount.

PER SERVING: CALORIES 284; PROTEIN 42G; CARBOHYDRATES 137G; SUGARS 12G; TOTAL FAT 12.8G; SATURATED FAT 1G; SODIUM 9MG; FIBER 9.6G; BETA-CAROTENE 225MCG; VITAMIN C 3MG; CALCIUM 82MG; IRON 69MG; FOLATE 15MCG; MAGNESIUM 69MG; POTASSIUM 288MG; ZINC 1.1MG; SELENIUM 0.6MCG

Carrot Cake Bites

SERVES 20

½ cup chopped dried pineapple

¼ cup water

1 cup pecans

1 cup walnuts

1 cup pitted dates

1 cup unsweetened shredded coconut

½ cup grated carrots

1 teaspoon cinnamon

½ teaspoon ground ginger

¼ teaspoon nutmeg

Place chopped pineapple in sealed container or plastic bag, add ¼ cup water, and let it soak overnight or for a few hours, mixing and turning occasionally. Pulse pecans and walnuts in a food processor until finely chopped. Remove nuts from food processor. Place dates and soaked dried pineapple with soaking liquid in the food processor and process until the mixture is like a paste and starts to form a ball. Return nuts to the food processor along with coconut, grated carrots, and spices, and pulse until mixture is well-combined. Form into balls. Makes about 20.

PER SERVING: CALORIES 131; PROTEIN 2G; CARBOHYDRATES 11G; SUGARS 8G; TOTAL FAT 9.8G; SATURATED FAT 3.2G; SODIUM 4MG; FIBER 2.4G; BETA-CAROTENE 230MCG; CALCIUM 17MG; IRON 0.6MG; FOLATE 9MCG; MAGNESIUM 23MG; POTASSIUM 153MG; ZINC 0.5MG; SELENIUM 1.5MCG

Cherry Apricot Oatmeal Cookies

SERVES 12

1 large ripe banana

½ cup raw almond butter

¼ cup chopped dates

1 teaspoon pure vanilla bean powder or alcohol-free vanilla extract

1 teaspoon almond extract

½ cup chopped unsulfured dried apricots

½ cup chopped unsweetened dried cherries

2 cups old-fashioned rolled oats

2 tablespoons ground flaxseeds

Preheat oven to 335°F. Place banana in a medium bowl and mash well; stir in the almond butter; mash in the dates, vanilla, and almond extract until well-combined. Add the apricots, cherries, oats, and flaxseed. Divide mixture into a lightly oiled muffin pan, using only enough dough to make cookies that are about ¾ inch in height. Bake for 12 minutes or until golden. Using a knife, loosen cookies and place on a cooling rack. Mixture can also be poured onto a nonstick baking sheet. Bake for 15 minutes or until golden, cool, then cut into squares.

PER SERVING: CALORIES 195; PROTEIN 5G; CARBOHYDRATES 31G; SUGARS 13G; TOTAL FAT 7.4G; SATURATED FAT 0.7G; SODIUM 2MG; FIBER 4.5G; BETA-CAROTENE 131MCG; VITAMIN C 2MG; CALCIUM 50MG; IRON 4MG; FOLATE 11MCG; MAGNESIUM 45MG; POTASSIUM 269MG; ZINC 0.5MG; SELENIUM 0.8MCG

Chocolate Almond Pudding

SERVES 4

3 ripe bananas
½ medium avocado
¼ cup unsweetened natural cocoa powder

¼ cup raw almond butter
1 Medjool or 2 regular dates, pitted
1 teaspoon pure vanilla bean powder or alcohol-free vanilla extract

Add all ingredients to a high-powered blender and blend until smooth. Refrigerate until ready to serve.

PER SERVING: CALORIES 233; PROTEIN 6G; CARBOHYDRATES 32G; SUGARS 16G; TOTAL FAT 12.3G; SATURATED FAT 1.5G; SODIUM 5MG; FIBER 7.3G; BETA-CAROTENE 39MCG; VITAMIN C 9MG; CALCIUM 72MG; IRON 1.7MG; FOLATE 44MCG; MAGNESIUM 102MG; POTASSIUM 644MG; ZINC 1.2MG; SELENIUM 2.1MCG

Creamy Pumpkin Chia Pudding

SERVES 2

1 cup unsweetened soy, hemp, or almond milk
3 Medjool or 6 regular dates, pitted and chopped
¼ cup chia seeds
1 teaspoon vanilla bean powder or alcohol-free vanilla extract

½ teaspoon cinnamon
½ cup pumpkin purée
2 tablespoons unsweetened shredded coconut, toasted
2 tablespoons chopped pecans, toasted

Blend nondairy milk and dates in a high-powered blender until well-blended. Add chia seeds, vanilla, and cinnamon and blend for 30 seconds. Divide pumpkin purée between two serving dishes, pour blended chia mixture over pumpkin, cover, and chill overnight or for at least 4 hours. Before serving, top with coconut and pecans. This will keep refrigerated for up to 5 days.

PER SERVING: CALORIES 314; PROTEIN 10G; CARBOHYDRATES 34G; SUGARS 17G; TOTAL FAT 17.5G; SATURATED FAT 4.9G; SODIUM 24MG; FIBER 11.7G; BETA-CAROTENE 1,308MCG; VITAMIN C 3MG; CALCIUM 166MG; IRON 3.3MG; FOLATE 33MCG; MAGNESIUM 131MG; POTASSIUM 453MG; ZINC 1.8MG; SELENIUM 12.7MCG

Dark Chocolate Mousse

SERVES 3

3 ½ ounces 100% cocoa chocolate, roughly chopped

12 ounces silken tofu, drained and brought to room temperature

3 Medjool or 6 regular dates, soaked in warm water until soft, then drained, pitted, and chopped

1 teaspoon pure vanilla bean powder or alcohol-free vanilla extract

½ teaspoon agar powder (see Note)

OPTIONAL TOPPING INGREDIENTS

100% cocoa chocolate shavings

Raspberries or blackberries

Fresh mint leaves

Raw almonds, toasted and chopped

Melt chocolate on medium power in the microwave in 30-second intervals, stirring each time, until melted. Set aside to cool. Meanwhile, place silken tofu, dates, vanilla, and agar in a high-powered blender and blend until smooth. Add melted chocolate and blend to combine. Transfer the mixture into three ramekins or small dishes and refrigerate for 30 minutes. Garnish with your choice of toppings. Will keep in the refrigerator for 4–5 days.

Note: Agar is a vegetarian gelatin made from seaweed. You can use it as a thickening agent in a variety of recipes. If using agar flakes instead of powder, double the amount.

PER SERVING: CALORIES 299; PROTEIN 10G; CARBOHYDRATES 32G; SUGARS 18G; TOTAL FAT 20.4G; SATURATED FAT 11.1G; SODIUM 15MG; FIBER 7.3G; BETA-CAROTENE 21MCG; CALCIUM 92MG; IRON 7.2MG; FOLATE 21MCG; MAGNESIUM 164MG; POTASSIUM 661MG; ZINC 4MG; SELENIUM 2.8MCG

No-Bake Brownies

SERVES 16

½ cup almonds
½ cup cashews
¼ cup hemp seeds
¾ cup walnuts, divided
2 cups Medjool dates, pitted

½ cup unsweetened cocoa powder
⅓ cup no-oil-added, unsalted peanut butter
2 tablespoons cocoa nibs

Line an 8-by-8-inch square pan with parchment paper. Place the almonds, cashews, hemp seeds, and ½ cup of the walnuts in a food processor. Process until finely ground. Add the dates and process until mixture is like a paste and begins to form a ball. Add the cocoa powder, peanut butter, and cocoa nibs; process and knead together until well-combined. Press mixture evenly into pan. Coarsely chop the remaining ¼ cup of walnuts. Sprinkle over the top of brownies. Refrigerate for 1 hour. Cut into 16 pieces.

PER SERVING: CALORIES 213; PROTEIN 6G; CARBOHYDRATES 21G; SUGARS 13G; TOTAL FAT 13.8G; SATURATED FAT 2.2G; SODIUM 3MG; FIBER 4G; BETA-CAROTENE 2MCG; CALCIUM 35MG; IRON 1.8MG; FOLATE 24MCG; MAGNESIUM 92MG; POTASSIUM 334MG; ZINC 1.4MG; SELENIUM 3.4MCG

Pistachio Gelato

SERVES 8

¾ cup unsalted shelled pistachios, divided
2 cups water
½ cup raw cashews
⅔ cup silken tofu

8 Medjool or 16 regular dates, pitted
1 ½ cups frozen mango chunks
1 small or ½ large avocado
1 handful raw spinach
¼ teaspoon almond extract

Coarsely chop ¼ cup of the pistachios and set aside. Blend the other ½ cup of pistachios along with remaining ingredients in a high-powered blender until smooth and creamy. Stir in chopped pistachios. Freeze in an ice-cream maker or just freeze.

PER SERVING: CALORIES 229; PROTEIN 6G; CARBOHYDRATES 29G; SUGARS 20G; TOTAL FAT 12.2G; SATURATED FAT 1.7G; SODIUM 9MG; FIBER 4.7G; BETA-CAROTENE 404MCG; VITAMIN C 11MG; CALCIUM 55MG; IRON 1.6MG; FOLATE 43MCG; MAGNESIUM 62MG; POTASSIUM 483MG; ZINC 1MG; SELENIUM 2.7MCG

Strawberry Banana Ice Cream

SERVES 4

3 ripe bananas, frozen (see Note)
⅓ cup unsweetened soy, hemp, or almond milk
2 cups frozen strawberries

2 tablespoons chopped walnuts
1 tablespoon ground flaxseeds
½ teaspoon pure vanilla bean powder or alcohol-free vanilla extract

Blend ingredients in a high-powered blender until smooth and creamy. Add additional nondairy milk if needed to adjust consistency.

For Blueberry Banana Ice Cream, substitute frozen blueberries.

Note: Freeze ripe bananas at least 8 hours in advance. Peel bananas and seal in a plastic bag before freezing.

PER SERVING: CALORIES 146; PROTEIN 3G; CARBOHYDRATES 29G; SUGARS 15G; TOTAL FAT 3.4G; SATURATED FAT 0.4G; SODIUM 13MG; FIBER 4.6G; BETA-CAROTENE 44MCG; VITAMIN C 38MG; CALCIUM 29MG; IRON 1.1MG; FOLATE 39MCG; MAGNESIUM 49MG; POTASSIUM 479MG; ZINC 0.4MG; SELENIUM 3MCG

Summer Fruit Pie with
Simple Almond Coconut Pie Crust

SERVES 8

FOR THE PIE CRUST
2/3 cup raw almonds
2/3 cup shredded unsweetened coconut
2/3 cup old-fashioned rolled oats
4 Medjool or 8 regular dates, pitted
1/4 cup water

FOR THE FILLING
2–3 bananas, sliced
1 teaspoon fresh lemon juice
2 kiwis, peeled and sliced
1 quart organic strawberries, sliced
1 pint blueberries and/or blackberries
1/2 cup unsweetened vanilla soy, hemp, or almond milk
1 1/4 cups frozen strawberries (or an additional pint fresh organic strawberries)
2 Medjool or 4 regular dates, pitted
2 teaspoons chopped fresh mint leaves, if desired

To make the pie crust, place almonds in a food processor or high-powered blender and process until finely chopped. Add remaining ingredients and process until well-blended. Remove from food processor, knead into a ball, and then flatten and press into a 9-inch pie plate to form crust.

To make the filling, spread bananas on the crust, pressing down slightly. Sprinkle lemon juice over the bananas. Place kiwis, strawberries, and blueberries over the bananas. If desired, reserve some fruit to decorate the top of the pie. In a high-powered blender, blend nondairy milk, frozen strawberries, dates, and mint until smooth. Pour mixture over the fruit. Decorate with additional fruit as desired. Cover and freeze for at least 2 hours before serving.

PER SERVING: CALORIES 277; PROTEIN 6G; CARBOHYDRATES 45G; SUGARS 29G; TOTAL FAT 10.7G; SATURATED FAT 1.6G; SODIUM 12MG; FIBER 8.8G; BETA-CAROTENE 47MCG; VITAMIN C 91MG; CALCIUM 90MG; IRON 1.9MG; FOLATE 56MCG; MAGNESIUM 91MG; POTASSIUM 666MG; ZINC 1MG; SELENIUM 3MCG

Vanilla or Chocolate Nice Cream

SERVES 4

¼ cup walnuts (or macadamia nuts)

2 ripe bananas, frozen (see Note)

⅓ cup unsweetened soy, hemp, or almond milk (frozen ahead of time)

1 teaspoon pure vanilla bean powder or alcohol-free vanilla extract

Using a high-powered blender, blend walnuts to a fine powder. Add remaining ingredients and blend on high speed until smooth and creamy. Serve immediately or store in freezer for later use.

To make Chocolate Nice Cream, add 2 tablespoons of natural, unsweetened cocoa powder and 2 pitted Medjool or 4 regular dates.

Note: Freeze ripe bananas at least 8 hours in advance. Peel bananas and seal in a plastic bag before freezing.

PER SERVING: CALORIES 138; PROTEIN 2G; CARBOHYDRATES 25G; SUGARS 16G; TOTAL FAT 4.6G; SATURATED FAT 0.5G; SODIUM 11MG; FIBER 2.9G; BETA-CAROTENE 27MCG; VITAMIN C 5MG; CALCIUM 22MG; IRON 0.6MG; FOLATE 23MCG; MAGNESIUM 37MG; POTASSIUM 346MG; ZINC 0.4MG; SELENIUM 1.9MCG

In Closing

A wide array of opportunities regarding our health are available to us in the modern world. We can find dangerous foods for sale on almost every street corner, which enable us to eat ourselves to death; but at the same time, we have the opportunity to buy a wide variety of healthful produce, which had not been an option for prior generations. We now have wide access all year long to a large assortment of fresh leafy vegetables, mushrooms, seeds high in omega-3 fatty acids, and berries. Excellent health is now a choice that is truly in our hands.

Many factors play a role in determining the quality of our health, and our lives in general. Having a rich purpose in life, enjoying our work and leisure time, feeling gratitude, meditating, caring for the Earth, having goodwill toward others—all of these enhance our life's journey. When we learn to appreciate the beauty that surrounds us, when we laugh and have fun, we learn to cherish—and value—every minute we have.

The food we eat is the most critical factor that determines our health and happiness. We have one body and must protect, nurture, and cherish it, for that is the foundation of a pleasurable life.

I wish you great health, much fun, and happiness always. They can be yours.

Acknowledgments

I appreciate my skilled, supportive team at DrFuhrman.com. Specifically, Deana Ferrari, PhD, who assisted me with research collection and interpretation; Linda Popescu, RD, who helped with recipes, nutritional analysis, and menus; Lauren Russell and Tim Shay, who provided diagrams and graphs; Eileen Murphy, who proofread; Mary Becker, who helped with preparing, testing, and tweaking recipes to my liking; Doris Wellfield, who helped contact and interview people for testimonials; and Lisa Fuhrman, who supplied important feedback and editing. I am also appreciative of the wonderful team at HarperOne, especially Gideon Weil, for his strong support of my work and his early vision for my series of books to spearhead nutrition-based healthcare, and Senior Director of Publicity Melinda Mullin.

Notes

Introduction: The Nutritarian Diet

1 Fuhrman J, Singer M. Improved cardiovascular parameter with a nutrient-dense, plant-rich diet-style: a patient survey with illustrative cases. *Am J Lifestyle Med.* 2015;11(3)264–73, doi:10.1177/1559827615611024.

2 Bertoia ML, Mukamai KJ, Cahill LE et al. Changes in intake of fruits and vegetables and weight change in United States men and women followed for up to 24 years: analysis from three prospective cohort studies. *PLOS Med.* 2015;12(9):e1001878.

3 Sartorelli DS, Franco LJ, Cardosa MA. High intake of fruits and vegetables predicts weight loss in Brazilian overweight adults. *Nutr Res.* 2008;28(4):233–38.

4 Jenkins DJ, Kendall CW, Faulkner D et al. A dietary portfolio approach to cholesterol reduction: combined effects of plant sterols, vegetable proteins, and viscous fibers in hypercholesterolemia. *Metabolism.* 2002;51(12):1596–604.

5 Fuhrman, Singer. Improved cardiovascular parameter.

6 Dunaief DM, Fuhrman J, Dunaief JL, Ying G. Glycemic and cardiovascular parameters improved in type 2 diabetes with the high nutrient density (HND) diet. *Open J Prev Med.* 2012;2(3):364–71.

7 Fuhrman J, Sarter B, Glaser D, Accocella S. Changing perceptions of hunger on a high nutrient density diet. *Nutr J.* 2010;9:51, doi:10.1186/1475-2891-9-51.

8 Fuhrman et al. https://nutritionj.biomedcentral.com/articles/10.1186/1475-2891-9-51.

Chapter One: The Science of Longevity

1 Sergiev PV, Dontsova OA, Berezkin GV. Theories of aging: an ever-evolving field. *Acta Naturae* 2015;7(1):9–18.

2 Jin K. Modern biological theories of aging. *Aging Dis.* 2010;1(2):72–74; Madison HE. Theories of aging. In: Lueckenotte AG, ed., *Gerontologic Nursing.* St. Louis: Mosby, 2002.

3 Barzilai N, Cuervo AM, Austad S. Aging as a biological target for prevention and therapy. *JAMA.* 2018;320(13):1321–22; Tchkonia T, Kirkland JL. Aging, cell senescence, and chronic disease: emerging therapeutic strategies. *JAMA.* 2018;320(13):1319–20.

4 Ruiz-Canela M, Bes-Rastrollo M, Martinez-Gonzalez MA. The role of dietary inflammatory index in cardiovascular disease, metabolic syndrome and mortality. *Int J Mol Sci.* 2016;17(8), E1265, doi:10.3390/ijms17081265; Federico A, Morgillo F, Tuccillo C et al. Chronic inflammation and oxidative stress in human carcinogenesis. *Int J Cancer.* 2007;121:2381–86; Fowler ME, Akinyemiju TF. Meta-analysis of the association between dietary inflammatory index (DII) and cancer outcomes. *Int J Cancer.* 2017;141:2215–27; Higashi Y, Noma K, Yoshizumi M, Kihara Y. Endothelial function and oxidative stress in cardiovascular diseases. *Circ J.* 2009;73:411–18.

5 Marcon F, Siniscalchi E, Crebelli R et al. Diet-related telomere shortening and chromosome stability. *Mutagenesis.* 2012;27:49–57; Houben JM, Moonen HJ, van Schooten FJ, Hageman GJ. Telomere length assessment: biomarker of chronic oxidative stress? *Free Radic Biol Med.* 2008;44:235–46; von Zglinicki T, Martin-Ruiz CM. Telomeres as biomarkers for ageing and age-related diseases. *Curr Mol Med.* 2005 Mar;5(2):197–203.

6 Cassidy A, De Vivo I, Liu Y et al. Associations between diet, lifestyle factors, and telomere length in women. *Am J Clin Nutr.* 2010;91:1273–80; Ludlow AT, Ludlow LW, Roth SM. Do telomeres adapt to physiological stress? Exploring the effect of exercise on telomere length and telomere-related proteins. *Biomed Res Int.* 2013; 2013:601368; Tiainen AM, Mannisto S, Blomstedt PA et al. Leukocyte telomere length and its relation to food and nutrient intake in an elderly population. *Eur J Clin Nutr.* 2012;66:1290–94; LaRocca TJ, Seals DR, Pierce GL. Leukocyte telomere length is preserved with aging in endurance exercise-trained adults and related to maximal aerobic capacity. *Mech Ageing Dev.* 2010;131:165–67; Marcon F, Siniscalchi E, Crebelli R et al. Diet-related telomere shortening and chromosome stability. *Mutagenesis.* 2012;27:49–57; Min KB, Min JY. Association between leukocyte telomere length and serum carotenoid in US adults. *Eur J Nutr.* 2017;56(3):1045–52; Cherkas LF, Hunkin JL, Kato BS et al. The association between physical activity in leisure time and leukocyte telomere length. *Arch Intern Med.* 2008;168:154–58.

7 Quach A, Levine ME, Tanaka T et al. Epigenetic clock analysis of diet, exercise, education, and lifestyle factors. *Aging* (Albany NY). 2017;9:419–46.

8 Astrup A, Gotzsche PC, van de Werken K et al. Meta-analysis of resting metabolic rate in formerly obese subjects. *Am J Clin Nutr.* 1999;69:1117–22; Ravussin E, Lillioja S, Knowler WC et al. Reduced rate of energy expenditure as a risk factor for body-weight gain. *N Engl J Med.* 1988;318:467–72.

9 Joseph JA, Denisova N, Fisher D et al. Age-related neurodegeneration and oxidative stress: putative nutritional intervention. *Neurol Clin.* 1998;16:747–55.

10 Hulbert AJ, Pamplona R, Buffenstein R et al. Life and death: metabolic rate, membrane composition, and life span of animals. *Physiol Rev.* 2007;87:1175–213; Farooqui T, Farooqui AA. Aging: an important factor for the pathogenesis of neurodegenerative diseases. *Mech Ageing Dev.* 2009;130:203–15.

11 Speakman JR, Selman C, McLaren JS et al. Living fast, dying when? The link between aging and energetics. *J Nutr.* 2002;132:1583S–97S.

12 Jumpertz R, Hanson RL, Sievers ML et al. Higher energy expenditure in humans predicts natural mortality. *J Clin Endocrinol Metab.* 2011;96(6):E972–76.

13 Bouchard C, Tremblay A, Nadeau A et al. Genetic effect in resting and exercise metabolic rates. *Metabolism.* 1989;38:364–70.

14 Martin CK, Heilbronn LK, de Jonge L et al. Effect of calorie restriction on resting metabolic rate and spontaneous physical activity. *Obesity.* 2007;15:2964–73; Roberts SB, Fuss P, Evans WJ et al. Energy expenditure, aging and body composition. *J Nutr.* 1993;123:474–80.

15 Fontana L. The scientific basis of caloric restriction leading to longer life. *Curr Opin Gastroenterol.* 2009;25:144–50.

16 Broeder CE, Burrhus KA, Svanevik LS et al. The effects of aerobic fitness on resting metabolic rate. *Am J Clin Nutr.* 1992;55:795–801.

17 Manini TM, Everhart JE, Patel KV et al. Daily activity energy expenditure and mortality among older adults. *JAMA.* 2006;296:171–79.

18 Pontzer H, Raichlen DA, Godon AD et al. Primate energy expenditure and life history. *PNAS.* 2014;111(4):1433–37. https://doi.org/10.1073/pnas.1316940111; Brown JH, Gillooly JF, Allen A et al. Toward a metabolic theory of ecology. *ESA: Ecology,* 1 July 2004. https://doi .org/10.1890/03-9000.

19 Rizzo MR, Mari D, Barbieri M et al. Resting metabolic rate and respiratory quotient in human longevity. *J Clin Endocrinol Metab.* 2005;90(1):409–13.

20 Bano A, Khana K, Chaker L et al. Association of thyroid function with life expectancy with and without cardiovascular disease: the Rotterdam Study. *JAMA Intern Med.* 2017;177(11):1650–57.

21 Chaker L, van den Berg ME, Neimeijer MN et al. Thyroid function and sudden cardiac death. *Circulation*. 2016;134:713–22.

22 Baumgartner C, de Costa BR, Collet TH et al. Thyroid function within the normal range, subclinical hypothyroidism and the risk of atrial fibrillation. *Circulation*. 2017;136:2100–116.

23 Ruggierro C, Metter EJ, Melenovsky V et al. High basal metabolic fate is a risk factor for mortality: the Baltimore Longitudinal Study of Aging. *J Gerontol A Biol Sci Med Sci*. 2008;63(7):698–706.

24 US Centers for Disease Control and Prevention. National Center for Health Statistics. Health, United States, 2017. Table 14. Life expectancy at birth and at age 65, by sex: Organisation for Economic Co-operation and Development (OECD) countries, selected years 1980–2015. https://www.cdc.gov/nchs/hus/contents2017.htm?search=Life_expectancy.

25 World Health Organization. Global Health Observatory data repository. Healthy life expectancy (HALE): Data by country. http://apps.who.int/gho/data/view.main.HALEXv?lang=en.

26 Aune D, Giovannucci E, Boffetta P et al. Fruit and vegetable intake and the risk of cardiovascular disease, total cancer and all-cause mortality: a systematic review and dose-response meta-analysis of prospective studies. *Int J Epidemiol*. 2017 Jun 1;46(3):1029–56; Kwok CS, Gulati M, Michos ED et al. Dietary components and risk of cardiovascular disease and all-cause mortality: a review of evidence from meta-analyses. *Eur J Prev Cardiol*. 2019;26(13):1415–29; Mori N, Shimazu T, Charvat H et al. JPHC Study Group. Cruciferous vegetable intake and mortality in middle-aged adults: a prospective cohort study. *Clin Nutr*. 2019 Apr;38(2):631–43; Zhang X, Shu XO, Xiang YB et al. Cruciferous vegetable consumption is associated with a reduced risk of total and cardiovascular disease mortality. *Am J Clin Nutr*. 2011 Jul;94(1):240–46.

27 Ames BN. Prolonging healthy aging: longevity vitamins and proteins. *Proc Natl Acad Sci USA*. 2018;115:10836–44.

Chapter Two: Your Hormones and Your Health

1 Shanmugalingam T, Bosco C, Ridley AJ, Van Hemelrijck MV. Is there a role for IGF-1 in the development of second primary cancers? *Cancer Med*. 2016;5(11):3353–67.

2 Cleary MP, Grossmann ME. Minireview: Obesity and breast cancer: the estrogen connection. *Endocrinology*. 2009;150:2537–42; Nelles JL, Hu WY, Prins GS et al. Estrogen action and prostate cancer. *Expert Rev Endocrinol Metab*. 2011;6(3):437–51.

3 Cleary MP, Grossmann ME. Obesity and breast cancer: the estrogen connection. *Endocrinology*. 2009;150(6):2537–42.

4 Iyengar NM, Hudis CA, Danenberg AJ. Obesity and inflammation: new insights into breast cancer development and progression. *Am Soc Clin Oncol Educ Book*. 2013;33:46–51; Poloz Y, Stambolic V. Obesity and cancer, a case for insulin signaling. *Cell Death Dis*. 2015;6:e2037; Rausch LK, Netzer NC, Hoegel J et al. The linkage between breast cancer, hypoxia, and adipose tissue. *Front Oncol*. 2017;7:211.

5 Gallagher EJ, LeRoith D. The proliferating role of insulin and insulin-like growth factors in cancer. *Trends Endocrinol Metab*. 2010;21(10):610–18.

6 Sluijs I, van der Schouw YT, van der A DL et al. Carbohydrate quantity and quality and risk of type 2 diabetes in the European Prospective Investigation into Cancer and Nutrition–Netherlands (EPIC-NL) study. *Am J Clin Nutr*. 2010;92(4):905–11; Barclay AW, Petocz P, McMillan-Price J et al. Glycemic index, glycemic load, and chronic disease risk: a meta-analysis of observational studies. *Am J Clin Nutr*. 2008;87(3):627–37; Gnagnarella P, Gandini S, La Vecchia C et al. Glycemic index, glycemic load, and cancer risk: a meta-analysis. *Am J Clin Nutr*. 2008;87:1793–801; Sieri S, Krogh V, Berrino F et al. Dietary glycemic load and index and risk of coronary heart disease in a large Italian cohort: the EPICOR study. *Arch Intern Med*. 2010;170:640–47; Kaushik S, Wang JJ, Flood V et al. Dietary glycemic index and the risk of age-related macular degeneration. *Am J Clin Nutr*. 2008;88(4):1104–10.

7 Gnagnarella P, Gandini S, La Vecchia C et al. Glycemic index, glycemic load, and cancer risk: a meta-analysis. *Am J Clin Nutr*. 2008;87:1793–801; Dong JY, Qin LQ. Dietary glycemic index, glycemic load, and risk of breast cancer: meta-analysis of prospective cohort studies. *Breast*

Cancer Res Treat. 2011;126(2):287–94; Sieri S, Pala V, Brighenti F et al. Dietary glycemic index, glycemic load, and the risk of breast cancer in an Italian prospective cohort study. *Am J Clin Nutr.* 2007;86(4):1160–66.

8 Yun SH, Kim K, Nam SJ et al. The association of carbohydrate intake, glycemic load, glycemic index, and selected rice foods with breast cancer risk: a case-control study in South Korea. *Asia Pac J Clin Nutr.* 2010;19(3):383–92.

9 Wolpert HA, Atakov-Castillo A, Smith SA, Steil GM. Dietary fat acutely increases glucose concentrations and insulin requirements in patients with type 1 diabetes. *Diabetes Care.* 2013;36(4):810–16; von Frankenberg AD, Marina A, Song X et al. A high-fat, high-saturated fat diet decreases insulin sensitivity without changing intra-abdominal fat in weight-stable overweight and obese adults. *Eur J Nutr.* 2017;56(1):431–43.

10 Mattes RD, Dreher ML. Nuts and healthy body weight maintenance mechanisms. *Asia Pac J Clin Nutr.* 2010;19(1):137–41.

11 Tsugane S, Inoue M. Insulin resistance and cancer: epidemiological evidence. *Cancer Sci.* 2010;101:1073–79.

12 Van Wyk H, Daniels M. The use of very low calorie diets in the management of type 2 diabetes mellitus. *South African J Clin Nutr.* 2016;29(2):96–102.

13 Gallagher EJ, LeRoith D. The proliferating role of insulin and insulin-like growth factors in cancer. *Trend Endocrinol Metab.* 2010;21(10):610–18.

14 Arcidiacono B, Iiritano S, Nocera A et al. Insulin resistance and cancer risk: an overview of the pathogenetic mechanisms. *Exp Diabetes Res.* 2012;2012:789174, doi:10.1155/2012/789174.

15 Atay K, Canbakan B, Koroglu E et al. Apoptosis and disease severity is associated with insulin resistance in non-alcoholic fatty liver disease. *Acta Gastroenterol Belg.* 2017;80:271–77; Civera M, Urios A, Garcia-Torres ML et al. Relationship between insulin resistance, inflammation and liver cell apoptosis in patients with severe obesity. *Diabetes Metab Res Rev.* 2010;26:187–92.

16 Arcidiacono B, Iiritano S, Nocera A et al. Insulin resistance and cancer risk: an overview of the pathogenetic mechanisms. *Exp Diabetes Res.* 2012;2012:789174, doi:10.1155/2012/789174; Gallagher EJ, LeRoith D. The proliferating role of insulin and insulin-like growth factors in cancer. *Trend Endocrinol Metab.* 2010;21(10):610–18.

17 Djiogue S, Kamdje AHN, Vecchio L et al. Insulin resistance and cancer: the role of IGFs in endocrine-related cancer. *Endocrinology.* 2013:20(1):R1–R17; Key TJ, Appleby PN, Reeves GK, Roddam AW. Insulin-like growth factor 1 (IGF1), IGF binding protein 3 (IGFBP3), and breast cancer risk: pooled individual data analysis of 17 prospective studies. *Lancet Oncology.* 2010;11:530–42; Rowlands MA, Gunnell D, Harris R et al. Circulating insulin-like growth factor peptides and prostate cancer risk: a systematic review and meta-analysis. *Int J Cancer.* 2009;124:2416–29.

18 Jakobson MU, Dethlefsen C, Joensen AM et al. Intake of carbohydrates compared with intake of saturated fatty acids and risk of myocardial infarction: importance of the glycemic index. *Am J Clin Nutr.* 2010;91:1764–68; Liu S, Willett WC, Stampfer MJ et al. A prospective study of dietary glycemic load, carbohydrate intake, and risk of coronary heart disease in US women. *Am J Clin Nutr.* 2000;71:1455–61; Hu FB. Are refined carbohydrates worse than saturated fat? *Am J Clin Nutr.* 2010;91:1541–42.

19 Halton TL, Willett WC, Liu S et al. Potato and French fry consumption and risk of type 2 diabetes in women. *Am J Clin Nutr.* 2006;83(2):284–90.

20 Williams CD, Satia JA, Adair LS et al. Dietary patterns, food groups, and rectal cancer risk in Whites and African-Americans. *Cancer Epidemiol Biomarkers Prev.* 2009;18(5):1552–61.

21 Mozaffarian D, Hao T, Rimm EB et al. Changes in diet and lifestyle and long-term weight gain in women and men. *N Engl J Med.* 2011;364(25):2392–404.

22 Sonnenburg ED, Sonnenburg JL. Starving our microbial self: the deleterious consequences of a diet deficient in microbiota-accessible carbohydrates. *Cell Metab.* 2014;20:779–86; Higdon J, Drake VJ. Fiber. In: *An Evidence-Based Approach to Phytochemicals and Other Dietary Factors.* New York: Thieme, 2013: 133–48.

23 Aune D, Chan DS, Greenwood DC et al. Dietary fiber and breast cancer risk: a systematic review and meta-analysis of prospective studies. *Ann Oncol.* 2012;23(6):1394–2402; Aune D,

Chan DS, Lau R et al. Dietary fibre, whole grains, and risk of colorectal cancer: systematic review and dose-response meta-analysis of prospective studies. *BMJ.* 2011;343:d6617; Sun L, Zhang Z, Xu J et al. Dietary fiber intake reduces risk for Barrett's esophagus and esophageal cancer. *Crit Rev Food Sci Nutr.* 2017;57:2749–57; Mao QQ, Lin YW, Chen H et al. Dietary fiber intake is inversely associated with risk of pancreatic cancer: a meta-analysis. *Asia Pac J Clin Nutr.* 2017;26:89–96; Ang CH, Qiao C, Wang RC, Zhou WP. Dietary fiber intake and pancreatic cancer risk: a meta-analysis of epidemiologic studies. *Sci Rep.* 2015;5:10834; Zhang Z, Xu G, Ma M et al. Dietary fiber intake reduces risk for gastric cancer: a meta-analysis. *Gastroenterology.* 2013;145:113–20; Bandera EV, Kushi LH, Moore DF et al. Association between dietary fiber and endometrial cancer: a dose-response meta-analysis. *Am J Clin Nutr.* 2007;86:1730–37; Park Y, Hunter DJ, Spiegelman D et al. Dietary fiber intake and risk of colorectal cancer: a pooled analysis of prospective cohort studies. *JAMA.* 2005;294:2849–57.

24 Aubertin-Leheudre M, Gorbach S, Woods M et al. Fat/fiber intakes and sex hormones in healthy premenopausal women in USA. *J Steroid Biochem Mol Biol.* 2008;112:32–39; Aubertin-Leheudre M, Hamalainen E, Adlercreutz H. Diets and hormonal levels in postmenopausal women with or without breast cancer. *Nutr Cancer.* 2011;63:514–24; Goldin BR, Adlercreutz H, Gorbach SL et al. Estrogen excretion patterns and plasma levels in vegetarian and omnivorous women. *N Engl J Med.* 1982;307:1542–47.

25 Reynolds A, Mann J, Cummings J et al. Carbohydrate quality and human health: a series of systemic reviews and meta-analysis. *Lancet.* 2019;393(10170):P434–45. https://doi.org/10.1016/S0140-6736(18)31809-9.

26 World Health Organization, Food and Agriculture Organization of the United Nations. *Carbohydrates in Human Nutrition:* Report of a Joint FAO/WHO Expert Consultation. FAO Food and Nutrition Paper 66. World Health Organization, 1998.

27 Jenkins DJ, Kendall CW, Augustin LS et al. Effect of legumes as part of a low glycemic index diet on glycemic control and cardiovascular risk factors in type 2 diabetes mellitus: a randomized controlled trial. *Arch Intern Med.* 2012;172(21):1653–60.

28 Mollard RC, Wong CL, Luhovyy BL et al. First and second meal effect of pulses on blood glucose appetite, and food intake at a later meal. *Appl Physio Nutr Metab.* 2011;36(5):634–42; Wolever TM, Jenkins DJ, Ocana AM et al. Second-meal effect: low-glycemic-index foods eaten at dinner improve subsequent breakfast glycemic response. *Am J Clin Nutr.* 1988;48(4):1041–47; Brighenti F, Benini L, Del Rio D et al. Colonic fermentation of indigestible carbohydrates contributes to the second-meal effect. *Am J Clin Nutr.* 2006;83(4):817–22.

29 Darmadi-Blackberry I, Wahlqvist ML, Kouris-Blazos A et al. Legumes: the most important dietary predictor of survival in older people of different ethnicities. *Asia Pac J Clin Nutr.* 2004;13(2):217–20.

30 Maffucci T, Piccolo E, Cumashi A et al. Inhibition of the phosphatidylinositol 3-kinase/Akt pathway by inositol pentakisphosphate results in antiangiogenic and antitumor effects. *Cancer Res.* 2005 Sept 15;65(18):8339–49; Singh J, Basu PS. Non-nutritive bioactive compounds in pulses and their impact on human health: an overview. *Food Nutr Sci.* 2012;3(12):1664–72; Zhang Z, Song Y, Wang XL. Inositol hexaphosphate-induced enhancement of natural killer cell activity correlates with suppression of colon carcinogenesis in rats. *World J Gastroenterol.* 2005;11(32):5044–46.

31 Mollard RC, Luhovyy BL, Panahi S et al. Regular consumption of pulses for 8 weeks reduces metabolic syndrome risk factors in overweight and obese adults. *Br J Nutr.* 2012;108(suppl 1):S111–22.

32 Thissen JP, Ketelslegers JM, Underwood LE. Nutritional regulation of the insulin-like growth factors. *Endocr Rev.* 1994;15:80–101; Clemmons DR, Seek MM, Underwood LE. Supplemental essential amino acids augment the somatomedin-C/insulin-like growth factor I response to refeeding after fasting. *Metabolism.* 1985;34:391–95; Crowe FL, Key TJ, Allen NE et al. The association between diet and serum concentrations of IGF-I, IGFBP-1, IGFBP-2, and IGFBP-3 in the European Prospective Investigation into Cancer and Nutrition. *Cancer Epidemiol Biomarkers Prev.* 2009;18:1333–40.

33 Runchey SS, Pollak MN, Valsta LM et al. Glycemic load effect on fasting and post-prandial serum glucose, insulin, IGF-1 and IGFBP-3 in a randomized, controlled feeding study. *Eur*

J Clin Nutr. 2012;66:1146–52; Brand-Miller JC, Liu V, Petocz P, Baxter RC. The glycemic index of foods influences postprandial insulin-like growth factor–binding protein responses in lean young subjects. Am J Clin Nutr. 2005;82:350–54; Biddinger SB, Ludwig DS. The insulin-like growth factor axis: a potential link between glycemic index and cancer. Am J Clin Nutr. 2005;82:277–78.

34 Kaaks R. Nutrition, insulin, IGF-1 metabolism and cancer risk: a summary of epidemiological evidence. Novartis Found Symp. 2004;262:247–60, discussion 260–68; Fung TT, van Dam RM, Hankinson SE et al. Low-carbohydrate diets and all-cause and cause-specific mortality: two cohort studies. Ann Intern Med. 2010;153(5):289–98.

35 Levine ME, Suarez JA, Brandhorst S et al. Low protein intake is associated with a major reduction in IGF-1, cancer, and overall mortality in the 65 and younger but not older population. Cell Metab. 2014;19(3):407–17.

36 Fung TT, van Dam RM, Hankinson SE et al. Low-carbohydrate diets and all-cause and cause-specific mortality: two cohort studies. Ann Intern Med. 2010;153(5):289–98; Lagiou P, Sandin S, Lof M et al. Low carbohydrate–high protein diet and incidence of cardiovascular diseases in Swedish women: prospective cohort study. BMJ. 2012;344:e4026; Endogenous Hormones and Breast Cancer Collaborative Group, Key TJ, Appleby PN, Reeves GK, Roddam AW. Insulin-like growth factor 1 (IGF1), IGF binding protein 3 (IGFBP3), and breast cancer risk: pooled individual data analysis of 17 prospective studies. Lancet Oncol. 2010;6:530–42; Grant WB. A multicountry ecological study of cancer incidence rates in 2008 with respect to various risk-modifying factors. Nutrients. 2013;6(1):163–89; Chitnis MM, Yuen JS, Protheroe AS et al. The type 1 insulin-like growth factor receptor pathway. Clin Cancer Res. 2008;14:6364–70; Werner H, Bruchim I. The insulin-like growth factor-I receptor as an oncogene. Arch Physiol Biochem. 2009;115:58–71; Davies M, Gupta S, Goldspink G, Winslet M. The insulin-like growth factor system and colorectal cancer: clinical and experimental evidence. Int J Colorectal Dis. 2006;21:201–8; Sandhu MS, Dunger DB, Giovannucci EL. Insulin, insulin-like growth factor-I (IGF-I), IGF binding proteins, their biologic interactions, and colorectal cancer. J Natl Cancer Inst. 2002;94:972–80; Kaaks R. Nutrition, insulin, IGF-1 metabolism and cancer risk: a summary of epidemiological evidence. Novartis Found Symp. 2004;262:247–60, discussion 260–68.

37 Liang Z, Diepstra A, Xu C et al. Insulin-like growth factor 1 receptor is a prognostic factor in classical Hodgkin lymphoma. PLOS One. 2014;9(1):e87474; Lann D, LeRoith D. The role of endocrine insulin-like growth factor-I and insulin in breast cancer. J Mammary Gland Biol Neoplasia. 2008;13:371–79; Allen NE, Roddam AW, Allen DS et al. A prospective study of serum insulin-like growth factor-I (IGF-I), IGF-II, IGF-binding protein-3 and breast cancer risk. Br J Cancer. 2005;92:1283–87; Fletcher O, Gibson L, Johnson N et al. Polymorphisms and circulating levels in the insulin-like growth factor system and risk of breast cancer: a systematic review. Cancer Epidemiol Biomarkers Prev. 2005;14:2–19; Renehan AG, Zwahlen M, Minder C et al. Insulin-like growth factor (IGF)-I, IGF binding protein-3, and cancer risk: systematic review and meta-regression analysis. Lancet. 2004;363:1346–53; Shi R, Yu H, McLarty J, Glass J. IGF-I and breast cancer: a meta-analysis. Int J Cancer. 2004;111:418–23; Sugumar A, Liu YC, Xia Q et al. Insulin-like growth factor (IGF)-I and IGF-binding protein 3 and the risk of premenopausal breast cancer: a meta-analysis of literature. Int J Cancer. 2004;111:293–97; Baglietto L, English DR, Hopper JL et al. Circulating insulin-like growth factor-I and binding protein-3 and the risk of breast cancer. Cancer Epidemiol Biomarkers Prev. 2007;16:763–68; Key TJ, Appleby PN, Reeves GK, Roddam AW. Insulin-like growth factor 1 (IGF1), IGF binding protein 3 (IGFBP3), and breast cancer risk: pooled individual data analysis of 17 prospective studies. Lancet Oncol. 2010;11:530–42; Rowlands MA, Gunnell D, Harris R et al. Circulating insulin-like growth factor peptides and prostate cancer risk: a systematic review and meta-analysis. Int J Cancer. 2009;124:2416–29; Davies M, Gupta S, Goldspink G, Winslet M. The insulin-like growth factor system and colorectal cancer: clinical and experimental evidence. Int J Colorectal Dis. 2006;21:201–8; Dziadziuszko R, Camidge DR, Hirsch FR. The insulin-like growth factor pathway in lung cancer. J Thorac Oncol. 2008;3:815–18.

38 Thissen JP, Ketelslegers JM, Underwood LE. Nutritional regulation of the insulin-like growth factors. Endocr Rev. 1994;15:80–101.

39 Aune D, Navarro Rosenblatt DA, Chan DS et al. Dairy products, calcium, and prostate cancer risk: a systematic review and meta-analysis of cohort studies. *Am J Clin Nutr.* 2015;101:87–117.

40 Abid Z, Cross AJ, Sinha R. Meat, dairy, and cancer. *Am J Clin Nutr.* 2014;100(suppl 1):386S–93S.

41 de Lorgeril M, Salen P. New insights into the health effects of dietary saturated and omega-6 and omega-3 polyunsaturated fatty acids. *BMC Med.* 2012;10:50.

42 Bastide NM, Pierre FHF, Corpet DE. Heme iron from meat and risk of colorectal cancer: a meta-analysis and a review of the mechanisms involved. *Cancer Prev Res (Phila).* 2011;4(2):177–84, doi:10.1158/1940-6207.

43 Tang WH, Wang Z, Levison BS et al. Intestinal microbial metabolism of phosphatidylcholine and cardiovascular risk. *N Engl J Med.* 2013;368:1575–84; Wang Z, Klipfell E, Bennett BJ et al. Gut flora metabolism of phosphatidylcholine promotes cardiovascular disease. *Nature.* 2011;472:57–63; Velasquez MT, Ramezani A, Manai A, Raj DS. Trimethylamine N-oxide: the good, the bad and the unknown. *Toxins (Basel).* 2016;8(11):326; Richman EL, Kenfield SA, Stampfer MJ et al. Choline intake and risk of lethal prostate cancer: incidence and survival. *Am J Clin Nutr.* 2012;96:855–63.

44 National Cancer Institute. "Chemicals in meat cooked at high temperatures and cancer risk." http://www.cancer.gov/cancertopics/factsheet/Risk/cooked-meats. Reviewed 11 July 2017; Thomson B. Heterocyclic amine levels in cooked meat and the implication for New Zealanders. *Eur J Cancer Prev.* 1999;8:201–6; Zheng W, Lee S-A. Well-done meat intake, heterocyclic amine exposure, and cancer risk. *Nutr Cancer.* 2009;61:437–46; Herrmann SS, Granby K, Duedahl-Olesen L. Formation and mitigation of N-nitrosamines in nitrite preserved cooked sausages. *Food Chem.* 2015;174:516–26.

45 *Red Meat and Processed Meat.* IARC Monographs on the Evaluation of Carcinogenic Risks to Humans, No. 114. Lyon: International Agency for Research on Cancer, 2018.

46 Alshahrani SM, Fraser GE, Sabate J et al. Red and processed meat and mortality in a low meat intake population. *Nutrients.* 2019;11(3):622.

47 Snowdon DA, Phillips RL, Choi W. Diet, obesity, and risk of fatal prostate cancer. *Am J Epidemiol.* 1984;120:244–50; Richman EL, Kenfield SA, Stampfer MJ et al. Egg, red meat, and poultry intake and risk of lethal prostate cancer in the prostate-specific antigen-era: incidence and survival. *Cancer Prev Res (Phila).* 2011;4:2110–21.

48 Tse G, Eslick GD. Egg consumption and risk of GI neoplasms: dose-response meta-analysis and systematic review. *Eur J Nutr.* 2014;53(7):1581–90.

49 Johansson M, Van Guelpen B, Vollset SE et al. One-carbon metabolism and prostate cancer risk: prospective investigation of seven circulating B vitamins and metabolites. *Cancer Epidemiol Biomarkers Prev.* 2009;18:1538–43; Platz EA, Clinton SK, Giovannucci E. Association between plasma cholesterol and prostate cancer in the PSA era. *Int J Cancer.* 2008;123:1693–98; Pelton K, Freeman MR, Solomon KR. Cholesterol and prostate cancer. *Curr Opin Pharmacol.* 2012;12:751–59.

50 Cruz PM, Mo H, McConathy WJ et al. The role of cholesterol metabolism and cholesterol transport in carcinogenesis: a review of scientific findings, relevant to future cancer therapeutics. *Front Pharmacol.* 2013;4:119; Steinmetz KA, Potter JD. Egg consumption and cancer of the colon and rectum. *Eur J Cancer Prev.* 1994;3:237–45; Cruse P, Lewin M, Clark CG. Dietary cholesterol is co-carcinogenic for human colon cancer. *Lancet.* 1979;1:752–55.

51 Tang WH, Wang Z, Levison BS et al. Intestinal microbial metabolism of phosphatidylcholine and cardiovascular risk. *N Engl J Med.* 2013;368:1575–84; Wang Z, Klipfell E, Bennett BJ et al. Gut flora metabolism of phosphatidylcholine promotes cardiovascular disease. *Nature.* 2011;472:57–63; Velasquez MT, Ramezani A, Manai A, Raj DS. Trimethylamine N-oxide: the good, the bad and the unknown. *Toxins (Basel).* 2016;8(11):326; Richman EL, Kenfield SA, Stampfer MJ et al. Choline intake and risk of lethal prostate cancer: incidence and survival. *Am J Clin Nutr.* 2012;96:855–63.

52 Schiattarella GG, Sannino A, Toscano E et al. Gut microbe–generated metabolite trimethylamine-N-oxide as cardiovascular risk biomarker: a systematic review and dose-response meta-analysis. *Eur Heart J.* 2017;38(39):2948–56.

53 European Society of Cardiology. "Low carbohydrate diets are unsafe and should be avoided, study suggests." 28 Aug 2018. https://www.escardio.org/The-ESC/Press-Office/Press-releases/Low-carbohydrate-diets-are-unsafe-and-should-be-avoided.

54 European Society of Cardiology.

55 Mazidi M, Katsiki N, Mikhailidis DP, Banach M: Low-carbohydrate diets and all-cause and cause-specific mortality: a population-based cohort study and pooling prospective studies in European Society of Cardiology Congress. *Eur Heart J.* 2018;(39S):1112–13.

56 Levine ME, Suarez JA, Brandhorst S et al. Low protein intake is associated with a major reduction in IGF-1, cancer, and overall mortality in the 65 and younger but not older population. *Cell Metab.* 2014;19:407–17.

57 Fontana L, Klein S, Holloszy JO. Long-term low-protein, low-calorie diet and endurance exercise modulate metabolic factors associated with cancer risk. *Am J Clin Nutr.* 2006;84:1456–62; Fontana L, Weiss EP, Villareal DT et al. Long-term effects of calorie or protein restriction on serum IGF-1 and IGFBP-3 concentration in humans. *Aging Cell.* 2008;7:681–87.

58 Hankinson SE, Willett WC, Colditz GA et al. Circulating concentrations of insulin-like growth factor-I and risk of breast cancer. *Lancet.* 1998;351:1393–96.

59 Chan JM, Stampfer MJ, Giovannucci E et al. Plasma insulin-like growth factor-I and prostate cancer risk: a prospective study. *Science.* 1998;279:563–66.

60 Burgers AM, Biermasz NR, Schoones JW et al. Meta-analysis and dose-response metaregression: circulating insulin-like growth factor I (IGF-I) and mortality. *J Clin Endocrinol Metab.* 2011;96:2912–20.

61 Bidlingmaier M, Friedrich N, Emeny RT et al. Reference intervals for insulin-like growth factor-1 (IGF-I) from birth to senescence: results from a multicenter study using a new automated chemiluminescence IGF-I immunoassay conforming to recent international recommendations. *J Clin Endocrinol Metab.* 2014;99:1712–21; Brabant G, von zur Muhlen A, Wuster C et al. Serum insulin-like growth factor I reference values for an automated chemiluminescence immunoassay system: results from a multicenter study. *Horm Res.* 2003;60:53–60; Ranke MB, Osterziel KJ, Schweizer R et al. Reference levels of insulin-like growth factor I in the serum of healthy adults: comparison of four immunoassays. *Clin Chem Lab Med.* 2003;41:1329–34.

62 Crowe FL, Key TJ, Allen NE et al. The association between diet and serum concentrations of IGF-I, IGFBP-1, IGFBP-2, and IGFBP-3 in the European Prospective Investigation into Cancer and Nutrition. *Cancer Epidemiol Biomarkers Prev.* 2009;18:1333–40.

63 Witard OC, McGlory C, Hamilton DL, Phillips SM. Growing older with health and vitality: a nexus of physical activity, exercise and nutrition. *Biogerontology.* 2016;17:529–46.

64 Campbell WW, Crim MC, Dallal GE et al. Increased protein requirements in elderly people: new data and retrospective reassessments. *Am J Clin Nutr.* 1994;60:501–9; Campbell WW, Trappe TA, Wolfe RR, Evans WJ. The recommended dietary allowance for protein may not be adequate for older people to maintain skeletal muscle. *J Gerontol A Biol Sci Med Sci.* 2001;56:M373–80.

65 Bauer J, Biolo G, Cederholm T et al. Evidence-based recommendations for optimal dietary protein intake in older people: a position paper from the PROT-AGE Study Group. *J Am Med Dir Assoc.* 2013;14:542–59.

66 Mustafa J, Ellison RC, Singer MR et al. Dietary protein and preservation of physical functioning among middle-aged and older adults in the Framingham Offspring Study. *Am J Epidemiol.* 2018;187:1411–19.

67 Lamberts SW, van den Beld AW, van der Lely AJ. The endocrinology of aging. *Science.* 1997;278:419–24; Doi T, Shimada H, Makizako H et al. Association of insulin-like growth factor-1 with mild cognitive impairment and slow gait speed. *Neurobiol Aging* 2015;36:942–47; Calvo D, Gunstad J, Miller LA et al. Higher serum insulin-like growth factor-1 is associated with better cognitive performance in persons with mild cognitive impairment. *Psychogeriatrics.* 2013;13:170–74.

68 Johnsen SP, Hundborg HH, Sorensen HT et al. Insulin-like growth factor (IGF) I, -II, and IGF binding protein-3 and risk of ischemic stroke. *J Clin Endocrinol Metab.* 2005;90:5937–41; Friedrich N, Haring R, Nauck M et al. Mortality and serum insulin-like growth factor (IGF)-I and IGF binding protein 3 concentrations. *J Clin Endocrinol Metab.* 2009;94:1732–39; Carlzon D, Svensson J, Petzold M et al. Both low and high serum IGF-1 levels associate with increased risk of cardiovascular events in elderly men. *J Clin Endocrinol Metab.* 2014;99:E2308–16;

Svensson J, Carlzon D, Petzold M et al. Both low and high serum IGF-I levels associate with cancer mortality in older men. *J Clin Endocrinol Metab.* 2012;97:4623–30; van Bunderen CC, van Nieuwpoort IC, van Schoor NM et al. The association of serum insulin-like growth factor-I with mortality, cardiovascular disease, and cancer in the elderly: a population-based study. *J Clin Endocrinol Metab.* 2010;95(10):4616–24; Arai Y, Takayama M, Gondo Y et al. Adipose endocrine function, insulin-like growth factor-1 axis, and exceptional survival beyond 100 years of age. *J Gerontol A Biol Sci Med Sci.* 2008;63:1209–18; Doi T, Shimada H, Makizako H et al. Association of insulin-like growth factor-1 with mild cognitive impairment and slow gait speed. *Neurobiol Aging* 2015;36:942–47.

69 Friedrich N, Haring R, Nauck M et al. Mortality and serum insulin-like growth factor (IGF)-I and IGF binding protein 3 concentrations. *J Clin Endocrinol Metab.* 2009;94:1732–39; van Bunderen CC, van Nieuwpoort IC, van Schoor NM et al. The association of serum insulin-like growth factor-I with mortality, cardiovascular disease, and cancer in the elderly: a population-based study. *J Clin Endocrinol Metab.* 2010;95(10):4616–24.

Chapter Three: It's All About the Plants

1 Fenech M. The Genome Health Clinic and Genome Health Nutrigenomics concepts: diagnosis and nutritional treatment of genome and epigenome damage on an individual basis. *Mutagenesis.* 2005;20(4):255–69.

2 Minnet C, Koc A, Aycicek A, Kocyigit A. Vitamin B-12 treatment reduces mononuclear DNA damage. *Pediatr Int.* 2011 Dec;53(6):1023–27.

3 Blount BC, Mack MM, Wehr CM et al. Folate deficiency causes uracil misincorporation into human DNA and chromosome breakage: implications for cancer and neuronal damage. *Proc Natl Acad Sci USA.* 1997;94(7):3290–95.

4 Ames BN. Low micronutrient intake may accelerate the degenerative diseases of aging through allocation of scarce micronutrients by triage. *Proc Natl Acad Sci USA.* 2006;103:17589–94.

5 Ames BN. Prevention of mutation, cancer, and other age-associated diseases by optimizing micronutrient intake. *J Nucleic Acids.* 2010 Sep 22;2010, doi:10.4061/2010/725071.

6 Ames BN. Prolonging healthy aging: longevity vitamins and proteins. *Proc Natl Acad Sci USA.* 2018;115(43):10836–44.

7 McCann JC, Ames BN. Vitamin K, an example of triage theory: is micronutrient inadequacy linked to diseases of aging? *Am J Clin Nutr.* 2009;90(4):889–907.

8 National Center for Health Statistics. *Health, United States, 2014. With Special Feature on Adults Aged 55–64.* Hyattsville, MD: National Center for Health Statistics, 2015. https://www.cdc.gov/nchs/data/hus/hus14.pdf.

9 Barnett JB, Hamer DH, Meydani SN. Low zinc status: a new risk factor for pneumonia in the elderly? *Nutr Rev.* 2010;68:30–37; Mocchegiani E, Romeo J, Malavolta M et al. Zinc: dietary intake and impact of supplementation on immune function in elderly. *Age (Dordr).* 2013;35:839–60.

10 Prasad AS, Fitzgerald JT, Hess JW et al. Zinc deficiency in elderly patients. *Nutrition.* 1993;9:218–24; Pepersack T, Rotsaert P, Benoit F et al. Prevalence of zinc deficiency and its clinical relevance among hospitalised elderly. *Arch Gerontol Geriatr.* 2001;33:243–53; Briefel RR, Bialostosky K, Kennedy-Stephenson J et al. Zinc intake of the US population: findings from the Third National Health and Nutrition Examination Survey, 1988–1994. *J Nutr.* 2000;130:1367S–73S.

11 King JC. Zinc: An essential but elusive nutrient. *Am J Clin Nutr.* 2011;94:679S–84S; Prasad AS. Zinc in human health: effect of zinc on immune cells. *Mol Med.* 2008;14:353–57.

12 Prasad AS. Zinc in human health: effect of zinc on immune cells. *Mol Med.* 2008;14:353–57; Office of Dietary Supplements, National Institutes of Health. "Zinc." http://ods.od.nih.gov/factsheets/Zinc-HealthProfessional/. Updated 10 July 2019.

13 Office of Dietary Supplements, National Institutes of Health. "Zinc"; Reddy NR, Pierson MD, Sathe SK, Salunkhe DK. *Phytates in Cereals and Legumes.* CRC Press, 1989: 88–91; Foster M, Chu A, Petocz P, Samman S. Effect of vegetarian diets on zinc status: a systematic review and meta-analysis of studies in humans. *J Sci Food Agric.* 2013;93:2362–71; Hunt JR. Bioavailability of iron, zinc, and other trace minerals from vegetarian diets. *Am J Clin Nutr.*

2003;78:633S–39S; Frassinetti S, Bronzetti G, Caltavuturo L et al. The role of zinc in life: a review. *J Environ Pathol Toxicol Oncol.* 2006;25:597–610; Miller LV, Krebs NF, Hambidge KM. A mathematical model of zinc absorption in humans as a function of dietary zinc and phytate. *J Nutr.* 2007;137:135–41.

14 Madej D, Borowska K, Bylinowska J et al. Dietary intakes of iron and zinc assessed in a selected group of the elderly: are they adequate? *Rocz Panstw Zakl Hig.* 2013;64(2):97–104.

15 Meydani SN, Barnett JB, Dallal GE et al. Serum zinc and pneumonia in nursing home elderly. *Am J Clin Nutr.* 2007;86:1167–73.

16 Prasad AS, Beck FW, Bao B et al. Zinc supplementation decreases incidence of infections in the elderly: effect of zinc on generation of cytokines and oxidative stress. *Am J Clin Nutr.* 2007;85:837–44.

17 Swardfager W, Herrmann N, Mazereeuw G et al. Zinc in depression: a meta-analysis. *Biol Psychiatry.* 2013;74(12):872–78; Maserejian NN, Hall SA, McKinlay JB. Low dietary or supplemental zinc is associated with depression symptoms among women, but not men, in a population-based epidemiological survey. *J Affect Disord.* 2012 Feb;136(3):781–88; Nowak G, Siwek M, Dudek D et al. Effect of zinc supplementation on antidepressant therapy in unipolar depression: a preliminary placebo-controlled study. *Pol J Pharmacol.* 2003;55(6):1143–47.

18 Lai HTM, de Oliveira Otto MC, Lemaitre RN et al. Serial circulating omega 3 polyunsaturated fatty acids and healthy ageing among older adults in the Cardiovascular Health Study: prospective cohort study. *BMJ.* 2018;363:k4067.

19 Rizos EC, Ntzani EE, Bika E et al. Association between omega-3 fatty acid supplementation and risk of major cardiovascular disease events: a systematic review and meta-analysis. *JAMA.* 2012;308(10):1024–33.

20 Aung T, Halsey J, Kromhout D et al. Associations of omega-3 fatty acid supplement use with cardiovascular disease risks: meta-analysis of 10 trials involving 77,917 individuals. *JAMA Cardiol.* 2018;3(3):225–33.

21 Manson JE, Cook NR, Lee IM et al., on behalf of the VITAL Research Group. Vitamin D supplements and prevention of cancer and cardiovascular disease. *N Engl J Med.* 2019;380:33–44; Manson JE, Cook NR, Lee IM et al., on behalf of the VITAL Research Group. Marine n-3 fatty acids and prevention of cardiovascular disease and cancer. *N Engl J Med.* 2019;380:23–32.

22 Burr ML, Ashfield-Watt PA, Dunstan FD et al. Lack of benefit of dietary advice to men with angina: results of a controlled trial. *Eur J Clin Nutr.* 2003;57(2):193–200.

23 Mozaffarian D, Lemaitre RN, King IB et al. Plasma phospholipid long-chain omega-3 fatty acids and total and cause-specific mortality in older adults: the Cardiovascular Health Study. *Ann Intern Med.* 2013;158(7):515–25.

24 de Oliveira Otto MC, Wu JH, Baylin A et al. Circulating and dietary omega-3 and omega-6 polyunsaturated fatty acids and incidence of CVD in the Multi-Ethnic Study of Atherosclerosis. *J Am Heart Assoc.* 2013;2(6):e000506, doi:10.1161/JAHA.113.000506.

25 Grosso G, Galvano F, Marventano S et al. Omega-3 fatty acids and depression: scientific evidence and biological mechanisms. *Oxid Med Cell Longev.* 2014;2014:313570.

26 Baydoun MA, Kaufman JS, Satia JA et al. Plasma n-3 fatty acids and the risk of cognitive decline in older adults: the Atherosclerosis Risk in Communities Study. *Am J Clin Nutr.* 2007;85(4):1103–11; Connor WE, Connor SL. The importance of fish and docosahexaenoic acid in Alzheimer disease. *Am J Clin Nutr.* 2007;85(4):929–30; Cole GM, Ma QL, Frautschy SA et al. Omega-3 fatty acids and dementia. *Prostaglandins Leukot Essent Fatty Acids.* 2009;81(2–3):213–21; van Gelder BM, Tijhuis M, Kalmijn S, Kromhout D. Fish consumption, n-3 fatty acids, and subsequent 5-y cognitive decline in elderly men: the Zutphen Elderly Study. *Am J Clin Nutr.* 2007;85(4):1142–47.

27 Pottala JV, Yaff K, Robinson JG et al. Higher RBC EPA + DHA corresponds with larger total brain and hippocampal volumes. *Neurology.* 2014;82(5):435–42.

28 Sarter B, Kelsey KS, Schwartz TA, Harris WS. Blood docosahexaenoic acid and eicosapentaenoic acid in vegans: associations with age and gender and effects of an algal-derived omega-3 fatty acid supplement. *Clin Nutr.* 2015;34(2):212–18.

29 McNamara RK, Strawn JR. Role of long-chain omega-3 fatty acids in psychiatric practice. *PharmaNutrition*. 2013 Apr;1(2):41–49; Grosso G, Galvano F, Marventano S et al. Omega-3 fatty acids and depression: scientific evidence and biological mechanisms. *Oxid Med Cellular Longev*. 2014;2014:313570, doi:10.1155/2014/313570; Grosso G, Pajak A, Marventano S et al. Role of omega-3 fatty acids in the treatment of depressive disorders: a comprehensive meta-analysis of randomized clinical trials. *PLOS One*. 2014;9(5):e96905, doi:10.1371/journal.pone.0096905.

30 Kiecolt-Glaser JK, Belury MA, Porter K et al. Depressive symptoms, omega-6:omega-3 fatty acids, and inflammation in older adults. *Psychosom Med*. 2007;69(3):217–24; Mazza M, Pomponi M, Janiri L et al. Omega-3 fatty acids and antioxidants in neurological and psychiatric diseases: an overview. *Prog Neuropsychopharmacol Biol Psychiatry*. 2007;31(1):12–26.

31 McNamara RK, Strimpfel J, Jandacek R, Rider T et al. Detection and treatment of long-chain omega-3 fatty acid deficiency in adolescents with SSRI-resistant major depressive disorder. *PharmaNutrition*. 2014;2(2):38–46; Kraguljac NV, Montori VM, Pavuluri M et al. Efficacy of omega-3 fatty acids in mood disorders: a systematic review and meta analysis. *Psychopharmacol Bull*. 2009;42(3):39–54; Lin PY, Su KP. A meta-analytic review of double-blind, placebo-controlled trials of antidepressant efficacy of omega-3 fatty acids. *J Clin Psychiatry*. 2007;68(7):1056–61; Frangou S, Lewis M, McCrone P. Efficacy of ethyl-eicosapentaenoic acid in bipolar depression: randomised double-blind placebo-controlled study. *Br J Psychiatry*. 2006;188:46–50.

32 Wilson VK, Houston DK, Kilpatrick L et al. Relationship between 25-hydroxyvitamin D and cognitive function in older adults: the Health, Aging and Body Composition Study. *J Am Geriatr Soc*. 2014;62(4):636–41; Toffanello ED, Coin A, Egle Perissinotto E et al. Vitamin D deficiency predicts cognitive decline in older men and women: the Pro.V.A. Study. *Neurology*. 2014;83(24):2292–98.

Chapter Four: Your Health Is in Your Hands

1 Niedernhofer LJ, Daniels JS, Rouzer CA et al. Malondialdehyde, a product of lipid peroxidase is mutagenetic in human cells. *J Biol Chem*. 2003;278(33):31426–33; National Cancer Institute. "Chemicals in Meat Cooked at High Temperature and Cancer Risk." https://www.cancer.gov/about-cancer/causes-prevention/risk/diet/cooked-meats-fact-sheet. Reviewed 11 July 2017.

2 US Food and Drug Administration. "Survey Data on Acrylamide in Food: Individual Food Products." http://www.fda.gov/Food/FoodborneIllnessContaminants/ChemicalContaminants/ucm053549.htm. Last updated July 2006; current as of 25 Jan 2018.

3 Michalak J, Gujska E, Klepacka J. The effect of domestic preparation of some potato products on acrylamide content. *Plant Foods Hum Nutr*. 2011;66:307–12.

4 Je Y. Dietary acrylamide intake and risk of endometrial cancer in prospective cohort studies. *Arch Gynecol Obstet*. 2015;291:1395–401; Pelucchi C, Bosetti C, Galeone C, La Vecchia C. Dietary acrylamide and cancer risk: an updated meta-analysis. *Int J Cancer*. 2015;136:2912–22; Virk-Baker MK, Nagy TR, Barnes S, Groopman J. Dietary acrylamide and human cancer: a systematic review of literature. *Nutr Cancer*. 2014;66:774–90.

5 Di Marco E, Gray SP, Jandeileit-Dahm K. Diabetes alters activation and repression of pro- and anti-inflammatory signaling pathways in the vasculature. *Front Endocrinol (Lausanne)*. 2013;4:68; Nowotny K, Jung T, Höhn A et al. Advanced glycation end products and oxidative stress in type 2 diabetes mellitus. *Biomolecules*. 2015 Mar 16;5(1):194–222; Del Turco S, Basta G. An update on advanced glycation endproducts and atherosclerosis. *Biofactors*. 2012 July–Aug;38(4):266–74.

6 Goldberg T, Cai W, Peppa M et al. Advanced glycoxidation end products in commonly consumed foods. *J Am Diet Assoc*. 2004;104:1287–91; Uribarri J, Woodruff S, Goodman S et al. Advanced glycation end products in foods and a practical guide to their reduction in the diet. *J Am Diet Assoc*. 2010;110:911–16.

7 Hirose A, Tanikawa T, Mori H et al. Advanced glycation end products increase endothelial permeability through the RAGE/Rho signaling pathway. *FEBS Lett*. 2010;584(1):61–66.

8 Di Pino A, Currenti W, Urbano F et al. High intake of dietary advanced glycation end-products is associated with increased arterial stiffness and inflammation in subjects with type 2 diabetes. *Nutr Metab Cardiovasc Dis.* 2017;27(11):978–84; McNulty M, Mahmud A, Feely J. Advanced glycation end-products and arterial stiffness in hypertension. *Am J Hypertens.* 2007;20(3):242–47.

9 Sobal G, Menzel J, Sinzinger H. Why is glycated LDL more sensitive to oxidation than native LDL? A comparative study. *Prostaglandins Leukot Essent Fatty Acids.* 2000;63(4):177–86; Del Turco S, Basta G. An update on advanced glycation endproducts and atherosclerosis. *Biofactors.* 2012;38(4):266–74.

10 Nowotny K, Jung T, Höhn A et al. Advanced glycation end products and oxidative stress in type 2 diabetes mellitus. *Biomolecules.* 2015;5(1):194–222.

11 Peppa M, Raptis SA. Glycoxidation and wound healing in diabetes: an interesting relationship. *Curr Diabetes Rev.* 2011;7(6):416–25; Peppa M, Stavroulakis P, Raptis SA. Advanced glycoxidation products and impaired diabetic wound healing. *Wound Repair Regen.* 2009;17:461–72; Goldin A, Beckman JA, Schmidt AM, Creager MA. Advanced glycation end products: sparking the development of diabetic vascular injury. *Circulation.* 2006;114:597–605; Yamagishi S, Matsui T. Advanced glycation end products, oxidative stress and diabetic nephropathy. *Oxid Med Cell Longev.* 2010;3:101–8.

12 Crinnion WJ. Polychlorinated biphenyls: persistent pollutants with immunological, neurological, and endocrinological consequences. *Altern Med Rev.* 2011;16:5–13; Carpenter DO. Polychlorinated biphenyls (PCBs): routes of exposure and effects on human health. *Rev Environ Health.* 2006;21:1–23.

13 Environmental Working Group. "PCBs in Farmed Salmon." 31 July 2003. http://www.ewg.org/research/pcbs-farmed-salmon.

14 Karagas MR, Choi AL, Oken E et al. Evidence on the human health effects of low-level methylmercury exposure. *Environ Health Perspect.* 2012;120:799–806.

15 US Food and Drug Administration, US Environmental Protection Agency. "Advice About Eating Fish: For Women Who Are or Might Become Pregnant, Breastfeeding Mothers, and Young Children." Content current as of 2 July 2019. https://www.fda.gov/food/consumers/advice-about-eating-fish.

16 Fisher DJ, Yonkos LT, Staver KW. Environmental concerns of roxarsone in broiler poultry feed and litter in Maryland, USA. *Environ Sci Technol.* 2015;49:1999–2012.

17 Rao CV, Pal S, Mohammed A et al. Biological effects and epidemiological consequences of arsenic exposure, and reagents that can ameliorate arsenic damage in vivo. *Oncotarget.* 2017;8:57605–21.

18 US Food and Drug Administration. "Questions & Answers: Apple Juice and Arsenic." 15 July 2013. http://wayback.archive-it.org/7993/20170111224422/http://www.fda.gov/Food/FoodborneIllnessContaminants/Metals/ucm271595.htm; "Arsenic in Your Food." *Consumer Reports.* Nov 2012. https://www.consumerreports.org/cro/magazine/2012/11/arsenic-in-your-food/index.htm.

19 Tchounwou PB, Yedjou CG, Patlolla AK, Sutton DJ. Heavy metal toxicity and the environment. *Exp Suppl.* 2012;101:133–64; US Environmental Protection Agency. "Lead." https://www.epa.gov/lead. Last updated 9 July 2019.

20 US Environmental Protection Agency. "What Are the Health Effects of Lead?" In: "Learn About Lead." https://www.epa.gov/lead/learn-about-lead#effects. Last updated 12 Aug 2019.

21 Codex Alimentarius Commission. Joint FAO/WHO Food Standards Programme. Codex Committee on Contaminants in Foods, 9th session, New Delhi, India, 16–20 March 2015. Proposed draft maximum levels for cadmium in chocolate and cocoa-derived products, 2014.

22 Satarug S, Garrett SH, Sens MA, Sens DA. Cadmium, environmental exposure, and health outcomes. *Environ Health Perspect.* 2010;118:182–90.

23 Kim K, Melough MM, Vance TM et al. Dietary cadmium intake and sources in the US. *Nutrients.* 2019 Jan;11(1):2.

24 Satarug S, Garrett SH, Sens MA, Sens DA. Cadmium, environmental exposure, and health outcomes. *Environ Health Perspect.* 2010;118:182–90.

25 Abt E, Fong SJ, Gray P et al. Cadmium and lead in cocoa powder and chocolate products in the US market. *Food Addit Contam Part B Surveill.* 2018;11:92–102.

26 US Environmental Protection Agency. Office of Pollution Prevention and Toxics. "Fight Lead Poisoning with a Healthy Diet." EPA-747-F-01-004, Nov 2001. https://www.epa.gov/sites /production/files/2014-02/documents/fight_lead_poisoning_with_a_healthy_diet.pdf.

27 Boonprasert K, Kongiam P, Limpatanachote P et al. Urinary and blood cadmium levels in relation to types of food and water intake and smoking status in a Thai population residing in cadmium-contaminated areas in Mae Sot. *Southeast Asian J Trop Med Public Health.* 2011;42(6):1521–30; Zhai Q, Narbad A, Chen W. Dietary strategies for the treatment of cadmium and lead toxicity. *Nutrients.* 2015;7(1):552–71.

28 Bai SH, Ogbourne SM. Glyphosate: environmental contamination, toxicity and potential risks to human health via food contamination. *Environ Sci Pollut Res Int.* 2016;23:18988– 9001; Vandenberg LN, Blumberg B, Antoniou MN et al. Is it time to reassess current safety standards for glyphosate-based herbicides? *J Epidemiol Community Health.* 2017;71:613–18.

29 Schinasi L, Leon ME. Non-Hodgkin lymphoma and occupational exposure to agricultural pesticide chemical groups and active ingredients: a systematic review and meta-analysis. *Int J Environ Res Public Health.* 2014;11:4449–4527; Bohn T, Cuhra M, Traavik T et al. Compositional differences in soybeans on the market: glyphosate accumulates in Roundup Ready GM soybeans. *Food Chem.* 2014;153:207–15.

30 Guyton KZ, Loomis D, Grosse Y et al. Carcinogenicity of tetrachlorvinphos, parathion, malathion, diazinon, and glyphosate. *Lancet Oncol.* 2015;16:490–91.

31 IARC Director. IARC response to criticisms of the *Monographs* and the glyphosate evaluation. Jan 2018. https://www.iarc.fr/wp-content/uploads/2018/07/IARC_response_to_criticisms _of_the_Monographs_and_the_glyphosate_evaluation.pdf.

32 United States Department of Agriculture. 2012 agricultural chemical use survey: Wheat. *NASS Highlights.* May 2013, no. 2013-2. http://www.nass.usda.gov/Surveys/Guide_to_NASS _Surveys/Chemical_Use/ChemUseHighlights-Wheat-2012.pdf.

33 Herrmann SS, Granby K, Duedahl-Olesen L. Formation and mitigation of N-nitrosamines in nitrite preserved cooked sausages. *Food Chem.* 2015;174:516–26; Santarelli R, Pierre F, Corpet D. Processed meat and colorectal cancer: a review of epidemiologic and experimental evidence. *Nutr Cancer,* 2008;60:131–44; Chan DS, Lau R, Aune D et al. Red and processed meat and colorectal cancer incidence: meta-analysis of prospective studies. *PLOS One.* 2011;6:e20456; Hu J, La Vecchia C, Morrison H et al. Salt, processed meat and the risk of cancer. *Eur J Cancer Prev.* 2011;20:132–39; International Agency for Research on Cancer, World Health Organization. IARC Monographs evaluate consumption of red meat and processed meat. 26 Oct 2015, press release no. 240. http://www.iarc.fr/en/media-centre /pr/2015/pdfs/pr240_E.pdf.

34 Dubrow R, Darefsky AS, Park Y et al. Dietary components related to N-nitroso compound formation: a prospective study of adult glioma. *Cancer Epidemiol Biomarkers Prev.* 2010;19:1709–22; Hord NG, Tang Y, Bryan NS. Food sources of nitrates and nitrites: the physiologic context for potential health benefits. *Am J Clin Nutr.* 2009;90:1–10.

35 Puangsombat K, Gadgil P, Houser TA et al. Occurrence of heterocyclic amines in cooked meat products. *Meat Sci.* 2012;90:739–46; Zheng W, Lee S-A. Well-done meat intake, heterocyclic amine exposure, and cancer risk. *Nutr Cancer.* 2009;61:437–46; Chan DS, Lau R, Aune D et al. Red and processed meat and colorectal cancer incidence: meta-analysis of prospective studies. *PLOS One.* 2011;6:e20456; International Agency for Research on Cancer, World Health Organization. IARC Monographs evaluate consumption of red meat and processed meat. 26 Oct 2015, press release no. 240. http://www.iarc.fr/en/media-centre/pr/2015/pdfs/pr240_E.pdf.

36 Byrne C, Sinha R, Platz EA et al. Predictors of dietary heterocyclic amine intake in three prospective cohorts. *Cancer Epidemiol Biomarkers Prev.* 1998;7:523–29; Sullivan KM, Erickson MA, Sandusky CB, Barnard ND. Detection of PhIP in grilled chicken entrees at popular chain restaurants throughout California. *Nutr Cancer.* 2008;60:592–602; Thomson B. Heterocyclic amine levels in cooked meat and the implication for New Zealanders. *Eur J Cancer Prev.* 1999;8:201–6.

37 Abid Z, Cross AJ, Sinha R. Meat, dairy, and cancer. *Am J Clin Nutr*. 2014;100(suppl 1):386S–93S; National Cancer Institute. "Chemicals in Meat Cooked at High Temperatures and Cancer Risk." http://www.cancer.gov/cancertopics/factsheet/Risk/cooked-meats. Reviewed 11 July 2017.

38 Viegas O, Amaro LF, Ferreira IM, Pinho O. Inhibitory effect of antioxidant-rich marinades on the formation of heterocyclic aromatic amines in pan-fried beef. *J Agric Food Chem*. 2012;60:6235–40; Smith JS, Ameri F, Gadgil P. Effect of marinades on the formation of heterocyclic amines in grilled beef steaks. *J Food Sci*. 2008;73:T100–105; Sugimura T. Nutrition and dietary carcinogens. *Carcinogenesis*. 2000;21:387–95; Murray S, Lake BG, Gray S et al. Effect of cruciferous vegetable consumption on heterocyclic aromatic amine metabolism in man. *Carcinogenesis*, 2001;22:1413–20.

39 Ranciere F, Lyons JG, Loh VH et al. Bisphenol A and the risk of cardiometabolic disorders: a systematic review with meta-analysis of the epidemiological evidence. *Environ Health*. 2015;14:46.

40 Bittner GD, Yang CZ, Stoner MA. Estrogenic chemicals often leach from BPA-free plastic products that are replacements for BPA-containing polycarbonate products. *Environ Health*. 2014;13:41; Yang CZ, Yaniger SI, Jordan VC et al. Most plastic products release estrogenic chemicals: a potential health problem that can be solved. *Environ Health Perspect*. 2011;119:989–96.

41 Centers for Disease Control and Prevention. Environmental Health. "Phthalates." Nov 2009. http://www.cdc.gov/biomonitoring/pdf/Phthalates_FactSheet.pdf; Rudel RA, Gray JM, Engel CL et al. Food packaging and bisphenol A and bis(2-ethyhexyl) phthalate exposure: findings from a dietary intervention. *Environ Health Perspect*. 2011;119:914–20.

42 Diamanti-Kandarakis E, Bourguignon JP, Giudice LC et al. Endocrine-disrupting chemicals: an Endocrine Society scientific statement. *Endocrine Rev*. 2009;30:293–342; Grindler NM, Allsworth JE, Macones GA et al. Persistent organic pollutants and early menopause in US women. *PLOS One*. 2015;10:e0116057.

43 Steingraber S. "The Falling Age of Puberty in US Girls: What We Know, What We Need to Know." Breast Cancer Fund, Aug 2007. http://gaylesulik.com/wp-content/uploads/2010/07 /falling-age-of-puberty.pdf; Natural Resources Defense Council. "Smarter living: Chemical Index. Phthalates." http://www.nrdc.org/living/chemicalindex/phthalates.asp; Breast Cancer Fund. "Phthalates." http://www.breastcancerfund.org/clear-science/radiation-chemicals-and -breast-cancer/phthalates.html.

44 Sharma S, Chatterjee S. Microplastic pollution, a threat to marine ecosystem and human health: a short review. *Environ Sci Pollut Res Int*. 2017;24(27):21530–47.

45 Baudry J, Assmann A, Touvier M. Association of frequency of organic food consumption with cancer risk: findings from the NutriNet-Santé Prospective Cohort Study. *JAMA Intern Med*. 2018;178(12):1597–606.

46 Mie A, Andersen HR, Gunnarsson S et al. Human health implications of organic food and organic agriculture: a comprehensive review. *Environ Health*. 2017;16:111; Baudry J, Assmann A, Touvier M. Association of frequency of organic food consumption with cancer risk: findings from the NutriNet-Santé Prospective Cohort Study. *JAMA Intern Med*. 2018;178(12):1597–606.

47 Bradbury KE, Balkwill A, Spencer EA et al. Organic food consumption and the incidence of cancer in a large prospective study of women in the United Kingdom. *Br J Cancer*. 2014;110:2321–26.

48 Baudry J, Lelong H, Adriouch S et al. Association between organic food consumption and metabolic syndrome: cross-sectional results from the NutriNet-Sante study. *Eur J Nutr*. 2018;57:2477–88; Kesse-Guyot E, Baudry J, Assmann KE et al. Prospective association between consumption frequency of organic food and body weight change, risk of overweight or obesity: results from the NutriNet-Sante Study. *Br J Nutr*. 2017;117:325–34.

49 Torjusen H, Brantsaeter AL, Haugen M et al. Reduced risk of pre-eclampsia with organic vegetable consumption: results from the prospective Norwegian Mother and Child Cohort Study. *BMJ Open*. 2014;4:e006143.

50 International Agency for Research on Cancer. IARC Monographs volume 112: evaluation of five organophosphate insecticides and herbicides, 20 Mar 2015. https://www.iarc .fr/en/media-centre/iarcnews/pdf/MonographVolume112.pdf; Hemler EC, Chavarro JE, Hu FB. Organic foods for cancer prevention—worth the investment? *JAMA Intern Med.* 2018;178(12):1606–7.

51 Brown TP, Rumsby PC, Capleton AC et al. Pesticides and Parkinson's disease—is there a link? *Environ Health Perspect.* 2006;114:156–64; Sanderson WT, Talaska G, Zaebst D et al. Pesticide prioritization for a brain cancer case-control study. *Environ Res.* 1997;74:133–44; Zahm SH, Blair A. Cancer among migrant and seasonal farmworkers: an epidemiologic review and research agenda. *Am J Ind Med.* 1993;24:753–66; Lewis-Mikhael AM, Bueno-Cavanillas A, Ofir Guiron T et al. Occupational exposure to pesticides and prostate cancer: a systematic review and meta-analysis. *Occup Environ Med.* 2016;73:134–44; Schinasi L, Leon ME. Non-Hodgkin lymphoma and occupational exposure to agricultural pesticide chemical groups and active ingredients: a systematic review and meta-analysis. *Int J Environ Res Public Health.* 2014;11:4449–4527.

52 Curl CL, Beresford SA, Fenske RA et al. Estimating pesticide exposure from dietary intake and organic food choices: the Multi-Ethnic Study of Atherosclerosis (MESA). *Environ Health Perspect.* 2015;123:475–83; Brantsaeter AL, Ydersbond TA, Hoppin JA et al. Organic food in the diet: exposure and health implications. *Annu Rev Public Health.* 2017;38:295–313; Oates L, Cohen M. Assessing diet as a modifiable risk factor for pesticide exposure. *Int J Environ Res Public Health.* 2011;8:1792–804; Oates L, Cohen M, Braun L et al. Reduction in urinary organophosphate pesticide metabolites in adults after a week-long organic diet. *Environ Res.* 2014;132:105–11; Bradman A, Quiros-Alcala L, Castorina R et al. Effect of organic diet intervention on pesticide exposures in young children living in low-income urban and agricultural communities. *Environ Health Perspect.* 2015;123:1086–93.

53 Munoz-Quezada MT, Lucero BA, Barr DB et al. Neurodevelopmental effects in children associated with exposure to organophosphate pesticides: a systematic review. *Neurotoxicology.* 2013;39:158–68.

54 Winter CK, Katz JM. Dietary exposure to pesticide residues from commodities alleged to contain the highest contamination levels. *J Toxicol.* 2011;2011:589674, doi: 10.1155/2011/589674.

55 Baudry J, Assmann A, Touvier M. Association of frequency of organic food consumption with cancer risk: findings from the NutriNet-Santé Prospective Cohort Study. *JAMA Intern Med.* 2018;178(12):1597–606.

56 Reganold JP, Wachter JM. Organic agriculture in the twenty-first century. *Nat Plants.* 2016;2:15221; Costa C, Garcia-Leston J, Costa S et al. Is organic farming safer to farmers' health? A comparison between organic and traditional farming. *Toxicol Lett.* 2014;230:166–76; Baranski M, Srednicka-Tober D, Volakakis N et al. Higher antioxidant and lower cadmium concentrations and lower incidence of pesticide residues in organically grown crops: a systematic literature review and meta-analyses. *Br J Nutr.* 2014:1–18.

57 Benz CC, Yau C. Ageing, oxidative stress and cancer: paradigms in parallax. *Nat Rev Cancer.* 2008;8:875–79.

58 Nowotny K, Jung T, Hohn A et al. Advanced glycation end products and oxidative stress in type 2 diabetes mellitus. *Biomolecules.* 2015;5:194–222.

59 Betteridge DJ. What is oxidative stress? *Metabolism.* 2000;49:3–8.

60 Gordon MH. Significance of dietary antioxidants for health. *Int J Mol Sci.* 2012;13:173–79; Reuter S, Gupta SC, Chaturvedi MM, Aggarwal BB. Oxidative stress, inflammation, and cancer: how are they linked? *Free Radic Biol Med.* 2010;49:1603–16.

61 Vallejo MJ, Salazar L, Grijalva M. Oxidative stress modulation and ROS-mediated toxicity in cancer: a review on in vitro models for plant-derived compounds. *Oxid Med Cell Longev.* 2017;2017:4586068; Saha SK, Lee SB, Won J et al. Correlation between oxidative stress, nutrition, and cancer initiation. *Int J Mol Sci.* 2017 July 17;18(7), doi:10.3390/ijms18071544.

62 Himbert C, Thompson H, Ulrich CM. Effects of intentional weight loss on markers of oxidative stress, DNA repair and telomere length: a systematic review. *Obes Facts.* 2017;10:648–65.

63 Goldberg T, Cai W, Peppa M et al. Advanced glycoxidation end products in commonly consumed foods. *J Am Diet Assoc.* 2004;104:1287–91; Pruser KN, Flynn NE. Acrylamide in

health and disease. *Front Biosci (Schol Ed)*. 2011;3:41–51; Uribarri J, Woodruff S, Goodman S et al. Advanced glycation end products in foods and a practical guide to their reduction in the diet. *J Am Diet Assoc*. 2010;110:911–16.

64 Nowotny K, Jung T, Hohn A et al. Advanced glycation end products and oxidative stress in type 2 diabetes mellitus. *Biomolecules*. 2015;5:194–222; Peppa M, Raptis SA. Glycoxidation and wound healing in diabetes: an interesting relationship. *Curr Diabetes Rev*. 2011;7(6):416–25; Peppa M, Stavroulakis P, Raptis SA. Advanced glycoxidation products and impaired diabetic wound healing. *Wound Repair Regen*. 2009;17:461–72; Goldin A, Beckman JA, Schmidt AM, Creager MA. Advanced glycation end products: sparking the development of diabetic vascular injury. *Circulation*. 2006;114:597–605; Yamagishi S, Matsui T. Advanced glycation end products, oxidative stress and diabetic nephropathy. *Oxid Med Cell Longev*. 2010;3:101–8.

65 Saha SK, Lee SB, Won J et al. Correlation between oxidative stress, nutrition, and cancer initiation. *Int J Mol Sci*. 2017 July 17;18(7), doi:10.3390/ijms18071544.

66 Reuter S, Gupta SC, Chaturvedi MM, Aggarwal BB. Oxidative stress, inflammation, and cancer: how are they linked? *Free Radic Biol Med*. 2010;49:1603–16.

67 Vallejo MJ, Salazar L, Grijalva M. Oxidative stress modulation and ROS-mediated toxicity in cancer: a review on in vitro models for plant-derived compounds. *Oxid Med Cell Longev*. 2017;2017:4586068.

68 Tsai WC, Li YH, Lin CC et al. Effects of oxidative stress on endothelial function after a high-fat meal. *Clin Sci (Lond)*. 2004;106:315–19; Lacroix S, Rosiers CD, Tardif JC, Nigam A. The role of oxidative stress in postprandial endothelial dysfunction. *Nutr Res Rev*. 2012;25:288–301.

69 Le NA. Lipoprotein-associated oxidative stress: a new twist to the postprandial hypothesis. *Int J Mol Sci*. 2014;16:401–19; Betteridge DJ. What is oxidative stress? *Metabolism*. 2000;49:3–8.

70 Kudryavtseva AV, Krasnov GS, Dmitriev AA et al. Mitochondrial dysfunction and oxidative stress in aging and cancer. *Oncotarget*. 2016;7:44879–905; Liu Z, Zhou T, Ziegler AC et al. Oxidative stress in neurodegenerative diseases: from molecular mechanisms to clinical applications. *Oxid Med Cell Longev*. 2017;2017:2525967.

71 "Carotenoids." Oregon State University, Linus Pauling Institute. Micronutrient Information Center. http://lpi.oregonstate.edu/mic/dietary-factors/phytochemicals/carotenoids. Last updated July 2016; "Vitamin A." Oregon State University, Linus Pauling Institute, Micronutrient Information Center. http://lpi.oregonstate.edu/infocenter/vitamins/vitaminA/. Last updated January 2015.

72 Shardell MD, Alley DE, Hicks GE et al. Low-serum carotenoid concentrations and carotenoid interactions predict mortality in US adults: the Third National Health and Nutrition Examination Survey. *Nutr Res*. 2011;31:178–89.

73 Li C, Ford ES, Zhao G et al. Serum α-carotene concentrations and risk of death among US adults: the Third National Health and Nutrition Examination Survey Follow-up Study. *Arch Intern Med*. 2011;171(6):507–15.

74 Shardell MD, Alley DE, Hicks GE et al. Low-serum carotenoid concentrations and carotenoid interactions predict mortality in US adults: the Third National Health and Nutrition Examination Survey. *Nutr Res*. 2011;31:178–89.

75 Min KB, Min JY. Association between leukocyte telomere length and serum carotenoid in US adults. *Eur J Nutr*. 2017;56(3):1045–52.

76 "Carotenoids." Oregon State University, Linus Pauling Institute, Micronutrient Information Center. http://lpi.oregonstate.edu/mic/dietary-factors/phytochemicals/carotenoids. Last updated July 2016.

77 Stringham JM, Bovier ER, Wong JC, Hammond BR Jr. The influence of dietary lutein and zeaxanthin on visual performance. *J Food Sci*. 2010;75:R24–29; Abdel-Aal el SM, Akhtar H, Zaheer K, Ali R. Dietary sources of lutein and zeaxanthin carotenoids and their role in eye health. *Nutrients*. 2013;5:1169–85; Koushan K, Rusovici R, Li W et al. The role of lutein in eye-related disease. *Nutrients*. 2013;5:1823–39; Widomska J, Subczynski WK. Why has nature chosen lutein and zeaxanthin to protect the retina? *J Clin Exp Ophthalmol*. 2014;5:326.

78 Koushan K, Rusovici R, Li W et al. The role of lutein in eye-related disease. *Nutrients*. 2013;5:1823–39; Schleicher M, Weikel K, Garber C, Taylor A. Diminishing risk for age-

related macular degeneration with nutrition: a current view. *Nutrients*. 2013;5:2405–56; Seddon JM, Ajani UA, Sperduto RD et al. Dietary carotenoids, vitamins A, C, and E, and advanced age-related macular degeneration. Eye Disease Case-Control Study Group. *JAMA*. 1994;272:1413–20; Ma L, Dou HL, Wu YQ et al. Lutein and zeaxanthin intake and the risk of age-related macular degeneration: a systematic review and meta-analysis. *Br J Nutr*. 2012;107:350–59.

79 Age-Related Eye Disease Study 2 Research Group; Chew EY, Clemons TE et al. Secondary analyses of the effects of lutein/zeaxanthin on age-related macular degeneration progression: AREDS2 report No. 3. *JAMA Ophthalmol*. 2014;132:142–49.

80 USDA Agricultural Research Service. "USDA Food Composition Databases." https://ndb.nal .usda.gov/ndb/search/list.

81 Rizwan M, Rodriguez-Blanco I, Harbottle A et al. Tomato paste rich in lycopene protects against cutaneous photodamage in humans in vivo: a randomized controlled trial. *Br J Dermatol*. 2011;164(1):154–62; Evans JA, Johnson EJ. The role of phytonutrients in skin health. *Nutrients*. 2010;2:903–28; Kopcke W, Krutmann J. Protection from sunburn with beta-carotene—a meta-analysis. *Photochem Photobiol*. 2008;84:284–88.

82 Voutilainen S, Nurmi T, Mursu J, Rissanen TH. Carotenoids and cardiovascular health. *Am J Clin Nutr*. 2006;83:1265–71.

83 Xaplanteris P, Vlachopoulos C, Pietri P et al. Tomato paste supplementation improves endothelial dynamics and reduces plasma total oxidative status in healthy subjects. *Nutr Res*. 2012;32:390–94.

84 Lycopene. Monograph. *Altern Med Rev*. 2003;8(3):336–42.

85 Ried K, Fakler P. Protective effect of lycopene on serum cholesterol and blood pressure: meta-analyses of intervention trials. *Maturitas*. 2011;68:299–310.

86 Palozza P, Parrone N, Catalano A, Simone R. Tomato lycopene and inflammatory cascade: basic interactions and clinical implications. *Curr Med Chem*. 2010;17:2547–63; Palozza P, Parrone N, Simone RE, Catalano A. Lycopene in atherosclerosis prevention: an integrated scheme of the potential mechanisms of action from cell culture studies. *Arch Biochem Biophys*. 2010;504:26–33.

87 Voutilainen S, Nurmi T, Mursu J, Rissanen TH. Carotenoids and cardiovascular health. *Am J Clin Nutr*. 2006;83:1265–71.

88 Mansuri ML, Parihar P, Solanki I, Parihar MS. Flavonoids in modulation of cell survival signalling pathways. *Genes Nutr*. 2014;9:400.

89 Wang LS, Arnold M, Huang YW et al. Modulation of genetic and epigenetic biomarkers of colorectal cancer in humans by black raspberries: a phase I pilot study. *Clin Cancer Res*. 2011;17:598–610; Wang LS, Burke CA, Hasson H et al. A phase Ib study of the effects of black raspberries on rectal polyps in patients with familial adenomatous polyposis. *Cancer Prev Res (Phila)*. 2014;7:666–74; Chen T, Yan F, Qian J et al. Randomized phase II trial of lyophilized strawberries in patients with dysplastic precancerous lesions of the esophagus. *Cancer Prev Res (Phila)*. 2012;5:41–50; Mallery SR, Tong M, Shumway BS et al. Topical application of a mucoadhesive freeze-dried black raspberry gel induces clinical and histologic regression and reduces loss of heterozygosity events in premalignant oral intraepithelial lesions: results from a multicentered, placebo-controlled clinical trial. *Clin Cancer Res*. 2014;20:1910–24.

90 Mladenka P, Zatloukalova L, Filipsky T, Hrdina R. Cardiovascular effects of flavonoids are not caused only by direct antioxidant activity. *Free Radic Biol Med*. 2010;49:963–75; Reis JF, Monteiro VV, de Souza Gomes R et al. Action mechanism and cardiovascular effect of anthocyanins: a systematic review of animal and human studies. *J Transl Med*. 2016;14:315.

91 Cassidy A, O'Reilly EJ, Kay C et al. Habitual intake of flavonoid subclasses and incident hypertension in adults. *Am J Clin Nutr*. 2011;93:338–47; Wang X, Ouyang YY, Liu J, Zhao G. Flavonoid intake and risk of CVD: a systematic review and meta-analysis of prospective cohort studies. *Br J Nutr*. 2014;111:1–11; Kim Y, Je Y. Flavonoid intake and mortality from cardiovascular disease and all causes: a meta-analysis of prospective cohort studies. *Clin Nutr ESPEN*. 2017;20:68–77.

92 Kawser Hossain M, Abdal Dayem A, Han J et al. Molecular mechanisms of the anti-obesity and anti-diabetic properties of flavonoids. *Int J Mol Sci*. 2016;17:569.

93 Flanagan E, Muller M, Hornberger M, Vauzour D. Impact of flavonoids on cellular and molecular mechanisms underlying age-related cognitive decline and neurodegeneration. *Curr Nutr Rep*. 2018;7:49–57; Shukitt-Hale B. Blueberries and neuronal aging. *Gerontology*. 2012;58:518–23; Devore EE, Kang JH, Breteler MM, Grodstein F. Dietary intakes of berries and flavonoids in relation to cognitive decline. *Ann Neurol*. 2012;72(1):135–43.

94 Krikorian R, Shidler MD, Nash TA et al. Blueberry supplementation improves memory in older adults. *J Agric Food Chem*. 2010;58:3996–4000.

95 Boespflug EL, Eliassen JC, Dudley JA et al. Enhanced neural activation with blueberry supplementation in mild cognitive impairment. *Nutr Neurosci*. 2018;21:297–305.

96 Socci V, Tempesta D, Desideri G et al. Enhancing human cognition with cocoa flavonoids. *Front Nutr*. 2017;4:19.

97 "Flavonoids." Oregon State University, Linus Pauling Institute, Micronutrient Information Center. https://lpi.oregonstate.edu/mic/dietary-factors/phytochemicals/flavonoids. Last updated Nov 2015.

98 "Vitamin C." Oregon State University, Linus Pauling Institute, Micronutrient Information Center. http://lpi.oregonstate.edu/infocenter/vitamins/vitaminC/index.html. Last updated July 2018.

99 "Vitamin E." Oregon State University, Linus Pauling Institute, Micronutrient Information Center. https://lpi.oregonstate.edu/mic/vitamins/vitamin-E. Last updated May 2015.

100 van Het Hof KH, West CE, Weststrate JA, Hautvast JG. Dietary factors that affect the bioavailability of carotenoids. *J Nutr*. 2000;130:503–6.

101 Brown MJ, Ferruzzi MG, Nguyen ML et al. Carotenoid bioavailability is higher from salads ingested with full-fat than with fat-reduced salad dressings as measured with electrochemical detection. *Am J Clin Nutr*. 2004;80:396–403.

102 Stefanson AL, Bakovic M. Dietary regulation of Keap1/Nrf2/ARE pathway: focus on plant-derived compounds and trace minerals. *Nutrients* 2014, 6:3777–801.

103 Khan F, Niaz K, Maqbool F et al. Molecular targets underlying the anticancer effects of quercetin: an update. *Nutrients*. 2016 Aug 29;8(9), doi:10.3390/nu8090529; Done AJ, Traustadottir T. Nrf2 mediates redox adaptations to exercise. *Redox Biol*. 2016;10:191–99; Erdman JW Jr., Balentine D, Arab L et al. Flavonoids and heart health: proceedings of the ILSI North America Flavonoids Workshop, May 31 – June 1, 2005, Washington, DC. *J Nutr*. 2007;137:718S–37S; Higdon J. Flavonoids. In: *An Evidence-Based Approach to Dietary Phytochemicals*. New York: Thieme, 2006: 114–26; Kang SY, Seeram NP, Nair MG, Bourquin LD. Tart cherry anthocyanins inhibit tumor development in Apc(Min) mice and reduce proliferation of human colon cancer cells. *Cancer Lett*. 2003;194:13–19; Stoner GD, Wang LS, Casto BC. Laboratory and clinical studies of cancer chemoprevention by antioxidants in berries. *Carcinogenesis*. 2008;29:1665–74.

104 Hybertson BM, Gao B, Bose SK, McCord JM. Oxidative stress in health and disease: the therapeutic potential of Nrf2 activation. *Mol Aspects Med*. 2011;32(4–6):234–46; Zhang Q, Pi J, Woods CG, Andersen ME. A systems biology perspective on Nrf2-mediated antioxidant response. *Toxicol Appl Pharmacol*. 2010;244:84–97.

105 Stefanson AL, Bakovic M. Dietary regulation of Keap1/Nrf2/ARE pathway: focus on plant-derived compounds and trace minerals. *Nutrients*. 2014;6:3777–801; Huang Y, Li W, Su ZY, Kong AN. The complexity of the Nrf2 pathway: beyond the antioxidant response. *J Nutr Biochem*. 2015;26:1401–13.

106 Scapagnini G, Vasto S, Abraham NG et al. Modulation of Nrf2/ARE pathway by food polyphenols: a nutritional neuroprotective strategy for cognitive and neurodegenerative disorders. *Mol Neurobiol*. 2011;44:192–201.

107 Smith RE, Tran K, Smith CC et al. The role of the Nrf2/ARE antioxidant system in preventing cardiovascular diseases. *Diseases* 2016 Dec;4(4):34, doi:10.3390/diseases4040034.

108 Bat-Chen W, Golan T, Peri I et al. Allicin purified from fresh garlic cloves induces apoptosis in colon cancer cells via Nrf2. *Nutr Cancer*. 2010;62:947–57; Kelsey NA, Wilkins HM, Linseman DA. Nutraceutical antioxidants as novel neuroprotective agents. *Molecules*. 2010;15:7792–814; Stefanson AL, Bakovic M. Dietary regulation of Keap1/Nrf2/ARE pathway: focus on plant-

derived compounds and trace minerals. *Nutrients*. 2014;6:3777–801; Ho CY, Weng CJ, Jhang JJ et al. Diallyl sulfide as a potential dietary agent to reduce TNF-alpha- and histamine-induced proinflammatory responses in A7r5 cells. *Mol Nutr Food Res*. 2014;58:1069–78.

109 Cimino F, Speciale A, Anwar S et al. Anthocyanins protect human endothelial cells from mild hyperoxia damage through modulation of Nrf2 pathway. *Genes Nutr*. 2013;8:391–99.

110 Kavitha K, Thiyagarajan P, Rathna Nandhini J et al. Chemopreventive effects of diverse dietary phytochemicals against DMBA-induced hamster buccal pouch carcinogenesis via the induction of Nrf2-mediated cytoprotective antioxidant, detoxification, and DNA repair enzymes. *Biochimie*. 2013;95:1629–39.

111 Yang Y, Li W, Li Y et al. Dietary *Lycium barbarum* polysaccharide induces Nrf2/ARE pathway and ameliorates insulin resistance induced by high-fat via activation of PI3K/AKT signaling. *Oxid Med Cell Longev*. 2014;2014. http://dx.doi.org/10.1155/2014/145641.

112 Leonardo CC, Dore S. Dietary flavonoids are neuroprotective through Nrf2-coordinated induction of endogenous cytoprotective proteins. *Nutri Neurosci*. 2011;14:226–36; Mann GE, Rowlands DJ, Li FY et al. Activation of endothelial nitric oxide synthase by dietary isoflavones: role of NO in Nrf2-mediated antioxidant gene expression. *Cardiovasc Res*. 2007;75:261–74.

113 Song Y, Huang L, Yu J. Effects of blueberry anthocyanins on retinal oxidative stress and inflammation in diabetes through Nrf2/HO-1 signaling. *J Neuroimmunol*. 2016;301:1–6; Lee SG, Kim B, Yang Y et al. Berry anthocyanins suppress the expression and secretion of proinflammatory mediators in macrophages by inhibiting nuclear translocation of NF-κB independent of NRF2-mediated mechanism. *J Nutr Biochem*. 2014;25:404–11.

114 Granado-Serrano AB, Martin MA, Bravo L et al. Quercetin modulates Nrf2 and glutathione-related defenses in HepG2 cells: involvement of p38. *Chem Biol Interact*. 2012 Jan 25;195(2):154–64.

115 Han SG, Han SS, Toborek M, Hennig B. EGCG protects endothelial cells against PCB 126-induced inflammation through inhibition of AhR and induction of Nrf2-regulated genes. *Toxicol Appl Pharmacol*. 2012;261:181–88.

116 Zhao J, Moore AN, Redell JB, Dash PK. Enhancing expression of Nrf2-driven genes protects the blood brain barrier after brain injury. *J Neurosci*. 2007;27:10240–48; Kelsey NA, Wilkins HM, Linseman DA. Nutraceutical antioxidants as novel neuroprotective agents. *Molecules*. 2010;15:7792–814; Stefanson AL, Bakovic M. Dietary regulation of Keap1/Nrf2/ARE pathway: focus on plant-derived compounds and trace minerals. *Nutrients*. 2014;6:3777–801.

117 Huang CS, Lin AH, Liu CT et al. Isothiocyanates protect against oxidized LDL-induced endothelial dysfunction by upregulating Nrf2-dependent antioxidation and suppressing NFκB activation. *Mol Nutr Food Res*. 2013;57:1918–30; Saw CL, Cintron M, Wu TY et al. Pharmacodynamics of dietary phytochemical indoles I3C and DIM: induction of Nrf2-mediated phase II drug metabolizing and antioxidant genes and synergism with isothiocyanates. *Biopharm Drug Dispos*. 2011 Jul;32(5):289–300; Stefanson AL, Bakovic M. Dietary regulation of Keap1/Nrf2/ARE pathway: focus on plant-derived compounds and trace minerals. *Nutrients*. 2014;6:3777–801.

118 Ungvari Z, Bagi Z, Feher A et al. Resveratrol confers endothelial protection via activation of the antioxidant transcription factor Nrf2. *Am J Physiol Heart Circ Physiol*. 2010;299:H18–24; Stefanson AL, Bakovic M. Dietary regulation of Keap1/Nrf2/ARE pathway: focus on plant-derived compounds and trace minerals. *Nutrients*. 2014;6:3777–801.

119 Stefanson AL, Bakovic M. Dietary regulation of Keap1/Nrf2/ARE pathway: focus on plant-derived compounds and trace minerals. *Nutrients*. 2014;6:3777–801.

120 Shehzad A, Lee YS. Molecular mechanisms of curcumin action: signal transduction. *Biofactors*. 2013;39:27–36.

121 Li S, Ding Y, Niu Q et al. Lutein has a protective effect on hepatotoxicity induced by arsenic via Nrf2 signaling. *Biomed Res Int*. 2015;2015:315205.

122 Pandurangan AK, Saadatdoust Z, Mohd Esa N et al. Dietary cocoa protects against colitis-associated cancer by activating the Nrf2/Keap1 pathway. *Biofactors*. 2015;41:1–14.

123 Stefanson AL, Bakovic M. Dietary regulation of Keap1/Nrf2/ARE pathway: focus on plant-derived compounds and trace minerals. *Nutrients*. 2014;6:3777–801; Ben-Dor A, Steiner

M, Gheber L. et al. Carotenoids activate the antioxidant response element transcription system. *Mol Cancer Ther.* 2005;4:177–86.

124 Stefanson AL, Bakovic M. Dietary regulation of Keap1/Nrf2/ARE pathway: focus on plant-derived compounds and trace minerals. *Nutrients.* 2014;6:3777–801.

125 Ibid.; Bishayee A, Bhatia D, Thoppil RJ et al. Pomegranate-mediated chemoprevention of experimental hepatocarcinogenesis involves Nrf2-regulated antioxidant mechanisms. *Carcinogenesis.* 2011;32:888–96.

126 Stefanson AL, Bakovic M. Dietary regulation of Keap1/Nrf2/ARE pathway: focus on plant-derived compounds and trace minerals. *Nutrients.* 2014;6:3777–801; Zou X, Feng Z, Li Y et al. Stimulation of GSH synthesis to prevent oxidative stress–induced apoptosis by hydroxytyrosol in human retinal pigment epithelial cells: activation of Nrf2 and JNK-p62/SQSTM1 pathways. *J Nutr Biochem.* 2012;23:994–1006.

127 Stefanson AL, Bakovic M. Dietary regulation of Keap1/Nrf2/ARE pathway: focus on plant-derived compounds and trace minerals. *Nutrients.* 2014;6:3777–801.

128 Stefanson, Bakovic.

129 Paredes-Gonzalez X, Fuentes F, Su ZY, Kong AN. Apigenin reactivates Nrf2 anti-oxidative stress signaling in mouse skin epidermal JB6 P + cells through epigenetics modifications. *AAPS J.* 2014;16:727–35.

130 Lou H, Jing X, Wei X et al. Naringenin protects against 6-OHDA–induced neurotoxicity via activation of the Nrf2/ARE signaling pathway. *Neuropharmacology.* 2014;79:380–88.

131 Tang X, Wang H, Fan L et al. Luteolin inhibits Nrf2 leading to negative regulation of the Nrf2/ARE pathway and sensitization of human lung carcinoma A549 cells to therapeutic drugs. *Free Radic Biol Med.* 2011;50:1599–609.

132 Zhang M, Wang S, Mao L et al. Omega-3 fatty acids protect the brain against ischemic injury by activating Nrf2 and upregulating heme oxygenase 1. *J Neurosci.* 2014;34:1903–15.

133 Mann GE, Bonacasa B, Ishii T, Siow RC. Targeting the redox sensitive Nrf2-Keap1 defense pathway in cardiovascular disease: protection afforded by dietary isoflavones. *Curr Opin Pharmacol.* 2009;9:139–45.

134 Thoppil RJ, Bhatia D, Barnes KF et al. Black currant anthocyanins abrogate oxidative stress through Nrf2-mediated antioxidant mechanisms in a rat model of hepatocellular carcinoma. *Curr Cancer Drug Targets.* 2012;12:1244–57.

135 Yaku K, Enami Y, Kurajyo C et al. The enhancement of phase 2 enzyme activities by sodium butyrate in normal intestinal epithelial cells is associated with Nrf2 and p53. *Mol Cell Biochem.* 2012;370:7–14; Vaziri ND, Liu SM, Lau WL et al. High amylose resistant starch diet ameliorates oxidative stress, inflammation, and progression of chronic kidney disease. *PLOS One.* 2014;9:e114881.

136 Huang Y, Khor TO, Shu L et al. A γ-tocopherol-rich mixture of tocopherols maintains Nrf2 expression in prostate tumors of TRAMP mice via epigenetic inhibition of CpG methylation. *J Nutr.* 2012;142:818–23.

137 Wondrak GT, Villeneuve NF, Lamore SD et al. The cinnamon-derived dietary factor cinnamic aldehyde activates the Nrf2-dependent antioxidant response in human epithelial colon cells. *Molecules.* 2010;15:3338–55.

138 Martin D, Rojo AI, Salinas M et al. Regulation of heme oxygenase-1 expression through the phosphatidylinositol 3-kinase/Akt pathway and the Nrf2 transcription factor in response to the antioxidant phytochemical carnosol. *J Biol Chem.* 2004;279:8919–29.

139 Nakamura Y, Yoshida C, Murakami A et al. Zerumbone, a tropical ginger sesquiterpene, activates phase II drug metabolizing enzymes. *FEBS Lett.* 2004;572:245–50.

140 Ma JQ, Ding J, Zhang L, Liu CM. Protective effects of ursolic acid in an experimental model of liver fibrosis through Nrf2/ARE pathway. *Clin Res Hepatol Gastroenterol.* 2015;39:188–97.

141 Wei M, Zheng Z, Shi L et al. Natural polyphenol chlorogenic acid protects against acetaminophen-induced hepatotoxicity by activating ERK/Nrf2 antioxidative pathway. *Toxicol Sci.* 2018;162:99–112.

142 Donovan EL, McCord JM, Reuland DJ et al. Phytochemical activation of Nrf2 protects human coronary artery endothelial cells against an oxidative challenge. *Oxid Med Cell Longev.* 2012;2012:132931.

143 Zakkar M, van der Heiden K, Luong le A et al. Activation of Nrf2 in endothelial cells protects arteries from exhibiting a proinflammatory state. *Arterioscler Thromb Vasc Biol.* 2009;29:1851–57.

144 Huang CS, Lin AH, Liu CT et al. Isothiocyanates protect against oxidized LDL-induced endothelial dysfunction by upregulating Nrf2-dependent antioxidation and suppressing NFkappaB activation. *Mol Nutr Food Res.* 2013;57:1918–30; Chen XL, Dodd G, Kunsch C. Sulforaphane inhibits TNF-alpha-induced activation of p38 MAP kinase and VCAM-1 and MCP-1 expression in endothelial cells. *Inflamm Res.* 2009;58:513–21; Bai Y, Wang X, Zhao S et al. Sulforaphane protects against cardiovascular disease via Nrf2 activation. *Oxid Med Cell Longev.* 2015;2015:407580; Huang CS, Lin AH, Liu CT et al. Isothiocyanates protect against oxidized LDL-induced endothelial dysfunction by upregulating Nrf2-dependent antioxidation and suppressing NFkappaB activation. *Mol Nutr Food Res.* 2013;57:1918–30.

145 Zhao J, Moore AN, Redell JB, Dash PK. Enhancing expression of Nrf2-driven genes protects the blood brain barrier after brain injury. *J Neurosci.* 2007;27:10240–48.

146 Bai Y, Wang X, Zhao S et al. Sulforaphane protects against cardiovascular disease via Nrf2 activation. *Oxid Med Cell Longev.* 2015;2015:407580.

Chapter Five: We Can Prevent Cancer

1 National Cancer Institute. Surveillance, Epidemiology, and End Results (SEER) Program. https://www.seer.cancer.gov.

2 Grant WB. A multicountry ecological study of cancer incidence rates in 2008 with respect to various risk-modifying factors. *Nutrients.* 2013;6(1):163–89; Buettner D, Skemp S. Blue Zones: lessons from the world's longest lived. *Am J Lifestyle Med.* 2016;10(5):318–21; Willcox DC, Scapagnini G, Willcox BJ. Healthy aging diets other than the Mediterranean: a focus on the Okinawan diet. *Mech Ageing Dev.* 2014 Mar–Apr;136–37:148–62; Campbell TC, Parpia B, Chen J. Diet, lifestyle, and the etiology of coronary artery disease: the Cornell China study. *Am J Cardiol.* 1998;82(10B):18T–21T; Vardavas CI, Linardakis MK, Hatzis CM et al. Cardiovascular disease risk factors and dietary habits of farmers from Crete 45 years after the first description of the Mediterranean diet. *Eur J Cardiovasc Prev Rehabil.* 2010;17(4):440–46.

3 Inoue-Choi M, Robien K, Lazovich D. Adherence to the WCRF/AICR guidelines for cancer prevention is associated with lower mortality among older female cancer survivors. *Cancer Epidemiol Biomarkers Prev.* 2013;22(5):792–802; Frattaroli J, Weidner G, Dnistrian AM et al. Clinical events in prostate cancer lifestyle trial: results from two years of follow-up. *Urology.* 2008;72:1319–23; Sansbury LB, Wanke K, Albert PS et al. The effect of strict adherence to a high-fiber, high-fruit and -vegetable, and low-fat eating pattern on adenoma recurrence. *Am J Epidemiol.* 2009;170:576–84; Lanza E, Hartman TJ, Albert PS et al. High dry bean intake and reduced risk of advanced colorectal adenoma recurrence among participants in the polyp prevention trial. *J Nutr.* 2006;136:1896–903; Bobe G, Sansbury LB, Albert PS et al. Dietary flavonoids and colorectal adenoma recurrence in the Polyp Prevention Trial. *Cancer Epidemiol Biomarkers Prev.* 2008;17:1344–53; Gold EB, Pierce JP, Natarajan L et al. Dietary pattern influences breast cancer prognosis in women without hot flashes: the women's healthy eating and living trial. *J Clin Oncol.* 2009;27:352–59; Thomson CA, Rock CL, Thompson PA et al. Vegetable intake is associated with reduced breast cancer recurrence in tamoxifen users: a secondary analysis from the Women's Healthy Eating and Living Study. *Breast Cancer Res Treat.* 2011;125:519–27; Pierce JP, Natarajan L, Caan BJ et al. Dietary change and reduced breast cancer events among women without hot flashes after treatment of early-stage breast cancer: subgroup analysis of the Women's Healthy Eating and Living Study. *Am J Clin Nutr.* 2009;89:1565S–71S; Traka M, Gasper AV, Melchini A et al. Broccoli consumption interacts with GSTM1 to perturb oncogenic signalling pathways in the prostate. *PLOS One.* 2008;3:e2568; Twardowski P, Kanaya N, Frankel P et al. A phase I trial of mushroom powder in patients with biochemically recurrent prostate cancer: roles of cytokines and myeloid-derived suppressor cells for Agaricus bisporus-induced prostate-specific antigen responses. *Cancer.* 2015;121:2942–50; Wang LS, Arnold M, Huang YW et al. Modulation of genetic and epigenetic biomarkers of colorectal cancer in humans by black raspberries: a phase I pilot study. *Clin Cancer Res.* 2011;17:598–610; Chen T, Yan F, Qian J et al. Randomized phase II trial of lyophilized strawberries in patients with dysplastic precancerous lesions of the

esophagus. *Cancer Prev Res (Phila)*. 2012;5:41–50; Pantuck AJ, Leppert JT, Zomorodian N et al. Phase II study of pomegranate juice for men with rising prostate-specific antigen following surgery or radiation for prostate cancer. *Clin Cancer Res*. 2006;12:4018–26; Thompson LU, Chen JM, Li T et al. Dietary flaxseed alters tumor biological markers in postmenopausal breast cancer. *Clin Cancer Res*. 2005;11:3828–35.

4 American Institute for Cancer Research. "Take Control of Your Cancer Risk: Nearly Fifty Percent of Common Cancers Are Preventable." February 2018. http://www.aicr.org/press /press-releases/2018/nearly-fifty-percent-of-common-cancers-are-preventable.html.

5 David AR, Zimmerman, MR. Cancer: an old disease, a new disease or something in between? *Nature Rev Cancer*. 2010;10:728–33.

6 Boffetta P, Couto E, Wichmann J et al. Fruit and vegetable intake and overall cancer risk in the European Prospective Investigation into Cancer and Nutrition (EPIC). *J Natl Cancer Inst*. 2010;102:529–37; Vieira AR, Abar L, Vingeliene S et al. Fruits, vegetables and lung cancer risk: a systematic review and meta-analysis. *Ann Oncol*. 2016;27:81–96; Liu H, Wang XC, Hu GH et al. Fruit and vegetable consumption and risk of bladder cancer: an updated meta-analysis of observational studies. *Eur J Cancer Prev*. 2015;24:508–16; Liu J, Wang J, Leng Y, Lv C. Intake of fruit and vegetables and risk of esophageal squamous cell carcinoma: a meta-analysis of observational studies. *Int J Cancer*. 2013;133:473–85; Zhou Y, Zhuang W, Hu W et al. Consumption of large amounts of Allium vegetables reduces risk for gastric cancer in a meta-analysis. *Gastroenterology*. 2011;141:80–89; Li B, Jiang G, Zhang G et al. Intake of vegetables and fruit and risk of esophageal adenocarcinoma: a meta-analysis of observational studies. *Eur J Nutr*. 2014;53:1511–21.

7 Heron M. Deaths: leading causes for 2014. *Natl Vital Stat Rep*. 2016;65:1–96.

8 Cronin KA, Lake AJ, Scott S et al. Annual Report to the Nation on the Status of Cancer, part I: National cancer statistics. *Cancer*. 2018;124(13):2785–800.

9 Mariotto AB, Etzioni R, Hurlbert M et al. Estimation of the number of women living with metastatic breast cancer in the United States. *Cancer Epidemiol Biomarkers Prev*. 2017;26:809–15.

10 Higdon J, Drake VJ. Cruciferous vegetables. In: *An Evidence-based Approach to Phytochemicals and Other Dietary Factors*. 2nd ed. New York: Thieme, 2013.

11 "Chlorophyll and Chlorophyllin." Oregon State University, Linus Pauling Institute, Micronutrient Information Center. https://lpi.oregonstate.edu/mic/dietary-factors/phytochemicals/chlorophyll -chlorophyllin. Last updated June 2009.

12 Cavell BE, Syed Alwi SS, Donlevy A, Packham G. Anti-angiogenic effects of dietary isothiocyanates: mechanisms of action and implications for human health. *Biochem Pharmacol*. 2011;81:327–36; Kunimasa K, Kobayashi T, Kaji K, Ohta T. Antiangiogenic effects of indole-3-carbinol and 3,3'-diindolylmethane are associated with their differential regulation of ERK1/2 and Akt in tube-forming HUVEC. *J Nutr*. 2010;140:1–6; Davis R, Singh KP, Kurzrock R, Shankar S. Sulforaphane inhibits angiogenesis through activation of FOXO transcription factors. *Oncol Rep*. 2009;22:1473–78; Kumar A, D'Souza SS, Tickoo S et al. Antiangiogenic and proapoptotic activities of allyl isothiocyanate inhibit ascites tumor growth in vivo. *Int Cancer Ther*. 2009;8:75–87; Higdon J, Delage B, Williams D, Dashwood R. Cruciferous vegetables and human cancer risk: epidemiologic evidence and mechanistic basis. *Pharmacol Res*. 2007;55:224–36; Higdon J, Drake VJ. Cruciferous vegetables. In: *An Evidence-Based Approach to Phytochemicals and Other Dietary Factors*. 2nd ed. New York: Thieme, 2013; Clarke JD, Dashwood RH, Ho E. Multi-targeted prevention of cancer by sulforaphane. *Cancer Lett*. 2008;269:291–304; Weng JR, Tsai CH, Kulp SK, Chen CS. Indole-3-carbinol as a chemopreventive and anti-cancer agent. *Cancer Lett*. 2008;262:153–63.

13 Yuan F, Chen DZ, Liu K et al. Anti-estrogenic activities of indole-3-carbinol in cervical cells: implication for prevention of cervical cancer. *Anticancer Res*. 1999;19:1673–80; Meng Q, Yuan F, Goldberg ID et al. Indole-3-carbinol is a negative regulator of estrogen receptor-alpha signaling in human tumor cells. *J Nutr*. 2000;130:2927–31; Ramirez MC, Singletary K. Regulation of estrogen receptor alpha expression in human breast cancer cells by sulforaphane. *J Nutr Biochem*. 2009;20:195–201.

14 Wu QJ, Yang Y, Vogtmann E et al. Cruciferous vegetables intake and the risk of colorectal cancer: a meta-analysis of observational studies. *Ann Oncol.* 2013;24:1079–87; Liu X, Lv K. Cruciferous vegetables intake is inversely associated with risk of breast cancer: a meta-analysis. *Breast.* 2013 June;22(3):309–13; Liu B, Mao Q, Cao M, Xie L. Cruciferous vegetables intake and risk of prostate cancer: a meta-analysis. *Int J Urol.* 2012;19:134–41.

15 Higdon J, Delage B, Williams D, Dashwood R. Cruciferous vegetables and human cancer risk: epidemiologic evidence and mechanistic basis. *Pharmacol Res.* 2007;55:224–36; "Cruciferous Vegetables." Oregon State University, Linus Pauling Institute, Micronutrient Information Center. http://lpi.oregonstate.edu/mic/food-beverages/cruciferous-vegetables. Last updated Dec 2016; Pollock RL. The effect of green leafy and cruciferous vegetable intake on the incidence of cardiovascular disease: a meta-analysis. *JRSM Cardiovasc Dis.* 2016 Aug 1;5:2048004016661435; Dufour V, Stahl M, Baysse C. The antibacterial properties of isothiocyanates. *Microbiology.* 2015;161:229–43.

16 Stefanson AL, Bakovic M. Dietary regulation of Keap1/Nrf2/ARE pathway: focus on plant-derived compounds and trace minerals. *Nutrients.* 2014;6:3777–801.

17 Zakkar M, van der Heiden K, Luong le A et al. Activation of Nrf2 in endothelial cells protects arteries from exhibiting a proinflammatory state. *Arterioscler Thromb Vasc Biol.* 2009;29:1851–57.

18 Huang CS, Lin AH, Liu CT et al. Isothiocyanates protect against oxidized LDL-induced endothelial dysfunction by upregulating Nrf2-dependent antioxidation and suppressing NFkappaB activation. *Mol Nutr Food Res.* 2013;57:1918–30.

19 Ahn J, Gammon MD, Santella RM et al. Effects of glutathione S-transferase A1 (GSTA1) genotype and potential modifiers on breast cancer risk. *Carcinogenesis.* 2006;27:1876–82; Lee SA, Fowke JH, Lu W et al. Cruciferous vegetables, the GSTP1 Ile105Val genetic polymorphism, and breast cancer risk. *Am J Clin Nutr.* 2008;87:753–60.

20 Alumkal JJ, Slottke R, Schwartzman J et al. A phase II study of sulforaphane-rich broccoli sprout extracts in men with recurrent prostate cancer. *Invest New Drugs.* 2015;33:480–89; Cipolla BG, Mandron E, Lefort JM et al. Effect of sulforaphane in men with biochemical recurrence after radical prostatectomy. *Cancer Prev Res (Phila).* 2015;8:712–19.

21 Traka M, Gasper AV, Melchini A et al. Broccoli consumption interacts with GSTM1 to perturb oncogenic signalling pathways in the prostate. *PLOS One.* 2008;3:e2568.

22 Wirth MD, Murphy EA, Hurley TG, Hebert JR. Effect of cruciferous vegetable intake on oxidative stress biomarkers: differences by breast cancer status. *Cancer Invest.* 2017;35:277–87.

23 Schlemmer U, Frolich W, Prieto RM, Grases F. Phytate in foods and significance for humans: food sources, intake, processing, bioavailability, protective role and analysis. *Mol Nutr Food Res.* 2009;53(suppl 2):S330–75.

24 Sonnenburg ED, Sonnenburg JL. Starving our microbial self: the deleterious consequences of a diet deficient in microbiota-accessible carbohydrates. *Cell Metab.* 2014;20:779–86; Fung KY, Cosgrove L, Lockett T et al. A review of the potential mechanisms for the lowering of colorectal oncogenesis by butyrate. *Br J Nutr.* 2012;108:820–31; Hamer HM, Jonkers D, Venema K et al. Review article: the role of butyrate on colonic function. *Aliment Pharmacol Ther.* 2008;27:104–19; Williams EA, Coxhead JM, Mathers JC. Anti-cancer effects of butyrate: use of micro-array technology to investigate mechanisms. *Proc Nutr Soc.* 2003;62:107–15; Faris MA, Takruri HR, Shomaf MS, Bustanji YK. Chemopreventive effect of raw and cooked lentils (*Lens culinaris L*) and soybeans (*Glycine max*) against azoxymethane-induced aberrant crypt foci. *Nutr Res.* 2009;29:355–62.

25 Zhu B, Sun Y, Qi L et al. Dietary legume consumption reduces risk of colorectal cancer: evidence from a meta-analysis of cohort studies. *Sci Rep.* 2015;5:8797.

26 Li J, Mao QQ. Legume intake and risk of prostate cancer: a meta-analysis of prospective cohort studies. *Oncotarget.* 2017;8:44776–84.

27 USPSTF, Grossman DC, Curry SJ et al. Screening for prostate cancer: US Preventive Services Task Force recommendation statement. *JAMA.* 2018;319:1901–13; Zhang GQ, Chen JL, Liu Q et al. Soy intake is associated with lower endometrial cancer risk: a systematic review and meta-analysis of observational studies. *Medicine (Baltimore)* 2015;94:e2281; Yan L, Spitznagel

EL. Soy consumption and prostate cancer risk in men: a revisit of a meta-analysis. *Am J Clin Nutr.* 2009;89:1155–63; Applegate CC, Rowles JL, Ranard KM et al. Soy consumption and the risk of prostate cancer: an updated systematic review and meta-analysis. *Nutrients.* 2018;10(1), doi:10.3390/nu10010040; Messina M. Soy and health update: evaluation of the clinical and epidemiologic literature. *Nutrients.* 2016 Nov 24;8(12), doi:10.3390/nu8120754.

28 Messina M. Soy foods, isoflavones, and the health of postmenopausal women. *Am J Clin Nutr.* 2014;100(suppl 1):423S–30S; Oseni T, Patel R, Pyle J, Jordan VC. Selective estrogen receptor modulators and phytoestrogens. *Planta Med.* 2008;74:1656–65.

29 Higdon J, Drake VJ. Soy isoflavones. In: *An Evidence-Based Approach to Dietary Phytochemicals and Other Dietary Factors,* 2nd ed. New York: Thieme, 2013, 96–108.

30 Yang WS, Va P, Wong MY et al. Soy intake is associated with lower lung cancer risk: results from a meta-analysis of epidemiologic studies. *Am J Clin Nutr.* 2011;94:1575–83; Kim J, Kang M, Lee JS et al. Fermented and non-fermented soy food consumption and gastric cancer in Japanese and Korean populations: a meta-analysis of observational studies. *Cancer Sci.* 2011;102:231–44; Yan L, Spitznagel EL, Bosland MC. Soy consumption and colorectal cancer risk in humans: a meta-analysis. *Cancer Epidemiol Biomarkers Prev.* 2010;19:148–58.

31 Young VR, Pellett PL. Plant proteins in relation to human protein and amino acid nutrition. *Am J Clin Nutr.* 1994;59:1203S–12S.

32 Zhou XF, Ding ZS, Liu NB. Allium vegetables and risk of prostate cancer: evidence from 132,192 subjects. *Asian Pac J Cancer Prev.* 2013;14:4131–34; Turati F, Pelucchi C, Guercio V et al. Allium vegetable intake and gastric cancer: a case-control study and meta-analysis. *Mol Nutr Food Res.* 2015;59:171–79; Galeone C, Pelucchi C, Dal Maso L et al. Allium vegetables intake and endometrial cancer risk. *Public Health Nutr.* 2009;12:1576–79; Galeone C, Pelucchi C, Levi F et al. Onion and garlic use and human cancer. *Am J Clin Nutr.* 2006;84:1027–32; Galeone C, Turati F, Zhang ZF et al. Relation of allium vegetables intake with head and neck cancers: evidence from the INHANCE consortium. *Mol Nutr Food Res.* 2015;59:1641–50.

33 Galeone C, Pelucchi C, Levi F et al. Onion and garlic use and human cancer. *Am J Clin Nutr.* 2006;84:1027–32.

34 Khan F, Niaz K, Maqbool F et al. Molecular targets underlying the anticancer effects of quercetin: an update. *Nutrients.* 2016;8(9), doi:10.3390/nu8090529.

35 Higdon J, Drake VJ. Organosulfur compounds from garlic. In: *An Evidence-Based Approach to Dietary Phytochemicals and Other Dietary Factors,* 2nd ed. New York: Thieme, 2012, 149–61.

36 Powolny A, Singh S. Multitargeted prevention and therapy of cancer by diallyl trisulfide and related Allium vegetable-derived organosulfur compounds. *Cancer Lett.* 2008;269:305–14; Modem S, Dicarlo SE, Reddy TR. Fresh garlic extract induces growth arrest and morphological differentiation of MCF7 breast cancer cells. *Genes Cancer.* 2012;3:177–86; Na HK, Kim EH, Choi MA et al. Diallyl trisulfide induces apoptosis in human breast cancer cells through ROS-mediated activation of JNK and AP-1. *Biochem Pharmacol.* 2012;84(10):1241–50; Malki A, El-Saadani M, Sultan AS. Garlic constituent diallyl trisulfide induced apoptosis in MCF7 human breast cancer cells. *Cancer Biol Ther.* 2009;8:2175–85.

37 Higdon J, Drake VJ. Organosulfur compounds from garlic. In: *An Evidence-Based Approach to Dietary Phytochemicals and Other Dietary Factors,* 2nd ed. New York: Thieme, 2012, 149–61; Galeone C, Tavani A, Pelucchi C et al. Allium vegetable intake and risk of acute myocardial infarction in Italy. *Eur J Nutr.* 2009;48:120–23; Rahman K, Lowe GM. Garlic and cardiovascular disease: a critical review. *J Nutr.* 2006;136:736S–40S; Bradley JM, Organ CL, Lefer DJ. Garlic-derived organic polysulfides and myocardial protection. *J Nutr.* 2016;146:403S–9S; Makheja AN, Bailey JM. Antiplatelet constituents of garlic and onion. *Agents Actions.* 1990;29:360–63.

38 Ried K. Garlic lowers blood pressure in hypertensive individuals, regulates serum cholesterol, and stimulates immunity: an updated meta-analysis and review. *J Nutr.* 2016;146:389S–96S.

39 Chan GC, Chan WK, Sze DM. The effects of beta-glucan on human immune and cancer cells. *J Hematol Oncol.* 2009;2:25.

40 Hara M, Hanaoka T, Kobayashi M et al. Cruciferous vegetables, mushrooms, and gastrointestinal cancer risks in a multicenter, hospital-based case-control study in Japan. *Nutr Cancer.* 2003;46:138–47; Zhang CX, Ho SC, Chen YM et al.: Greater vegetable and

fruit intake is associated with a lower risk of breast cancer among Chinese women. *Int J Cancer*. 2009;125:181–88; Martin KR, Brophy SK. Commonly consumed and specialty dietary mushrooms reduce cellular proliferation in MCF-7 human breast cancer cells. *Exp Biol Med*. 2010;235:1306–14; Fang N, Li Q, Yu S et al. Inhibition of growth and induction of apoptosis in human cancer cell lines by an ethyl acetate fraction from shiitake mushrooms. *J Altern Complement Med*. 2006;12:125–32; Ng ML, Yap AT. Inhibition of human colon carcinoma development by lentinan from shiitake mushrooms (*Lentinus edodes*). *J Altern Complement Med*. 2002;8:581–89; Adams LS, Phung S, Wu X et al. White button mushroom (*Agaricus bisporus*) exhibits antiproliferative and proapoptotic properties and inhibits prostate tumor growth in athymic mice. *Nutr Cancer*. 2008;60:744–56; Lakshmi B, Ajith TA, Sheena N et al. Antiperoxidative, anti-inflammatory, and antimutagenic activities of ethanol extract of the mycelium of *Ganoderma lucidum* occurring in South India. *Teratog Carcinog Mutagen*. 2003;suppl 1:85–97; Cao QZ, Lin ZB. Antitumor and anti-angiogenic activity of *Ganoderma lucidum* polysaccharides peptide. *Acta Pharmacol Sinica*. 2004;25:833–38; Lin ZB, Zhang HN. Anti-tumor and immunoregulatory activities of *Ganoderma lucidum* and its possible mechanisms. *Acta Pharmacol Sinica*. 2004;25:1387–95; Patel S, Goyal A. Recent developments in mushrooms as anti-cancer therapeutics: a review. 3 *Biotech*. 2012 Mar;2(1):1–15.

41 Lee JS, Park BC, Ko YJ et al. *Grifola frondosa* (maitake mushroom) water extract inhibits vascular endothelial growth factor-induced angiogenesis through inhibition of reactive oxygen species and extracellular signal-regulated kinase phosphorylation. *J Med Food*. 2008;11:643–51; Xu H, Zou S, Xu X, Zhang L. Anti-tumor effect of beta-glucan from *Lentinus edodes* and the underlying mechanism. *Sci Rep*. 2016;6:28802; Chen S, Yong T, Zhang Y et al. Anti-tumor and anti-angiogenic ergosterols from *Ganoderma lucidum*. *Front Chem*. 2017;5:85; Cao QZ, Lin ZB. Antitumor and anti-angiogenic activity of *Ganoderma lucidum* polysaccharides peptide. *Acta Pharmacol Sinica*. 2004;25:833–38; Chang HC, Yang HL, Pan JH et al. *Hericium erinaceus* inhibits TNF-alpha-induced angiogenesis and ROS generation through suppression of MMP-9/NF-kappaB signaling and activation of Nrf2-mediated antioxidant genes in human EA.hy926 endothelial cells. *Oxid Med Cell Longev*. 2016;2016:8257238; Chang HH, Hsieh KY, Yeh CH et al. Oral administration of an Enoki mushroom protein FVE activates innate and adaptive immunity and induces anti-tumor activity against murine hepatocellular carcinoma. *Int Immunopharmacol*. 2010;10:239–46; Ho JC, Konerding MA, Gaumann A et al. Fungal polysaccharopeptide inhibits tumor angiogenesis and tumor growth in mice. *Life Sci*. 2004;75:1343–56; Sliva D, Jedinak A, Kawasaki J et al. *Phellinus linteus* suppresses growth, angiogenesis and invasive behaviour of breast cancer cells through the inhibition of AKT signalling. *Br J Cancer*. 2008 Apr 22;98(8):1348–56.

42 Feeney M, Miller A, Roupas P. Mushrooms—biologically distinct and nutritionally unique: exploring a "third food kingdom." *Nutr Today*. 2014;49:301–7.

43 Jin X, Ruiz Beguerie J, Sze DM, Chan GC. *Ganoderma lucidum* (Reishi mushroom) for cancer treatment. *Cochrane Database Syst Rev*. 2016;4:CD007731; Ina K, Kataoka T, Ando T. The use of lentinan for treating gastric cancer. *Anticancer Agents Med Chem*. 2013;13:681–88; Standish LJ, Wenner CA, Sweet ES et al. Trametes versicolor mushroom immune therapy in breast cancer. *J Soc Integr Oncol*. 2008;6:122–28.

44 Yu L, Fernig DG, Smith JA et al. Reversible inhibition of proliferation of epithelial cell lines by *Agaricus bisporus* (edible mushroom) lectin. *Cancer Res*. 1993;53:4627–32; Carrizo ME, Capaldi S, Perduca M et al. The antineoplastic lectin of the common edible mushroom (*Agaricus bisporus*) has two binding sites, each specific for a different configuration at a single epimeric hydroxyl. *J Biol Chem*. 2005;280:10614–23; Borchers AT, Krishnamurthy A, Keen CL et al. The immunobiology of mushrooms. *Exp Biol Med*. 2008;233:259–76.

45 Martin KR, Brophy SK. Commonly consumed and specialty dietary mushrooms reduce cellular proliferation in MCF-7 human breast cancer cells. *Exp Biol Med*. 2010;235:1306–14; Fang N, Li Q, Yu S et al. Inhibition of growth and induction of apoptosis in human cancer cell lines by an ethyl acetate fraction from shiitake mushrooms. *J Altern Complement Med*. 2006;12:125–32; Ng ML, Yap AT. Inhibition of human colon carcinoma development by lentinan from shiitake mushrooms (*Lentinus edodes*). *J Altern Complement Med*. 2002;8:581–

89; Adams LS, Phung S, Wu X et al. White button mushroom (*Agaricus bisporus*) exhibits antiproliferative and proapoptotic properties and inhibits prostate tumor growth in athymic mice. *Nutr Cancer*. 2008;60:744–56; Lakshmi B, Ajith TA, Sheena N et al. Antiperoxidative, anti-inflammatory, and antimutagenic activities of ethanol extract of the mycelium of *Ganoderma lucidum* occurring in South India. *Teratog Carcinog Mutagen*. 2003;suppl 1:85–97; Cao QZ, Lin ZB. Antitumor and anti-angiogenic activity of *Ganoderma lucidum* polysaccharides peptide. *Acta Pharmacol Sinica*. 2004;25:833–38; Lin ZB, Zhang HN. Anti-tumor and immunoregulatory activities of *Ganoderma lucidum* and its possible mechanisms. *Acta Pharmacol Sinica*. 2004;25:1387–95; Patel S, Goyal A. Recent developments in mushrooms as anti-cancer therapeutics: a review. *3 Biotech*. 2012;2:1–15.

46 Grube BJ, Eng ET, Kao YC et al. White button mushroom phytochemicals inhibit aromatase activity and breast cancer cell proliferation. *J Nutr*. 2001;131:3288–93.

47 Zhang M, Huang J, Xie X, Holman CD. Dietary intakes of mushrooms and green tea combine to reduce the risk of breast cancer in Chinese women. *Int J Cancer*. 2009;124:1404–8.

48 Li J, Zou L, Chen W et al. Dietary mushroom intake may reduce the risk of breast cancer: evidence from a meta-analysis of observational studies. *PLOS One*. 2014;9:e93437.

49 Twardowski P, Kanaya N, Frankel P et al. A phase I trial of mushroom powder in patients with biochemically recurrent prostate cancer: roles of cytokines and myeloid-derived suppressor cells for *Agaricus bisporus*-induced prostate-specific antigen responses. *Cancer*. 2015;121:2942–50.

50 Borchers AT, Krishnamurthy A, Keen CL et al. The immunobiology of mushrooms. *Exp Biol Med*. 2008;233:259–76; Vannucci L, Krizan J, Sima P et al. Immunostimulatory properties and antitumor activities of glucans (review). *Int J Oncol*. 2013;43:357–64; Akramiene D, Kondrotas A, Didziapetriene J, Kevelaitis E. Effects of beta-glucans on the immune system. *Medicina (Kaunas)*. 2007;43:597–606.

51 Jeong SC, Koyyalamudi SR, Pang G. Dietary intake of *Agaricus bisporus* white button mushroom accelerates salivary immunoglobulin A secretion in healthy volunteers. *Nutrition*. 2012;28:527–31.

52 Toth B, Erickson J. Cancer induction in mice by feeding of the uncooked cultivated mushroom of commerce *Agaricus bisporus*. *Cancer Res*. 1986;46:4007–11; Schulzova V, Hajslova J, Peroutka R et al. Influence of storage and household processing on the agaritine content of the cultivated *Agaricus* mushroom. *Food Addit Contam*. 2002;19:853–62.

53 Erdman JW Jr., Balentine D, Arab L et al. Flavonoids and heart health: proceedings of the ILSI North America Flavonoids Workshop, May 31 – June 1, 2005, Washington, DC. *J Nutr*. 2007;137:718S–37S.

54 Kristo AS, Klimis-Zacas D, Sikalidis AK. Protective role of dietary berries in cancer. *Antioxidants (Basel)*. 2016 Dec;5(4):37, doi:10.3390/antiox5040037.

55 Chen T, Yan F, Qian J et al. Randomized phase II trial of lyophilized strawberries in patients with dysplastic precancerous lesions of the esophagus. *Cancer Prev Res (Phila)*. 2012;5:41–50.

56 Khan N, Afaq F, Kweon MH et al. Oral consumption of pomegranate fruit extract inhibits growth and progression of primary lung tumors in mice. *Cancer Res*. 2007;67:3475–82; Toi M, Bando H, Ramachandran C et al. Preliminary studies on the anti-angiogenic potential of pomegranate fractions in vitro and in vivo. *Angiogenesis*. 2003;6:121–28; Sartippour MR, Seeram NP, Rao JY et al. Ellagitannin-rich pomegranate extract inhibits angiogenesis in prostate cancer in vitro and in vivo. *Int J Oncol*. 2008;32:475–80; Adams LS, Zhang Y, Seeram NP et al. Pomegranate ellagitannin-derived compounds exhibit antiproliferative and antiaromatase activity in breast cancer cells in vitro. *Cancer Prev Res (Phila)*. 2010;3:108–13.

57 Pantuck AJ, Leppert JT, Zomorodian N et al. Phase II study of pomegranate juice for men with rising prostate-specific antigen following surgery or radiation for prostate cancer. *Clin Cancer Res*. 2006;12:4018–26.

58 Paller CJ, Ye X, Wozniak PJ et al. A randomized phase II study of pomegranate extract for men with rising PSA following initial therapy for localized prostate cancer. *Prostate Cancer Prostatic Dis*. 2013;16:50–55.

59 Jayaprakasha GK, Murthy KN, Demarais R, Patil BS. Inhibition of prostate cancer (LNCaP) cell proliferation by volatile components from Nagami kumquats. *Planta Med.* 2012;78(10):974–80; Hakim IA, Harris RB, Ritenbaugh C. Citrus peel use is associated with reduced risk of squamous cell carcinoma of the skin. *Nutr Cancer.* 2000;37(2):161–68.

60 Grosso G, Yang J, Marventano S et al. Nut consumption on all-cause, cardiovascular, and cancer mortality risk: a systematic review and meta-analysis of epidemiologic studies. *Am J Clin Nutr.* 2015;101:783–93; Wu L, Wang Z, Zhu J et al. Nut consumption and risk of cancer and type 2 diabetes: a systematic review and meta-analysis. *Nutr Rev.* 2015;73:409–25.

61 Liu Y, Colditz GA, Cotterchio M et al. Adolescent dietary fiber, vegetable fat, vegetable protein, and nut intakes and breast cancer risk. *Breast Cancer Res Treat.* 2014;145:461–70.

62 Hardman WE, Primerano DA, Legenza MT et al. Dietary walnut altered gene expressions related to tumor growth, survival, and metastasis in breast cancer patients: a pilot clinical trial. *Nutr Res.* 2019;66:82–94, doi:10.1016/j.nutres.2019.03.004.

63 Yang M, Hu FB, Giovannucci EL et al. Nut consumption and risk of colorectal cancer in women. *Eur J Clin Nutr.* 2016;70:333–37; Bao Y, Hu FB, Giovannucci EL et al. Nut consumption and risk of pancreatic cancer in women. *Br J Cancer.* 2013;109:2911–16; Jenab M, Ferrari P, Slimani N et al. Association of nut and seed intake with colorectal cancer risk in the European Prospective Investigation into Cancer and Nutrition. *Cancer Epidemiol Biomarkers Prev.* 2004;13:1595–603.

64 Sabate J. Nut consumption, vegetarian diets, ischemic heart disease risk, and all-cause mortality: evidence from epidemiologic studies. *Am J Clin Nutr.* 1999;70:500S–503S; Fraser GE, Shavlik DJ. Ten years of life: is it a matter of choice? *Arch Intern Med.* 2001;161:1645–52; Tharry M, Mariotti F, Mashchak A. Patterns of plant and animal protein intake are strongly associated with cardiovascular mortality: the Adventist Health Study-2 cohort. *Int J Epidemiol.* 2018;47(5):1603–12; Baer HJ, Glynn RJ, Hu FB et al. Risk factors for mortality in the Nurses' Health Study: a competing risks analysis. *Am J Epidemiol.* 2011;173:319–29; Hshieh TT, Petrone AB, Gaziano JM, Djousse L. Nut consumption and risk of mortality in the Physicians' Health Study. *Am J Clin Nutr.* 2015;101:407–12; Bao Y, Han J, Hu FB et al. Association of nut consumption with total and cause-specific mortality. *N Engl J Med.* 2013;369:2001–11; Luu HN, Blot WJ, Xiang YB et al. Prospective evaluation of the association of nut/peanut consumption with total and cause-specific mortality. *JAMA Intern Med.* 2015 May;175(5):755–66; van den Brandt PA, Schouten LJ. Relationship of tree nut, peanut and peanut butter intake with total and cause-specific mortality: a cohort study and meta-analysis. *Int J Epidemiol.* 2015;44(3):1038–49; Grosso G, Yang J, Marventano S et al. Nut consumption on all-cause, cardiovascular, and cancer mortality risk: a systematic review and meta-analysis of epidemiologic studies. *Am J Clin Nutr.* 2015;101:783–93; Mayhew AJ, de Souza RJ, Meyre D et al. A systematic review and meta-analysis of nut consumption and incident risk of CVD and all-cause mortality. *Br J Nutr.* 2016;115:212–25.

65 Adlercreutz H, Bannwart C, Wahala K et al. Inhibition of human aromatase by mammalian lignans and isoflavonoid phytoestrogens. *J Steroid Biochem Mol Biol.* 1993;44:147–53; Brooks JD, Thompson LU. Mammalian lignans and genistein decrease the activities of aromatase and 17beta-hydroxysteroid dehydrogenase in MCF-7 cells. *J Steroid Biochem Mol Biol.* 2005;94:461–67; Adlercreutz H, Mousavi Y, Clark J et al. Dietary phytoestrogens and cancer: in vitro and in vivo studies. *J Steroid Biochem Mol Biol.* 1992;41:331–37; Adlercreutz H, Hockerstedt K, Bannwart C et al. Effect of dietary components, including lignans and phytoestrogens, on enterohepatic circulation and liver metabolism of estrogens and on sex hormone binding globulin (SHBG). *J Steroid Biochem.* 1987;27:1135–44; Low YL, Dunning AM, Dowsett M et al. Phytoestrogen exposure is associated with circulating sex hormone levels in postmenopausal women and interact with ESR1 and NR1I2 gene variants. *Cancer Epidemiol Biomarkers Prev.* 2007;16:1009–16; Flower G, Fritz H, Balneaves LG et al. Flax and breast cancer: a systematic review. *Integr Cancer Ther.* 2014;13:181–92; Thompson LU, Chen JM, Li T et al. Dietary flaxseed alters tumor biological markers in postmenopausal breast cancer. *Clin Cancer Res.* 2005;11:3828–35; Demark-Wahnefried W, Polascik TJ, George SL et al. Flaxseed supplementation (not dietary fat restriction) reduces prostate cancer proliferation rates in men presurgery. *Cancer Epidemiol Biomarkers Prev.* 2008;17:3577–87.

66 Adlercreutz H, Bannwart C, Wahala K et al. Inhibition of human aromatase by mammalian lignans and isoflavonoid phytoestrogens. *J Steroid Biochem Mol Biol.* 1993;44:147–53; Brooks JD, Thompson LU. Mammalian lignans and genistein decrease the activities of aromatase and 17beta-hydroxysteroid dehydrogenase in MCF-7 cells. *J Steroid Biochem Mol Biol.* 2005;94:461–67; Adlercreutz H, Mousavi Y, Clark J et al. Dietary phytoestrogens and cancer: in vitro and in vivo studies. *J Steroid Biochem Mol Biol.* 1992;41:331–37; Adlercreutz H, Hockerstedt K, Bannwart C et al. Effect of dietary components, including lignans and phytoestrogens, on enterohepatic circulation and liver metabolism of estrogens and on sex hormone binding globulin (SHBG). *J Steroid Biochem.* 1987;27:1135–44; Low YL, Dunning AM, Dowsett M et al. Phytoestrogen exposure is associated with circulating sex hormone levels in postmenopausal women and interact with ESR1 and NR1I2 gene variants. *Cancer Epidemiol Biomarkers Prev.* 2007;16:1009–16.

67 Nemes SM, Orstat V. Evaluation of a microwave-assisted extraction method for lignan quantification in flaxseed cultivars and selected oil seeds. *Food Analytical Methods.* 2012;5:551–63; Coulman KD, Liu Z, Hum WQ et al. Whole sesame seed is as rich a source of mammalian lignan precursors as whole flaxseed. *Nutr Cancer.* 2005;52:156–65.

68 Flower G, Fritz H, Balneaves LG et al. Flax and breast cancer: a systematic review. *Integr Cancer Ther.* 2014;13:181–92; Fabian CJ, Kimler BF, Zalles CM et al. Reduction in Ki-67 in benign breast tissue of high-risk women with the lignan secoisolariciresinol diglucoside. *Cancer Prev Res (Phila).* 2010;3:1342–50.

69 Thompson LU, Chen JM, Li T et al. Dietary flaxseed alters tumor biological markers in postmenopausal breast cancer. *Clin Cancer Res.* 2005;11:3828–35.

70 McCann SE, Thompson LU, Nie J et al. Dietary lignan intakes in relation to survival among women with breast cancer: the Western New York Exposures and Breast Cancer (WEB) Study. *Breast Cancer Res Treat.* 2010;122(1):229–35.

71 Azrad M, Vollmer RT, Madden J et al. Flaxseed-derived enterolactone is inversely associated with tumor cell proliferation in men with localized prostate cancer. *J Med Food.* 2013;16:357–60; Demark-Wahnefried W, Polascik TJ, George SL et al. Flaxseed supplementation (not dietary fat restriction) reduces prostate cancer proliferation rates in men presurgery. *Cancer Epidemiol Biomarkers Prev.* 2008;17:3577–87.

72 Peterson J, Dwyer J, Adlercreutz H et al. Dietary lignans: physiology and potential for cardiovascular disease risk reduction. *Nutr Rev.* 2010;68:571–603; Ren GY, Chen CY, Chen GC et al. Effect of flaxseed intervention on inflammatory marker C-reactive protein: a systematic review and meta-analysis of randomized controlled trials. *Nutrients.* 2016;8:136.

73 Eliassen AH, Liao X, Rosner B et al. Plasma carotenoids and risk of breast cancer over 20 y of follow-up. *Am J Clin Nutr.* 2015;101:1197–205; Wang Y, Cui R, Xiao Y et al. Effect of carotene and lycopene on the risk of prostate cancer: a systematic review and dose-response meta-analysis of observational studies. *PLOS One.* 2015;10:e0137427; Yu N, Su X, Wang Z et al. Association of dietary vitamin A and beta-carotene intake with the risk of lung cancer: a meta-analysis of 19 publications. *Nutrients.* 2015;7:9309–24; Eliassen AH, Hendrickson SJ, Brinton LA et al. Circulating carotenoids and risk of breast cancer: pooled analysis of eight prospective studies. *J Natl Cancer Inst.* 2012;104:1905–16; Etminan M, Takkouche B, Caamano-Isorna F. The role of tomato products and lycopene in the prevention of prostate cancer: a meta-analysis of observational studies. *Cancer Epidemiol Biomarkers Prev.* 2004;13:340–45; Leoncini E, Nedovic D, Panic N et al. Carotenoid intake from natural sources and head and neck cancer: a systematic review and meta-analysis of epidemiological studies. *Cancer Epidemiol Biomarkers Prev.* 2015;24:1003–11.

74 Shardell MD, Alley DE, Hicks GE et al. Low-serum carotenoid concentrations and carotenoid interactions predict mortality in US adults: the Third National Health and Nutrition Examination Survey. *Nutr Res.* 2011;31:178–89.

75 Evans JA, Johnson EJ. The role of phytonutrients in skin health. *Nutrients.* 2010;2:903–28. Stahl W, Sies H. Beta-carotene and other carotenoids in protection from sunlight. *Am J Clin Nutr.* 2012;96(5):1179S–84S.

76 van Het Hof KH, West CE, Weststrate JA, Hautvast JG. Dietary factors that affect the bioavailability of carotenoids. *J Nutr.* 2000;130:503–6.

77 Garcia AL, Koebnick C, Dagnelie PC et al. Long-term strict raw food diet is associated with favourable plasma beta-carotene and low plasma lycopene concentrations in Germans. *Br J Nutr.* 2008;99:1293–300; Brown MJ, Ferruzzi MG, Nguyen ML et al. Carotenoid bioavailability is higher from salads ingested with full-fat than with fat-reduced salad dressings as measured with electrochemical detection. *Am J Clin Nutr.* 2004;80:396–403.

78 Canene-Adams K, Campbell JK, Zaripheh S et al. The tomato as a functional food. *J Nutr.* 2005;135:1226–30.

79 van Breemen RB, Pajkovic N. Multitargeted therapy of cancer by lycopene. *Cancer Lett.* 2008;269:339–51; Holzapfel NP, Holzapfel BM, Champ S et al. The potential role of lycopene for the prevention and therapy of prostate cancer: from molecular mechanisms to clinical evidence. *Int J Mol Sci.* 2013;14:14620–46.

80 Wang Y, Cui R, Xiao Y et al. Effect of carotene and lycopene on the risk of prostate cancer: a systematic review and dose-response meta-analysis of observational studies. *PLOS One.* 2015;10:e0137427.

81 Paur I, Lilleby W, Bohn SK et al. Tomato-based randomized controlled trial in prostate cancer patients: effect on PSA. *Clin Nutr.* 2017;36:672–79; Bowen P, Chen L, Stacewicz-Sapuntzakis M et al. Tomato sauce supplementation and prostate cancer: lycopene accumulation and modulation of biomarkers of carcinogenesis. *Exp Biol Med (Maywood).* 2002;227:886–93.

82 Eliassen AH, Hendrickson SJ, Brinton LA et al. Circulating carotenoids and risk of breast cancer: pooled analysis of eight prospective studies. *J Natl Cancer Inst.* 2012;104(24):1905–16.

83 Neilson HK, Farris MS, Stone CR et al. Moderate-vigorous recreational physical activity and breast cancer risk, stratified by menopause status: a systematic review and meta-analysis. *Menopause.* 2017;24:322–44; Psaltopoulou T, Ntanasis-Stathopoulos I, Tzanninis IG et al. Physical activity and gastric cancer risk: a systematic review and meta-analysis. *Clin J Sport Med.* 2016;26:445–64; Keum N, Ju W, Lee DH et al. Leisure-time physical activity and endometrial cancer risk: dose-response meta-analysis of epidemiological studies. *Int J Cancer.* 2014;135:682–94; Singh S, Devanna S, Edakkanambeth Varayil J et al. Physical activity is associated with reduced risk of esophageal cancer, particularly esophageal adenocarcinoma: a systematic review and meta-analysis. *BMC Gastroenterol.* 2014;14:101; Keimling M, Behrens G, Schmid D et al. The association between physical activity and bladder cancer: systematic review and meta-analysis. *Br J Cancer.* 2014;110:1862–70; Robsahm TE, Aagnes B, Hjartaker A et al. Body mass index, physical activity, and colorectal cancer by anatomical subsites: a systematic review and meta-analysis of cohort studies. *Eur J Cancer Prev.* 2013;22:492–505.

84 Wu W, Guo F, Ye J et al. Pre- and post-diagnosis physical activity is associated with survival benefits of colorectal cancer patients: a systematic review and meta-analysis. *Oncotarget.* 2016;7:52095–103; Buffart LM, Kalter J, Sweegers MG et al. Effects and moderators of exercise on quality of life and physical function in patients with cancer: an individual patient data meta-analysis of 34 RCTs. *Cancer Treat Rev.* 2017;52:91–104.

85 He C, Bassik MC, Moresi V et al. Exercise-induced BCL2-regulated autophagy I required for muscle glucose homeostasis. *Nature.* 2012;481:511–19.

86 National Cancer Institute. "Obesity and Cancer." https://www.cancer.gov/about-cancer/causes-prevention/risk/obesity/obesity-fact-sheet. Reviewed 17 Jan 2017; Kolb R, Sutterwala FS, Zhang W. Obesity and cancer: inflammation bridges the two. *Curr Opin Pharmacol.* 2016;29:77–89.

87 National Cancer Institute. "Obesity and Cancer." https://www.cancer.gov/about-cancer/causes-prevention/risk/obesity/obesity-fact-sheet. Reviewed 17 Jan 2017; Lee J, Meyerhardt JA, Giovannucci E, Jeon JY. Association between body mass index and prognosis of colorectal cancer: a meta-analysis of prospective cohort studies. *PLOS One.* 2015;10:e0120706; Jiralerspong S, Goodwin PJ. Obesity and breast cancer prognosis: evidence, challenges, and opportunities. *J Clin Oncol.* 2016;34:4203–16; Hu MB, Xu H, Bai PD et al. Obesity has multifaceted impact on biochemical recurrence of prostate cancer: a dose-response meta-analysis of 36,927 patients. *Med Oncol.* 2014;31:829; Cao Y, Giovannucci E. Obesity and prostate cancer. *Recent Results Cancer Res.* 2016;208:137–53.

88 Damasceno NR, Perez-Heras A, Serra M et al. Crossover study of diets enriched with virgin olive oil, walnuts or almonds: effects on lipids and other cardiovascular risk markers. *Nutr Metab Cardiovasc Dis.* 2011;suppl 1:S14–20.

89 Stokowski LA. "No Amount of Alcohol Is Safe." Medscape Oncology, 30 Apr 2014. http:// www.medscape.com/viewarticle/824237_1; Rehm J, Shield, K. Alcohol consumption. In: World Cancer Report 2014, chap. 2.3, 2014: 96–104. https://www.iarc.fr/en/media-centre /iarcnews/2016/WCR_2014_Chapter_2-3.pdf.

90 Baan R, Straif K, Grosse Y et al. Carcinogenicity of alcoholic beverages. Lancet Oncol. 2007;8:292–93; Connor J. Alcohol consumption as a cause of cancer. Addiction. 2017 Feb;112(2):222–28.

91 Hartman TJ, Sisti JS, Hankinson SE et al. Alcohol consumption and urinary estrogens and estrogen metabolites in premenopausal women. Horm Cancer. 2016;7:65–74.

92 Bagnardi V, Rota M, Botteri E et al. Alcohol consumption and site-specific cancer risk: a comprehensive dose-response meta-analysis. Br J Cancer. 2015;112(3):580–93; LoConte NK, Brewster AM, Kaur JS, Merrill JK, Alberg AJ. Alcohol and cancer: a statement of the American Society of Clinical Oncology. J Clin Oncol. 2018;36(1):83–93.

93 Metayer C, Dahl G, Wiemels J, Miller M. Childhood leukemia: a preventable disease. Pediatrics. 2016;138:S45–S55; Haberg SE, London SJ, Stigum H et al. Folic acid supplements in pregnancy and early childhood respiratory health. Arch Dis Child. 2009;94:180–84; Whitrow MJ, Moore VM, Rumbold AR, Davies MJ. Effect of supplemental folic acid in pregnancy on childhood asthma: a prospective birth cohort study. Am J Epidemiol. 2009;170:1486–93; Haberg SE, London SJ, Nafstad P et al. Maternal folate levels in pregnancy and asthma in children at age 3 years. J Allergy Clin Immunol. 2011;127:262–64; Kallen B. Congenital malformations in infants whose mothers reported the use of folic acid in early pregnancy in Sweden: a prospective population study. Congenit Anom (Kyoto). 2007;47:119–24.

94 Voutilainen S, Rissanen TH, Virtanen J et al. Low dietary folate intake is associated with an excess incidence of acute coronary events: the Kuopio Ischemic Heart Disease Risk Factor Study. Circulation. 2001;103:2674–80; Kennedy DO. B vitamins and the brain: mechanisms, dose and efficacy—a review. Nutrients. 2016 Jan 27;8(2):68; Higdon J. Folic acid. In: An Evidence-Based Approach to Vitamins and Minerals: Health Benefits and Intake Recommendations. New York: Theime, 2003, 6–14.

95 Pitkin RM. Folate and neural tube defects. Am J Clin Nutr. 2007;85:285S–88S.

96 Troen AM, Mitchell B, Sorensen B et al. Unmetabolized folic acid in plasma is associated with reduced natural killer cell cytotoxicity among postmenopausal women. J Nutr. 2006;136:189–94.

97 Mason JB, Dickstein A, Jacques PF et al. A temporal association between folic acid fortification and an increase in colorectal cancer rates may be illuminating important biological principles: a hypothesis. Cancer Epidemiol Biomarkers Prev. 2007;16:1325–29.

98 Baggott JE, Oster RA, Tamura T. Meta-analysis of cancer risk in folic acid supplementation trials. Cancer Epidemiol. 2012;36(1):78–81; Wien TN, Pike E, Wisloff T et al. Cancer risk with folic acid supplements: a systematic review and meta-analysis. BMJ Open. 2012;2:e000653; Smith AD, Kim YI, Refsum H. Is folic acid good for everyone? Am J Clin Nutr. 2008;87:517– 33; Figueiredo JC, Grau MV, Haile RW et al. Folic acid and risk of prostate cancer: results from a randomized clinical trial. J Natl Cancer Inst. 2009;101:432–35; Kim YI. Will mandatory folic acid fortification prevent or promote cancer? Am J Clin Nutr. 2004;80:1123–28; Mason JB. Folate, cancer risk, and the Greek god, Proteus: a tale of two chameleons. Nutr Rev. 2009;67:206–12.

99 Metayer C, Dahl G, Wiemels J, Miller M. Childhood leukemia: a preventable disease. Pediatrics. 2016;138:S45–S55.

100 Haberg SE, London SJ, Stigum H et al. Folic acid supplements in pregnancy and early childhood respiratory health. Arch Dis Child. 2009;94:180–84; Whitrow MJ, Moore VM, Rumbold AR, Davies MJ. Effect of supplemental folic acid in pregnancy on childhood asthma: a prospective birth cohort study. Am J Epidemiol. 2009;170:1486–93; Haberg SE, London SJ, Nafstad P et al. Maternal folate levels in pregnancy and asthma in children at age 3 years. J Allergy Clin Immunol. 2011;127:262–64; Kallen B. Congenital malformations in infants whose mothers reported the use of folic acid in early pregnancy in Sweden: a prospective population study. Congenit Anom (Kyoto). 2007;47:119–24.

101 Baggott JE, Oster RA, Tamura T. Meta-analysis of cancer risk in folic acid supplementation trials. *Cancer Epidemiol.* 2012;36(1):78–81.

102 Stolzenberg-Solomon RZ, Chang SC, Leitzmann MF et al. Folate intake, alcohol use, and postmenopausal breast cancer risk in the Prostate, Lung, Colorectal, and Ovarian Cancer Screening Trial. *Am J Clin Nutr.* 2006;83:895–904; Sanjoaquin MA, Allen N, Couto E et al. Folate intake and colorectal cancer risk: a meta-analytical approach. *Int J Cancer.* 2005;113:825–28.

103 Stolzenberg-Solomon RZ, Chang SC, Leitzmann MF et al. Folate intake, alcohol use, and postmenopausal breast cancer risk in the Prostate, Lung, Colorectal, and Ovarian Cancer Screening Trial. *Am J Clin Nutr.* 2006;83:895–904; Larsson SC, Akesson A, Bergkvist L, Wolk A. Multivitamin use and breast cancer incidence in a prospective cohort of Swedish women. *Am J Clin Nutr.* 2010;91:1268–72.

104 Bjelakovic G, Nikolova D, Gluud LL et al. Antioxidant supplements for prevention of mortality in healthy participants and patients with various diseases. *Cochrane Database Syst Rev.* 2012 Mar;14(3):CD007176.

Chapter Six: The Struggle to Lose Weight

1 Santos-Lozano A, Pareja-Galeano H. Implications of obesity in exceptional longevity. *Ann Transl Med.* 2016;4(20):416.

2 Centers for Disease Control and Prevention, National Institute for Occupational Safety and Health (NIOSH). "NIOSH Facts: NFL Mortality Study." 1993. NIOSHTIC No. 00232586.

3 Loprinzi PD, Branscum A, Hanke J, Smit E. Healthy lifestyle characteristics and their joint association with cardiovascular disease biomarkers in US adults. *Mayo Clin Proc.* 2016;91(4):432–42.

4 Cavalo DN, Horino M, McCarthy WJ. Adult intake of minimally processed fruits and vegetables: associations with cardiometabolic disease risk factors. *J Acad Nutr Diet.* 2016;116(9):1387–94.

5 Wise J. Being overweight at any point increases risk for death. *BMJ.* 2017;357:j1650.

6 Fuhrman J, Sarter B, Acocella S et al. Changing perceptions of hunger on a high nutrient density diet. *Nutr J.* 2010;9:51.

7 Chobotova K. Aging and cancer: converging routes to disease prevention. *Int Cancer Ther.* 2009;8:115–22; Devaraj S, Wang-Polagruto J, Polagruto J et al. High-fat, energy-dense, fast-food-style breakfast results in an increase in oxidative stress in metabolic syndrome. *Metab Clin Exp.* 2008;57:867–70; Egger G, Dixon J. Inflammatory effects of nutritional stimuli: further support for the need for a big picture approach to tackling obesity and chronic disease. *Obesity Rev.* 2010;11(2):137–49; Esmaillzadeh A, Azadbakht L. Major dietary patterns in relation to general obesity and central adiposity among Iranian women. *J Nutr.* 2008;138:358–63.

8 Devaraj S, Mathur S, Basu A et al. A dose-response study on the effects of purified lycopene supplementation on biomarkers of oxidative stress. *J Am Coll Nutr.* 2008;27:267–73; Esmaillzadeh A, Azadbakht L. Dietary flavonoid intake and cardiovascular mortality. *Br J Nutr.* 2008;100:695–97; Esmaillzadeh A, Kimiagar M, Mehrabi Y et al. Fruit and vegetable intakes, C-reactive protein, and the metabolic syndrome. *Am J Clin Nutr.* 2006;84:1489–97; O'Keefe JH, Gheewala NM, O'Keefe JO et al. Dietary strategies for improving post-prandial glucose, lipids, inflammation, and cardiovascular health. *J Am Coll Cardiol.* 2008;51:249–55; Bose KS, Agrawal BK. Effect of lycopene from tomatoes (cooked) on plasma antioxidant enzymes, lipid peroxidation rate and lipid profile in grade-I hypertension. *Ann Nutr Metab.* 2007;51:477–81; Thompson HJ, Heimendinger J, Haegele A et al. Effect of increased vegetable and fruit consumption on markers of oxidative cellular damage. *Carcinogenesis.* 1999;20:2261–66.

9 Parylak S, Koob GT, Zorrilla, EP. The dark side of food addiction. *Physio Behav.* 2011;104(1):149–56.

10 Thompson HJ, Heimendinger J, Gillette C et al. In vivo investigation of changes in biomarkers of oxidative stress induced by plant food rich diets. *J Agric Food Chem.* 2005;53:6126–32; Peairs AT, Rankin JW. Inflammatory response to a high-fat, low-carbohydrate weight loss diet: effect of antioxidants. *Obesity.* 2008;16:1573–78; Patel C, Ghanim H, Ravishankar S et al. Prolonged reactive oxygen species generation and nuclear factor-kappaB activation after a high-fat, high-carbohydrate meal in the obese. *J Clin Endocrinol Metab.* 2007;92:4476–79.

11 Bockowski L, Sobaniec W, Kulak W et al. Serum and intraerythrocyte antioxidant enzymes and lipid peroxides in children with migraine. *Pharmacol Rep.* 2008;60:542–48; Khansari N, Shakiba Y, Mahmoudi M. Chronic inflammation and oxidative stress as a major cause of age-related diseases and cancer. *Recent Pat Inflamm Allergy Drug Discov.* 2009;3:73–80; Federico A, Morgillo F, Tuccillo C et al. Chronic inflammation and oxidative stress in human carcinogenesis. *Int J Cancer.* 2007;121:2381–86; Guo W, Kong E, Meydani M. Dietary polyphenols, inflammation, and cancer. *Nutr Cancer.* 2009;61:807–10; Schulte EM, Smeal JK, Lewis J, Gearhardt AN. Development of the highly processed food withdrawal scale. *Appetite.* 2018;131:148–54.

12 Blumenthal DM, Gold MS. Neurobiology of food addiction. *Curr Opin Clin Nutr Metab Care.* 2010;13:359–65; Cohen DA. Neurophysiological pathways to obesity: below awareness and beyond individual control. *Diabetes.* 2008;57:1768–73; Corwin RL, Grigson PS. Symposium overview. Food addiction: fact or fiction? *J Nutr.* 2009;139:617–19; Dagher A. The neurobiology of appetite: hunger as addiction. *Int J Obesity.* 2009;33(suppl 2):S30–33; Davis C, Carter JC. Compulsive overeating as an addiction disorder: a review of theory and evidence. *Appetite.* 2009;53:1–8; Del Parigi A, Chen K, Salbe AD et al. Are we addicted to food? *Obesity.* 2003;11:493–95; Gosnell BA, Levine AS. Reward systems and food intake: role of opioids. *Int J Obesity.* 2009;33(suppl 2):S54–58; Ifland JR, Preuss HG, Marcus MT et al. Refined food addiction: a classic substance use disorder. *Med Hypotheses.* 2009;72:518–26; Johnson PM, Kenny PJ. Dopamine D2 receptors in addiction-like reward dysfunction and compulsive eating in obese rats. *Nat Neurosci.* 2010;13(8):1033; Liu Y, von Deneen KM, Kobeissy FH et al. Food addiction and obesity: evidence from bench to bedside. *J Psychoactive Drugs.* 2010;42:133–45; Pelchat ML. Food addiction in humans. *J Nutr.* 2009;139:620–22; Spring B, Schneider K, Smith M et al. Abuse potential of carbohydrates for overweight carbohydrate cravers. *Psychopharmacology.* 2008;197:637–47; Yanover T, Sacco WP. Eating beyond satiety and body mass index. *Eating Weight Dis.* 2008;13:119–28; Yeomans MR. Alcohol, appetite and energy balance: is alcohol intake a risk factor for obesity? *Physiol Behav.* 2010;100:82–89.

13 Peairs AT, Rankin JW. Inflammatory response to a high-fat, low-carbohydrate weight loss diet: effect of antioxidants. *Obesity.* 2008;16:1573–78; Patel C, Ghanim H, Ravishankar S et al. Prolonged reactive oxygen species generation and nuclear factor-kappaB activation after a high-fat, high-carbohydrate meal in the obese. *J Clin Endocrinol Metab.* 2007;92:4476–79.

14 Olusi SO. Obesity is an independent risk factor for plasma lipid peroxidation and depletion of erythrocyte cytoprotectic enzymes in humans. *Int J Obesity Relat Metab Disord.* 2002;26:1159–64.

15 Bosch G, Verbrugghe A, Hesta M et al. The effects of dietary fibre type on satiety-related hormones and voluntary food intake in dogs. *Br J Nutr.* 2009;102:318–25; Lyly M, Liukkonen KH, Salmenkallio-Marttila M et al. Fibre in beverages can enhance perceived satiety. *Eur J Nutr.* 2009;48:251–58; Flood-Obbagy JE, Rolls BJ. The effect of fruit in different forms on energy intake and satiety at a meal. *Appetite.* 2009;52:416–22.

16 Major GC, Doucet E, Jacqmain M et al. Multivitamin and dietary supplements, body weight and appetite: results from a cross-sectional and a randomized double-blind placebo-controlled study. *Br J Nutr.* 2008;99:1157–67.

17 Fuhrman J, Singer M. Improved cardiovascular parameters with a nutrient-dense, plant-rich diet-style: a patient survey with illustrative cases. *Am J Lifestyle Med.* 2015;11(3):264–73. https://doi.org/10.1177/1559827615611024.

18 Ames BN. Micronutrients prevent cancer and delay aging. *Toxicol Lett.* 1998;102–3:5–18; Astley SB, Elliott RM, Archer DB et al. Increased cellular carotenoid levels reduce the persistence of DNA single-strand breaks after oxidative challenge. *Nutr Cancer.* 2002;43:202–13; Aviram M, Kaplan M, Rosenblat M, Fuhrman B. Dietary antioxidants and paraoxonases against LDL oxidation and atherosclerosis development. *Handb Exp Pharmacol.* 2005;(170):263–300; Collins AR, Harrington V, Drew J et al. Nutritional modulation of DNA repair in a human intervention study. *Carcinogenesis.* 2003;24:511–15; Ferguson LR, Philpott M, Karunasinghe N et al. Dietary cancer and prevention using antimutagens. *Toxicology.* 2004;198:147–59; Joseph JA, Denisova NA, Bielinski D et al. Oxidative stress protection and vulnerability in aging: putative

nutritional implications for intervention. *Mech Ageing Dev.* 2000;116:141–53; Martin KR, Failla ML, Smith JC Jr. Beta-carotene and lutein protect HepG2 human liver cells against oxidant-induced damage. *J Nutr.* 1996;126:2098–106; O'Brien NM, Carpenter R, O'Callaghan YC et al. Modulatory effects of resveratrol, citroflavan-3-ol, and plant-derived extracts on oxidative stress in U937 cells. *J Med Food.* 2006;9:187–95; O'Brien NM, Woods JA, Aherne SA, O'Callaghan YC. Cytotoxicity, genotoxicity and oxidative reactions in cell-culture models: modulatory effects of phytochemicals. *Biochem Soc Trans.* 2000;28:22–26; Prior RL. Fruits and vegetables in the prevention of cellular oxidative damage. *Am J Clin Nutr.* 2003;78:570S–78S; Schaefer S, Baum M, Eisenbrand G et al. Modulation of oxidative cell damage by reconstituted mixtures of phenolic apple juice extracts in human colon cell lines. *Mol Nutr Food Res.* 2006;50:413–17; Singh M, Arseneault M, Sanderson T et al. Challenges for research on polyphenols from foods in Alzheimer's disease: bioavailability, metabolism, and cellular and molecular mechanisms. *J Agric Food Chem.* 2008;56:4855–73; Sudheer AR, Muthukumaran S, Devipriya N et al. Ellagic acid, a natural polyphenol protects rat peripheral blood lymphocytes against nicotine-induced cellular and DNA damage in vitro: with the comparison of N-acetylcysteine. *Toxicology.* 2007;230:11–21; Tarozzi A, Hrelia S, Angeloni C et al. Antioxidant effectiveness of organically and non-organically grown red oranges in cell culture systems. *Eur J Nutr.* 2006;45:152–58; Willcox JK, Ash SL, Catignani GL et al. Antioxidants and prevention of chronic disease. *Crit Rev Food Sci Nutr.* 2004;44:275–95; Guo W, Kong E, Meydani M et al. Dietary polyphenols, inflammation, and cancer. *Nutr Cancer.* 2009;61:807–10.

19 Wang Y, Wang QJ. The prevalence of prehypertension and hypertension among adults according to the new joint National Committee guidelines. *Arch Intern Med.* 2004;164(19):2126–34.

20 Bakris GL. "High Blood Pressure." Merck Manual Consumer Version, Heart and Blood Vessel Disorders. https://www.merckmanuals.com/home/heart-and-blood-vessel-disorders/high -blood-pressure/high-blood-pressure. Last revised Mar 2018.

21 Frohlich ED, Varagic J. The role of sodium in hypertension is more complex than simply elevating arterial pressure. *Nat Clin Pract Cardiovasc Med.* 2004;1(1):24–30.

22 Dickenson BD, Havas S. Reducing the population burden of cardiovascular disease by reducing sodium intake: a report of the Council on Science and Public Health. *Arch Intern Med.* 2007;167(14):1460–68.

23 Karppanen H, Mervaala E. Sodium intake and hypertension. *Prog Cardiovasc Dis.* 2006;49(2):59–75; Cutler JA, Roccell E. Salt reduction for preventing hypertension and cardiovascular disease. *Hypertension.* 2006;48(5):818–19.

24 Cook N, Cutler J, Obarzanek E et al. Long term effects of dietary sodium reduction on cardiovascular disease outcomes: observational follow-up of the Trails of Hypertension Prevention (TOHP). *BMJ.* 2007;334:885.

25 Havas S, Roccella EJ, Lenfant C. Reducing the public health burden from elevated blood pressure levels in the United States by lowering intake of dietary sodium. *Am J Public Health.* 2004;94(1):19–22.

26 Prospective Studies Collaboration. Age specific relevance of usual blood pressure to vascular mortality: a meta analysis of individual data for one million adults in 61 prospective studies. *Lancet.* 2002;360:1903–13.

27 Weinberger MH. Salt sensitivity is associated with an increased mortality in both normal and hypertensive humans. *J Clin Hypertens.* 2002;4(4):274–76; Tuomilehto J, Jousilahti P, Rastenyte D et al. Urinary sodium excretion and cardiovascular mortality in Finland: a prospective study. *Lancet.* 2001;357:848–51.

28 Luke R. President's address: salt—too much of a good thing? *Trans Am Clin Climatol Assoc.* 2007;118:1–22.

29 Freis E. The role of salt in hypertension. *Blood Pressure.* 1991;1:196–200.

30 National High Blood Pressure Education Program. *The Seventh Report of the Joint National Committee on Prevention, Detection, Evaluation, and Treatment of High Blood Pressure.* Bethesda, MD: National Heart, Lung, and Blood Institute, 2004. Available at http://www.ncbi.nlm.nih .gov/books/NBK9636/.

31 Mozaffarian D, Fahimi S, Singh GM et al. Global sodium consumption and death from cardiovascular causes. *N Engl J Med.* 2014;371:624–34.

32 Havas S, Dickinson B, Wilson M. The urgent need to reduce sodium consumption. *JAMA.* 2007;298(12):1439–41.

33 Nestle M. *What to Eat.* New York: North Star, 2006, 365.

34 Stranahan AM, Norman ED, Lee K et al. Diet-induced insulin resistance impairs hippocampal synaptic plasticity and cognition in middle-aged rats. *Hippocampus.* 2008;18(11):1085–88.

35 Johnson PM, Kenny PJ. Dopamine D2 receptors in addiction-like reward dysfunction and compulsive eating in obese rats. *Nat Neurosci.* 2010:13(5):635–41.

36 Murphy C. Nutrition and chemosensory perception in the elderly. *Crit Rev Food Sci Nutr.* 1993;33(1):3–15.

37 Myers WC, Vondruska MA. Murder, minors, selective serotonin reuptake inhibitors, and the involuntary intoxication defense. *J Am Acad Psychiatry Law.* 1998;26(3):487–96; Patrick RP, Ames BN. Vitamin D and the omega-3 fatty acids control serotonin synthesis and action, part 2: relevance for ADHD, bipolar disorder, schizophrenia, and impulsive behavior. *FASEB J.* 2015;29(6):2207–22.

38 Williams E, Stewart-Knox B, Helander A et al. Associations between whole-blood serotonin and subjective mood in healthy male volunteers. *Biol Psychol.* 2006;71(2):171–74.

39 Schulte EM, Smeal JK, Lewis J, Gearhardt AN. Development of the highly processed food withdrawal scale. *Appetite* 2018;131(1):148–54.

40 Sánchez-Villegas A, Toledo E, de Irala J et al. Fast-food and commercial baked goods consumption and the risk of depression. *Pub Health Nutr.* 2011;15(3):424.

41 Grosso G, Galvano F, Marventano S et al. Omega-3 fatty acids and depression: scientific evidence and biological mechanisms. *Oxid Med Cell Longev.* 2014;2014:313570; Kennedy DO. B vitamins and the brain: mechanisms, dose and efficacy—a review. *Nutrients.* 2016;8(2):68.

42 Gardner CD, Kiazand A, Alhassan S et al. Comparison of the Atkins, Zone, Ornish, and LEARN diets for change in weight and related risk factors among overweight premenopausal women: the A TO Z Weight Loss Study: a randomized trial. *JAMA.* 2007;277(9):969–77.

43 Akbaraly TN, Brunner EJ, Ferrie JE et al. Dietary pattern and depressive symptoms in middle age. *Br J Psychiatry.* 2009;195(5):408–13.

44 Mujcic R, Oswald AJ. Evolution of well-being and happiness after increases in consumption of fruit and vegetables. *Am J Public Health.* 2016;106(8):1504–10.

45 Golden RN, Gaynes BN, Ekstrom RD et al. The efficacy of light therapy in the treatment of mood disorders: a review and meta-analysis of the evidence. *Am J Psychiatry.* 2005;162:658–62.

46 Martins JG. EPA but not DHA appears to be responsible for the efficacy of omega-3 long chain polyunsaturated fatty acid supplementation in depression: evidence from a meta-analysis of randomized controlled trials. *Am Coll Nutr.* 2009;28:525–42.

47 Dwyer AV, Whitten DL, Hawrelack JA. Herbal medicines other than St. John's Wort, in the treatment of depression: a systemic review. *Altern Med Rev.* 2011;16:40–49.

48 Mischoulon D, Fava M. Role of S-adenosyl-L-methionine in the treatment of depression: a review of the evidence. *Am J Clin Nutr.* 2002;76:1158s–61s; Williams AL, Girard C, Jui D et al. S-adenosylmethionine (SAMe) as treatment for depression: a systematic review. *Clin Invest Med.* 2005;28(3):132–39.

49 Linde K, Berner MM, Kriston I. St John's wort for major depression. *Cochrane Database Syst Rev.* 2008 Oct 8;(4):CD000448.

50 Shaw K, Turner J, Del Mar C. Tryptophan and 5-hydroxytryptophan for depression. *Cochrane Database Syst Rev.* 2002;(1):CD003198.

51 Sansone RA. Cholesterol quandaries: relationship to depression and the suicidal experience. *Psychiatry.* 2008;5(3):22–34.

52 "Fruits and veg give you the feel-good factor." Warwick, News & Events. 8 July 2016. http://www2.warwick.ac.uk/newsandevents/news/fruit_and_veg/.

53 Gangwisch JE, Hale L, Garcia L et al. High glycemic index diet as a risk factor for depression: analyses from the Women's Health Initiative. *Am J Clin Nutr.* 2015;102(2):454–63, doi:10.3945/ajcn.114.103846.

54 Kodi CT, Seaquist ER. Cognitive dysfunction and diabetes mellitus. *Endocrin Rev.* 2008;29:494–511; Sommerfield AJ, Deary IJ, Frier BM et al. Acute hyperglycemia alters mood state and impairs cognitive performance in people with type 2 diabetes. *Diabetes Care.* 2004;27:2335–40.

55 Schopf V, Fischmeister FP, Windischberger C et al. Effects of individual glucose levels on the neuronal correlates of emotions. *Front Hum Neurosci.* 2013;7:212.

56 O'Keefe SJD, Li JV, Lahti L et al. Fat, fiber and cancer risk in African Americans and rural Africans. *Nat Commun.* 2015;5:6342.

57 van Niekerk G, Hattingh SM, Engelbrecht AM. Enhanced therapeutic efficacy in cancer patients by short-term fasting: the autophagy connection. *Front Oncol.* 2016;6:242; Mattson MP, Longo VD, Harvie M. Impact of intermittent fasting on health and disease processes. *Ageing Res Rev.* 2017;39:46–58.

58 Cheng CW, Adams GB, Perin L et al. Prolonged fasting reduces IGF-1/PKA to promote hematopoietic stem cell-based regeneration and reverse immunosuppression. *Cell Stem Cell.* 2014;14(6):810–23; Mihaylova MM, Cheng CW, Cao AQ et al. Fasting activates fatty acid oxidation to enhance intestinal stem cell function during homeostasis and aging. *Cell Stem Cell.* 2018;22(5):769–78; Vera E, Bernardes de Jesus B, Foronda M et al. Telomerase reverse transcriptase synergizes with calorie restriction to increase health span and extend mouse longevity. *PLOS One.* 2013;8(1):e53760.

59 Mattson MP, Longo VD, Harvie M. Impact of intermittent fasting on health and disease processes. *Ageing Res Rev.* 2017;39:46–58; Antoni R, Johnston KL, Collins AL, Robertson MD. Effects of intermittent fasting on glucose and lipid metabolism. *Proc Nutr Soc.* 2017;76:361–68; Tinsley GM, La Bounty PM. Effects of intermittent fasting on body composition and clinical health markers in humans. *Nutr Rev.* 2015;73:661–74.

60 Longo VD, Mattson MP. Fasting: molecular mechanisms and clinical applications. *Cell Metab.* 2014;19:181–92; Cheng CW, Adams GB, Perin L et al. Prolonged fasting reduces IGF-1/PKA to promote hematopoietic-stem-cell-based regeneration and reverse immunosuppression. *Cell Stem Cell.* 2014;14:810–23.

61 Mattson MP, Allison DB, Fontana L et al. Meal frequency and timing in health and disease. *Proc Natl Acad Sci USA.* 2014;111:16647–53.

62 Nas A, Mirza N, Hagele F et al. Impact of breakfast skipping compared with dinner skipping on regulation of energy balance and metabolic risk. *Am J Clin Nutr.* 2017, 105:1351–61. Jakubowicz D, Wainstein J, Ahren B et al. Fasting until noon triggers increased postprandial hyperglycemia and impaired insulin response after lunch and dinner in individuals with type 2 diabetes: a randomized clinical trial. *Diabetes Care.* 2015, 38:1820–26. Chowdhury EA, Richardson JD, Tsintzas K et al. Carbohydrate-rich breakfast attenuates glycaemic, insulinaemic and ghrelin response to ad libitum lunch relative to morning fasting in lean adults. *Br J Nutr.* 2015, 114:98–107. Betts JA, Richardson JD, Chowdhury EA et al. The causal role of breakfast in energy balance and health: a randomized controlled trial in lean adults. *Am J Clin Nutr.* 2014, 100:539–47. Thomas EA, Higgins J, Bessesen DH et al. Usual breakfast eating habits affect response to breakfast skipping in overweight women. *Obesity* (Silver Spring). 2015, 23:750–59. Moro T, Tinsley G, Bianco A et al. Effects of eight weeks of time-restricted feeding (16/8) on basal metabolism, maximal strength, body composition, inflammation, and cardiovascular risk factors in resistance-trained males. *J Transl Med.* 2016;14:290.

63 St-Onge MP, Ard J, Baskin ML et al. Meal Timing and Frequency: Implications for Cardiovascular Disease Prevention: A Scientific Statement From the American Heart Association. *Circulation.* 2017, 135:e96–e121. Marinac CR, Sears DD, Natarajan L et al. Frequency and Circadian Timing of Eating May Influence Biomarkers of Inflammation and Insulin Resistance Associated with Breast Cancer Risk. *PLoS One.* 2015, 10:e0136240. Jakubowicz D, Barnea M, Wainstein J, Froy O. High caloric intake at breakfast vs. dinner differentially influences weight loss of overweight and obese women. *Obesity* (Silver Spring). 2013, 21:2504–12. Garaulet M, Gomez-Abellan P, Alburquerque-Bejar JJ et al. Timing of food intake predicts weight loss effectiveness. *Int J Obes* (Lond). 2013, 37:604–11. Gabel K,

Hoddy KK, Haggerty N et al. Effects of 8-hour time restricted feeding on body weight and metabolic disease risk factors in obese adults: A pilot study. *Nutrition and Healthy Aging.* 2018;4(4):345–53.

64 Marinac CR, Nelson SH, Breen CI et al. Prolonged Nightly Fasting and Breast Cancer Prognosis. *JAMA Oncol.* 2016.

65 Saad A, Dalla Man C, Nandy DK et al. Diurnal pattern to insulin secretion and insulin action in healthy individuals. *Diabetes.* 2012, 61:2691–700.

Chapter Seven: We Can Reverse Disease

1 Heslop CL, Frohlich JJ, Hill JS. Myeloperoxidase and C-reactive protein have combined utility for long-term prediction of cardiovascular mortality after coronary angiography. *Cardiology.* 2010;55(11):1102–9; Ames PRJ, Di Girolamo G, D'Andrea G et al. Predictive value of oxidized low-density lipoprotein/β_2-glycoprotein-I complexes (oxLDL/β_2GPI) in nonautoimmune atherothrombosis. *Clin Appl Thromb Hemost.* 2018;24(7):1050–55.

2 Mazidi M, Katsiki N, Mikhailidis DP et al. Lower carbohydrate diets and all-cause and cause-specific mortality: a population-based cohort study and pooling of prospective studies. *Eur Heart J.* 2019;40(34):2870–79.

3 Fung TT, van Dam RM, Hankinson SE et al. Low-carbohydrate diets and all-cause and cause-specific mortality: two cohort studies. *Ann Intern Med.* 2010;153:289–98.

4 Shang X, Scott D, Hodge AM et al. Dietary protein intake and risk of type 2 diabetes: results from the Melbourne Collaborative Cohort Study and a meta-analysis of prospective studies. *Am J Clin Nutr.* 2016;104:1352–65; Lagiou P, Sandin S, Weiderpass E et al. Low carbohydrate-high protein diet and mortality in a cohort of Swedish women. *J Intern Med.* 2007;261:366–74; Seidelmann SB, Claggett B, Cheng S et al. Dietary carbohydrate intake and mortality: a prospective cohort study and meta-analysis. *Lancet Pub Health.* 2018;3:e419–e28; Song M, Fung TT, Hu FB et al. Association of animal and plant protein intake with all-cause and cause-specific mortality. *JAMA Intern Med.* 2016;176(10):1453–63; Lagiou P, Sandin S, Lof M et al. Low carbohydrate-high protein diet and incidence of cardiovascular diseases in Swedish women: prospective cohort study. *BMJ.* 2012;344:e4026; Tharrey M, Mariotti F, Mashchak A et al. Patterns of plant and animal protein intake are strongly associated with cardiovascular mortality: the Adventist Health Study-2 cohort. *Int J Epidemiol.* 2018;47(5):1603–12; Li SS, Blanco Mejia S, Lytvyn L et al. Effect of plant protein on blood lipids: a systematic review and meta-analysis of randomized controlled trials. *J Am Heart Assoc.* 2017;6(12): pii:e006659; Tian S, Xu Q, Jiang R et al. Dietary protein consumption and the risk of type 2 diabetes: a systematic review and meta-analysis of cohort studies. *Nutrients.* 2017;9(9): pii:E982.

5 Song M, Fung TT, Hu FB et al. Association of animal and plant protein intake with all-cause and cause-specific mortality. *JAMA Intern Med.* 2016;176(10):1453–63.

6 Levine ME, Suarez JA, Brandhorst S et al. Low protein intake is associated with a major reduction in IGF-1, cancer, and overall mortality in the 65 and younger but not older population. *Cell Metab.* 2014;19:407–17; Key TJ, Appleby PN, Reeves GK, Roddam AW. Insulin-like growth factor 1 (IGF1), IGF binding protein 3 (IGFBP3), and breast cancer risk: pooled individual data analysis of 17 prospective studies. *Lancet Oncol.* 2010;11:530–42; Rowlands MA, Gunnell D, Harris R et al. Circulating insulin-like growth factor peptides and prostate cancer risk: a systematic review and meta-analysis. *Int J Cancer.* 2009;124:2416–29.

7 Koeth RA, Wang Z, Levison BS et al. Intestinal microbiota metabolism of L-carnitine, a nutrient in red meat, promotes atherosclerosis. *Nat Med.* 2013;19(5):576–85; Tang WH, Wang Z, Levison BS et al. Intestinal microbial metabolism of phosphatidylcholine and cardiovascular risk. *N Engl J Med.* 2013;368:1575–84.

8 Goldberg T, Cai W, Peppa M et al. Advanced glycoxidation end products in commonly consumed foods. *J Am Diet Assoc.* 2004;104:1287–91; Goldin A, Beckman JA, Schmidt AM, Creager MA. Advanced glycation end products: sparking the development of diabetic vascular injury. *Circulation.* 2006;114:597–605.

9 Brewer GJ. Iron and copper toxicity in diseases of aging, particularly atherosclerosis and Alzheimer's disease. *Exp Biol Med.* 2007;232:323–35.

10 de Lorgeril M, Salen P. New insights into the health effects of dietary saturated and omega-6 and omega-3 polyunsaturated fatty acids. *BMC Med.* 2012;10:50.

11 National Cancer Institute. "Chemicals in Meat Cooked at High Temperatures and Cancer Risk." http://www.cancer.gov/cancertopics/factsheet/Risk/cooked-meats. Reviewed 11 July 2017.

12 Nettleton JA, Brouwer IA, Gelenjinse JM, Hornstra G. Saturated fat consumption and risk of coronary heart disease and ischemic stroke: a science update. *Ann Nutr Metab.* 2017;70(1):26–33.

13 Bao Y, Han J, Hu FB et al. Association of nut consumption with total and cause-specific mortality. *N Engl J Med.* 2013;369:2001–11; Damasceno NR, Sala-Vial A, Cofan M et al. Mediterranean diet supplemented with nuts reduces waist circumference and shifts lipoprotein subfractions to a less atherogenic pattern in subjects at high cardiovascular risk. *Atherosclerosis.* 2013;230(2):347–53.

14 Pottala JV, Yaffe K, Robinson JG et al. Higher RBC EPA + DHA corresponds with larger total brain and hippocampal volumes: WHIMS-MRI study. *Neurology.* 2014;82(5):435–42.

15 Fraser GE, Sabate J, Beeson WL, Strahan TM. A possible protective effect of nut consumption on risk of coronary heart disease: the Adventist Health Study. *Arch Intern Med.* 1992;152:1416–24; Sabate J. Nut consumption, vegetarian diets, ischemic heart disease risk, and all-cause mortality: evidence from epidemiologic studies. *Am J Clin Nutr.* 1999;70:500S–503S; Fraser GE, Shavlik DJ. Ten years of life: is it a matter of choice? *Arch Intern Med.* 2001;161:1645–52; Tharrey M, Mariotti F, Mashchak A et al. Patterns of plant and animal protein intake are strongly associated with cardiovascular mortality: the Adventist Health Study-2 cohort. *Int J Epidemiol.* 2018;47(5):1603–12.

16 Ellsworth JL, Kushi LH, Folsom AR. Frequent nut intake and risk of death from coronary heart disease and all causes in postmenopausal women: the Iowa Women's Health Study. *Nutr Metab Cardiovasc Dis.* 2001;11:372–77.

17 Baer HJ, Glynn RJ, Hu FB et al. Risk factors for mortality in the Nurses' Health Study: a competing risks analysis. *Am J Epidemiol.* 2011;173:319–29; Bao Y, Han J, Hu FB et al. Association of nut consumption with total and cause-specific mortality. *N Engl J Med.* 2013;369:2001–11.

18 Bao Y, Han J, Hu FB et al. Association of nut consumption with total and cause-specific mortality. *N Engl J Med.* 2013;369:2001–11.

19 Kelly JH Jr., Sabate J. Nuts and coronary heart disease: an epidemiological perspective. *Br J Nutr.* 2006;96(suppl 2):S61–67.

20 Tharrey M, Mariotti F, Mashchak A et al. Patterns of plant and animal protein intake are strongly associated with cardiovascular mortality: the Adventist Health Study-2 cohort. *Int J Epidemiol.* 2018;47(5):1603–12.

21 Grosso G, Yang J, Marventano S et al. Nut consumption on all-cause, cardiovascular, and cancer mortality risk: a systematic review and meta-analysis of epidemiologic studies. *Am J Clin Nutr.* 2015;101(4):783–93.

22 Grosso G, Yang J, Marventano S et al. Nut consumption on all-cause, cardiovascular, and cancer mortality risk: a systematic review and meta-analysis of epidemiologic studies. *Am J Clin Nutr.* 2015;101:783–93; Aune D, Keum N, Giovannucci E et al. Nut consumption and risk of cardiovascular disease, total cancer, all-cause and cause-specific mortality: a systematic review and dose-response meta-analysis of prospective studies. *BMC Med.* 2016;14:207; van den Brandt PA, Schouten LJ. Relationship of tree nut, peanut and peanut butter intake with total and cause-specific mortality: a cohort study and meta-analysis. *Int J Epidemiol.* 2015 June;44(3):1038–49.

23 Liu G, Guasch-Ferre M, Hu Y et al. Nut consumption in relation to cardiovascular disease incidence and mortality among patients with diabetes mellitus. *Circ Res.* 2019;124:920–29.

24 Fraser GE, Shavlik DJ. Ten years of life: is it a matter of choice? *Arch Intern Med.* 2001;161:1645–52.

25 Kris-Etherton PM, Hu FB, Ros E, Sabate J. The role of tree nuts and peanuts in the prevention of coronary heart disease: multiple potential mechanisms. *J Nutr.* 2008;138:1746S–51S; Katz DL, Davidhi A, Ma Y et al. Effects of walnuts on endothelial function in overweight

adults with visceral obesity: a randomized, controlled, crossover trial. *J Am Coll Nutr.* 2012;31:415–23; Kris-Etherton PM. Walnuts decrease risk of cardiovascular disease: a summary of efficacy and biologic mechanisms. *J Nutr.* 2014;144:547S–54S; Bullo M, Juanola-Falgarona M, Hernandez-Alonso P, Salas-Salvado J. Nutrition attributes and health effects of pistachio nuts. *Br J Nutr.* 2015;113(suppl 2):S79–93; Rajaram S, Sabate J. Nuts, body weight and insulin resistance. *Br J Nutr.* 2006;96(suppl 2):S79–86; Khalesi S, Irwin C, Schubert M. Flaxseed consumption may reduce blood pressure: a systematic review and meta-analysis of controlled trials. *J Nutr.* 2015;145:758–65.

26 van Het Hof KH, West CE, Weststrate JA, Hautvast JG. Dietary factors that affect the bioavailability of carotenoids. *J Nutr.* 2000;130:503–6; Borel P, Desmarchelier C. Bioavailability of fat-soluble vitamins and phytochemicals in humans: effects of genetic variation. *Ann Rev Nutr.* 2018;38:69–96.

27 Fuhrman J, Singer M. Improved cardiovascular parameters with a nutrient-dense, plant-rich diet-style: a patient survey with illustrative cases. *Am J Lifestyle Med.* 2015;11(3):264–73. https://doi.org/10.1177/1559827615611024.

28 Janghorbani M, Dehghani M, Salehi-Marzijarani M. Systematic review and meta-analysis of insulin therapy and risk of cancer. *Horm Cancer.* 2012;3:137–46; Roumie CL, Min JY, D'Agostino McGowan L et al. Comparative safety of sulfonylurea and metformin monotherapy on the risk of heart failure: a cohort study. *J Am Heart Assoc.* 2017 Apr 19;6(4), doi:10.1161/JAHA.116.005379; Cao W, Ning J, Yang X, Liu Z. Excess exposure to insulin is the primary cause of insulin resistance and its associated atherosclerosis. *Curr Mol Pharmacol.* 2011;4:154–66.

29 Bray GA, Ryan DH. Medical therapy for the patient with obesity. *Circulation.* 2012;125:1695–703.

30 ACCORD Study Group; Gerstein HC, Miller ME, Byington RP et al. Effects of intensive glucose lowering in type 2 diabetes. *N Engl J Med.* 2008 June 12;358(24):2545–59.

31 Nissen SE, Wolski K. Effect of rosiglitazone on the risk of myocardial infarction and death from cardiovascular causes. *N Engl J Med.* 2007 June 14;356(24):2457–71.

32 Knop FK, Holst JJ, Vilsbøll T. Replacing SUs with incretin-based therapies for type 2 diabetes mellitus: challenges and feasibility. *IDrugs.* 2008 July;11(7):497–501; Takahashi A, Nagashima K, Hamasaki A et al. Sulfonylurea and glinide reduce insulin content, functional expression of K(ATP) channels, and accelerate apoptotic beta-cell death in the chronic phase. *Diabetes Res Clin Pract.* 2007 Sep;77(3):343–50; Del Prato S, Pulizzi N. The place of sulfonylureas in the therapy for type 2 diabetes mellitus. *Metabolism.* 2006 May;55(5 suppl 1):S20–7; Maedler K, Carr RD, Bosco D et al. Sulfonylurea induced beta-cell apoptosis in cultured human islets. *J Clin Endocrinol Metab.* 2005 Jan;90(1):501–6.

33 ACCORD Study Group; Gerstein HC, Miller ME, Byington RP et al. Effects of intensive glucose lowering in type 2 diabetes. *N Engl J Med.* 2008 June 12;358(24):2545–59.

34 Sunjaya AP, Sunjaya AF, Halim S, Ferdinal F. Risk and benefits of statins in glucose control management of type II diabetes. *Int J Angiol.* 2018 Sep;27(3):121–31; Casula M, Mozzanica F, Scotti L et al. Statin use and risk of new-onset diabetes: a meta-analysis of observational studies. *Nutr Metab Cardiovasc Dis.* 2017 May;27(5):396–406.

35 Dunaief DM, Fuhrman J, Dunaief JL, Ying G. Glycemic and cardiovascular parameters improved in type 2 diabetes with the high nutrient density (HND) diet. *Open Journal of Preventive Medicine.* 2012;2(3), doi:10.4236/ojpm.2012.23053.

36 Fuhrman J, Singer M. Improved cardiovascular parameter with a nutrient-dense, plant-rich diet-style: a patient survey with illustrative cases. *Am J Lifestyle Med.* 2015;11(3):264–73. https://doi.org/10.1177/1559827615611024.

37 Lagiou P, Sandin S, Lof M et al. Low carbohydrate–high protein diet and incidence of cardiovascular diseases in Swedish women: prospective cohort study. *BMJ.* 2012;344:e4026; Pan A, Sun Q, Bernstein AM et al. Red meat consumption and risk of type 2 diabetes: 3 cohorts of US adults and an updated meta-analysis. *Am J Clin Nutr.* 2011;94(4):1088–96; Vergnaud AC, Norat T, Romaguera D et al. Meat consumption and prospective weight change in participants of the EPIC-PANACEA study. *Am J Clin Nutr.* 2010;92:398–407; Brewer GJ. Iron and copper toxicity in diseases of aging, particularly atherosclerosis and Alzheimer's disease. *Exp Biol Med.* 2007;232:323–35; Barbaresko J, Koch M, Schulze MB, Nothlings U. Dietary

pattern analysis and biomarkers of low-grade inflammation: a systematic literature review. Nutr Rev. 2013;71:511–27.

38 Song M, Fung TT, Hu FB et al. Association of animal and plant protein intake with all-cause and cause-specific mortality. JAMA Intern Med. 2016;176(10):1453–63; Levine ME, Suarez JA, Brandhorst S et al. Low protein intake is associated with a major reduction in IGF-1, cancer, and overall mortality in the 65 and younger but not older population. Cell Metab. 2014;19:407–17.

39 Fuhrman J, Sarter B, Calabro DJ. Brief case reports of medically supervised, water-only fasting associated with remission of autoimmune disease. Altern Ther Health Med. 2002;8(4):112, 110–11; Maldonado-Puebla M, Price A, Gonzalez A, Fuhrman J et al. Efficacy of a plant-based anti-inflammatory diet as monotherapy in psoriasis. International Journal of Disease Reversal and Prevention. 2019;1(1). https://ijdrp.org/index.php/ijdrp/article/view/15.

40 Giat E, Ehrenfeld M, Shoenfeld Y. Cancer and autoimmune diseases. Autoimmun Rev. 2017;16(10):1049–57.

41 American Institute for Cancer Research. "Cancer Prevention Recommendations." https://www.aicr.org/reduce-your-cancer-risk/recommendations-for-cancer-prevention/.

42 Hastert TA, Beresford SA, Patterson RE et al. Adherence to WCRF/AICR cancer prevention recommendations and risk of postmenopausal breast cancer. Cancer Epidemiol Biomarkers Prev. 2013;22(9):1498–508.

43 Kabat GC, Matthews CE, Kamensky V et al. Adherence to cancer prevention guidelines and cancer incidence, cancer mortality, and total mortality: a prospective cohort study. Am J Clin Nutr. 2015;101(3):558–69.

44 Allen NE, Appleby PN, Key TJ et al. Macronutrient intake and risk of urothelial cell carcinoma in the European prospective investigation into cancer and nutrition. Int J Cancer. 2013;132(3):635–44.

45 Inoue-Choi M, Robien K, Lazovich D et al. Adherence to the WCRF/AICR guidelines for cancer prevention is associated with lower mortality among older female cancer survivors. Cancer Epidemiol Biomarkers Prev. 2013;22(5):792–802, doi:10.1158/1055-9965.

46 Ornish D, Weidner G, Fair WR et al. Intensive lifestyle changes may affect the progression of prostate cancer. J Urol. 2005;174:1065–69, discussion 1069–70.

47 Frattaroli J, Weidner G, Dnistrian AM et al. Clinical events in Prostate Cancer Lifestyle Trial: results from two years of follow-up. Urology. 2008;72:1319–23.

48 Sansbury LB, Wanke K, Albert PS et al. The effect of strict adherence to a high-fiber, high-fruit and -vegetable, and low-fat eating pattern on adenoma recurrence. Am J Epidemiol. 2009;170:576–84.

49 Lanza E, Hartman TJ, Albert PS et al. High dry bean intake and reduced risk of advanced colorectal adenoma recurrence among participants in the polyp prevention trial. J Nutr. 2006;136:1896–903.

50 Bobe G, Sansbury LB, Albert PS et al. Dietary flavonoids and colorectal adenoma recurrence in the Polyp Prevention Trial. Cancer Epidemiol Biomarkers Prev. 2008;17:1344–53.

51 Gold EB, Pierce JP, Natarajan L et al. Dietary pattern influences breast cancer prognosis in women without hot flashes: the women's healthy eating and living trial. J Clin Oncol. 2009;27:352–59; Thomson CA, Rock CL, Thompson PA et al. Vegetable intake is associated with reduced breast cancer recurrence in tamoxifen users: a secondary analysis from the Women's Healthy Eating and Living Study. Breast Cancer Res Treat. 2011;125:519–27; Pierce JP, Natarajan L, Caan BJ et al. Dietary change and reduced breast cancer events among women without hot flashes after treatment of early-stage breast cancer: subgroup analysis of the Women's Healthy Eating and Living Study. Am J Clin Nutr. 2009;89:1565S–71S.

52 Opp MR. Sleeping to fuel the immune system: mammalian sleep and resistance to parasites. BMC Evol Biol. 2009;9:8; Hakim F, Wang Y, Zhang SX et al. Fragmented sleep accelerates tumor growth and progression through recruitment of tumor-associated macrophages and TLR4 signaling. Cancer Res. 2014;74:1329–37.

53 Canaple L, Kakizawa T, Laudet V. The days and nights of cancer cells. Cancer Res. 2003;63:7545–52; Blask DE, Brainard GC, Dauchy RT et al. Melatonin-depleted blood from premenopausal

women exposed to light at night stimulates growth of human breast cancer xenografts in nude rats. *Cancer Res.* 2005;65:11174–84; Schernhammer ES, Schulmeister K. Melatonin and cancer risk: does light at night compromise physiologic cancer protection by lowering serum melatonin levels? *Br J Cancer.* 2004;90:941–43.

54 Yang WS, Deng Q, Fan WY et al. Light exposure at night, sleep duration, melatonin, and breast cancer: a dose-response analysis of observational studies. *Eur J Cancer Prev.* 2014;23:269–76; Stevens RG, Brainard GC, Blask DE et al. Breast cancer and circadian disruption from electric lighting in the modern world. *CA Cancer J Clin.* 2014;64:207–18.

55 Marinac CR, Nelson SH, Breen CI et al. Prolonged nightly fasting and breast cancer prognosis. *JAMA Oncol.* 2016;2(8):1049–55.

Index